The Seismicity of Central America

The Seismicity of Central America

A Descriptive Catalogue 1898–1995

N. N. Ambraseys
Imperial College, UK

R. D. Adams
International Seismological Centre, UK

ICP

Imperial College Press

Published by

Imperial College Press
57 Shelton Street
Covent Garden
London WC2H 9HE

Distributed by

World Scientific Publishing Co. Pte. Ltd.
P O Box 128, Farrer Road, Singapore 912805
USA office: Suite 1B, 1060 Main Street, River Edge, NJ 07661
UK office: 57 Shelton Street, Covent Garden, London WC2H 9HE

British Library Cataloguing-in-Publication Data
A catalogue record for this book is available from the British Library.

THE SEISMICITY OF CENTRAL AMERICA

ISBN 1-86094-244-X

Printed in Singapore by FuIsland Offset Printing

Preface

This study is the culmination of several decades of work. Our involvement in the region has grown over the years as more information and different types of analysis have become available.

We have developed and improved our own techniques in earthquake location, determination of size and the analysis of felt effects throughout this period, and because of this sometimes take the opportunity in the text of reviewing techniques previously used by others, and of tracing the development of these analyses. Modern workers do not always realise the limitations imposed on their early colleagues by inadequate instrumentation, the paucity of knowledge of basic earth structure and sometimes lack of appreciation of the physical bases of the parameters they were seeking. An example is our full discussion of the development of the magnitude scales, and of our own use of M_S. Similarly, our location of early events by using the travel time of the maximum of the surface wave train has proved most useful.

Better information is continually becoming available, enabling earlier work to be reassessed. We were helped in our final analysis by the late availability of the catalogue of Engdahl *et al.* (1998), which applies improved location techniques to the larger events in the ISC catalogues since 1964, producing much more reliable estimates of focal depth.

The most substantial parts of this work are the appendices containing the catalogue of earthquake parameters, including about 1,800 events, and the detailed description of macroseismic effects of major earthquakes and those of special interest. The earlier part of the text is intended to help in the understanding of these sections and to explain the methods used in their compilation.

We have made use of work of many previous workers, some of which we criticise and in which we find errors. Naturally we hope that our work is an improvement, but it would be presumptuous to assume that it is error free. For these errors we offer apologies, and hope that others will be able to carry out further improvements. We believe, however, that this work is the most reliable and complete compilation of earthquake activity in Central America that is currently available, and hope that it will serve as a basis for fuller investigations of seismicity.

Acknowledgements

We thank the following for help in collecting and assessing information:

Mrs M Laporte for field data from Costa Rica,

Dr L Feldman for retrieving unpublished information from repositories in Central America and USA,

Dr A Bodare for supplying reports of field work carried out in Central America by research students of the Royal Technical University of Stockholm,

Mr D White for assistance in the retrieval of data from the Public Records Office at Kew,

Mr J Vogt for newspaper data and unpublished reports kept in French and German repositories,

The Budesarchiv Abteilung, Potsdam, for copies of consular correspondence 1901–1910,

Libraries of the Royal Geographical Society, London, and the French Geographical Society, Paris,

Dr R White for providing us with copies of some early Wiechert seismograms from the Guatemala station,

Dr J J Bommer for information on earthquakes in El Salvador,

Dr H Bungum for advice on recent seismicity,

CEC National Scientific Cooperation Programme, Brussels, for support under contract CI1/CT94/1014, "Assessment of Seismic Hazard in El Salvador".

The computer re-location, and much of the text preparation has been carried out at the International Seismological Centre, in Berkshire, England over the last decade or so. We are grateful for the use of the Centre's facilities, which are funded by the Member organisations, including support from the U S National Science Foundation through grant EAR–9725096. Our sincere thanks are due the many staff members who helped with advice and practical assistance with computing and production of figures. Among those who made special contributions were Mr D M McGregor,

Mrs A-R Surguy, Mr J C Harris and Ms A L Bird. Ms Bird made a particular contribution in producing the final text for printing.

Contents

Chapter 1
Introduction

This work looks at the seismicity of Central America from the beginning of global instrumental recording in 1898 to the mid 1990s. Its main purpose is to produce a Catalogue of events, drawn from the best available sources. The Catalogue consists of a parametric listing of origin co-ordinates and estimates of size, and also a descriptive section giving details of macroseismic effects and comments on instrumental recordings. For many events we have re-evaluated the position and magnitude, using instrumental and macroseismic information, or a combination of both; otherwise we have chosen what we consider to be the best solution available from other sources.

Our area of concern is between latitudes 7°N and 17°N, and longitudes 93°W and 80°W. It comprises the whole of Central America, including the southeastern part of Mexico, parts of Panama and the Pacific Ocean and Caribbean Sea. This area is shown in Figure 1.1 We have already discussed the large-magnitude events in this region (Ambraseys and Adams, 1996) and the determination of magnitude there (Ambraseys, 1995). Our earlier studies were restricted to earthquakes of magnitude 7 or more; here we aim to reduce the threshold where possible by two orders of magnitude, to include events of magnitude greater than 5.

Naturally the availability of information varies with time. For the earliest part of our period we rely heavily on listings of the British Association for the Advancement of Science (BAAS), and standard catalogues such as those of Gutenberg (1958). There are also some events that we have identified and located from macroseismic information alone. Early instrumentally recorded events do not generally have magnitudes less than 6. Our Catalogue should contain all events in the region with magnitude more than 6, and those for which there are well-documented felt effects, even if their magnitude is small.

Several authors have made special studies of earthquakes in the region, often refining locations given by other agencies; these have been included. In recent years international listings include many events either with no magnitude specified, or given as less than 5. These have not been included unless they have been reported felt, but we have endeavoured to include all events for which the International

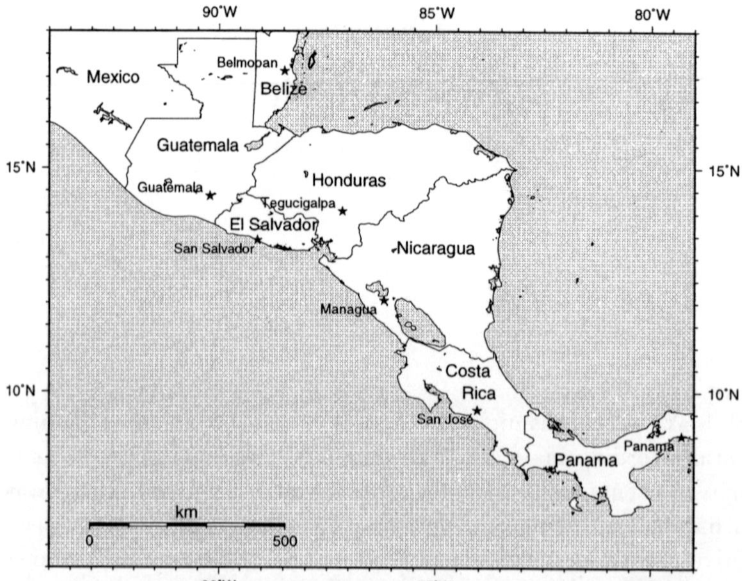

Fig. 1.1 **Map of area of study, extending one degree in each direction beyond main area of interest.**

Seismological Centre (ISC) gives a magnitude of 5 or more.

The information presented here should assist in the fuller understanding of the tectonic problems of the area, and in refining estimates of seismic hazard. It is not our intention to undertake an exhaustive analysis of these topics, but rather to provide a data base with which others can work. It is because of this that we sometimes give more detail in our Catalogue than might at first sight seem warranted.

1.1 Note on Transliteration

There is no accepted standard for the treatment of place names. Broadly speaking, our policy has been to leave uncompound names in their original forms, but to translate or adapt wherever possible the topographical appellatives of compound names. We cannot claim complete consistency; where consistency would have led us into absurdity, we have preferred inconsistency.

Where places are well known under English names, we have abandoned the local forms.

Many place names in the region have been officially changed, in some cases more than twice. We have decided not to use the latest name of places but rather retain as much as possible the contemporary names used in the references and maps at the time of the earthquake.

Chapter 2

Instrumental Earthquake Location

2.1 General Principles

The location of earthquakes in Central America is subject to the same difficulties as elsewhere, these being strongly influenced by the period concerned. The accuracy of earthquake location has improved steadily with time as problems of instrumentation, timing, earth models and station distribution are overcome. It is, however, quite wrong to disregard all early locations as being of little value.

Throughout all periods the same criteria for sound location apply. Foremost among these is an adequate coverage of recording stations in azimuth; without this, instrumentally determined solutions remain suspect. Readings from a large number of stations do not necessarily lead to a good solution, particularly if only *P* phases are being reported, unless there is an adequate distribution of stations in azimuth. It is true that clock errors and mis-reading of phase times and identification can lead to errors, but these are usually not difficult to detect. In all periods, properly interpreted felt information can confirm instrumental locations or highlight gross errors. For a reliable solution, it is also desirable to have readings from stations at a wide range of distance. In particular, close stations within a few degrees provide information that give good control of location and particularly depth. It is also important to accept that the use of a computer program to aid earthquake location does not necessarily improve the quality of the result. Poor station geometry cannot be overcome by faster or more complex computer programs, although modern techniques giving parameters such as slowness and azimuth can aid in deciding whether a given location is grossly in error.

An earthquake origin consists of four parameters; three spatial and one of time. The most difficult of the spatial parameters to determine is depth. If depth is being determined entirely geometrically, it is important to use rays leaving the focus at a wide range of angles downwards, that is, leaving to stations at a wide range of distances, or to use later phases such as surface or core reflections, that leave the focus at a different angle from that of the main *P* phase. Depth control is improved if the focus is deep enough for upward-going surface reflections to be recorded, but in this case it is sometimes difficult to decide whether such a phase is of type pP or sP . When an event is recorded mainly by distant stations, there may be a 'trade-off' between depth and origin time, with a shallower focus being associated with an earlier origin time.

Other, non-geometrical, methods may be used to help in depth determination. Deeper events in general excite smaller surface waves, and macroseismic observations can provide helpful evidence of depth, with deeper shocks being felt at a more uniform intensity over a wider area, compared with shallower ones, which tend to have a small area of higher intensity near the epicentre.

Difficulties in location can occur when the epicentre is located outside the area of the recording network. In particular, if P phases only are used, a trade-off between distance and origin time can occur, with an event being placed further away and earlier, and *vice versa*. The use of S phases can reduce such uncertainty.

When comparing earthquake locations determined instrumentally and macroseismically, it must be remembered that for large events, over magnitude about $6\frac{1}{2}$, the two need not necessarily agree within a few tens of kilometres. An instrumentally determined focus gives the point of initiation of rupture, whereas the macroseismic pattern defines the centre of maximum energy at the surface, and is therefore more closely related to the position given by the centroid moment tensor solution. For a symmetric rupture the two should be close, but if the focus is displaced from the centre of energy release, an apparent discrepancy may occur, bearing in mind that the linear dimension of an earthquake's source volume may be approximated by $logL = \frac{1}{2}(M - 4)$, where L is the linear source dimension in kilometres and M the magnitude, giving a dimension of about 30 km for a magnitude 7 event, and about 100 km for one of magnitude 8.

2.2 Agencies and Methods

2.2.1 *International Recording and Analysis*

The first global network of seismological recording stations was set up from 1898 by the Englishman John Milne. This network grew to comprise some 30 stations, widely spread, and including many island sites. The Milne instruments were undamped horizontal pendulums, of intermediate period (~12 s) and low magnification (~10). They were therefore ill suited to record body waves, and responded mainly to the longer period surface waves. Milne collected readings from the operators of his instruments, and published these from his observatory at Shide in the Isle of Wight off the south coast of England. These *Shide Circulars* contained no epicentres, but in 1911–1913 these were published for the years 1899–1910 by the Seismological Committee of the British Association for the Advancement of Science (BAAS).

In addition to Milne's network many other institutions and observatories set up further instruments, often of more advanced design, and took an interest in earthquake location. Among the earliest institutions to undertake global earthquake location was the Bureau Central International de Sismologie at Strasbourg (BCIS), which remained a major source of global earthquake locations until 1963. In the early part of the century, there were no reliable travel-time tables, and it was not until 1907 that the Zöppritz tables for a few main phases became available; these were incorporated in a modified form into BAAS analysis for earthquakes from 1914 onwards.

In 1922 the seismological work of BAAS was put under the control of the newly-formed International Union of Geodesy and Geophysics as the International Seismological Summary (ISS) which undertook analysis of earthquakes from 1918 onwards. Knowledge and techniques gradually improved, as did the number and distribution of reporting stations. In the early days, earthquakes were located manually at ISS on a large globe, but for events after 1930 a hand calculator was used to do a single least-squares iteration, with a preliminary version of the Jeffreys–Bullen travel-time tables.

About this time the U.S. government agencies also started a global earthquake location service, initially under the U.S. Coast and Geodetic Survey (USCGS), which started the production of a modest Preliminary Determination of Epicenters (PDE) service in 1928. This service has continued under different government agencies, and has become a major centre for global analysis and archiving. It is now operated by the National Earthquake Information Center of the U.S. Geological Survey (NEIC) at Boulder, Colorado.

Following the end of the Second World War there was an upsurge in interest in seismological location. Many new stations were opened, most equipped with new short-period instruments such as those designed by Benioff in the United States and Kirnos in the Soviet Union. A major advance came in the early 1960s, with the setting up under U.S. auspices of a World-Wide Standard Seismograph Network (WWSSN) of over 100 stations with identical instruments of both short and long period. Readings from these stations formed the basis of improved global locations by U.S. agencies, and provided a data base for further refinement and research by others.

Analysis techniques were also improving. In 1961 ISS started using a computer program to locate earthquakes from 1954 onwards. ISS was reorganised in 1964 as the International Seismological Centre (ISC), with broader international control, and improved analysis techniques. Both NEIC and ISC undertake global earthquake location, but their aims are somewhat different. NEIC provides a rapid service with as much detail as can be gathered within a reasonable time. The aim of ISC, on the other hand, is to provide a listing as complete as possible of all events and associated station readings, and it therefore delays the start of analysis for about two years. Locations found by NEIC and ISC are usually in good agreement, but the addition of later readings, particularly from close and regional stations, can result in ISC solutions being better controlled.

Unusual changes in the level of reported seismicity do not always have significance, but may reflect changes in the operation of stations. For instance, the work of the Central American Data Center in Guatemala in the early 1990s resulted in a significant increase in the number of small events reported in the ISC lists (Adams and Richardson, 1996). For example ISC listed 269 events in the Central American Region for the month of June 1992, after the beginning of CADCG reporting, compared with only 66 events for June 1991. This increase will not be evident in our Catalogue because most of the additional events are small, below magnitude 5, and will not therefore be included.

2.2.2 *Local and Regional Agencies*

With the development of local seismograph networks in seismically active areas, national agencies began to produce epicentres for earthquakes in their own region. Among the earliest such networks in Central America was that established in Mexico, centred at Tacubaya. Local agencies have also been set up in other Central American countries, particularly Guatemala and Costa Rica.

A concerted effort at regional co-operation was made in the early 1990s with the establishment of a Central American Data Center in Guatemala (CADCG) to carry out combined analysis of readings supplied from stations in Guatemala, El Salvador, Honduras, Nicaragua, Costa Rica and Panama. Originally, the determinations of the individual agencies often appeared separately, resulting in events being duplicated in catalogues. An example for an event on 3 September 1992, too small to be included in our Catalogue, is shown in Figure 2.1. Small stars show the positions given by three national agencies, several degrees apart, and the large star the position found by ISC after combining all the station readings (Adams and Richardson, 1996).

2.2.3 *Special Studies and Relocations*

Re-evaluation of earlier earthquakes has been carried out since the catalogues of Gutenberg (1958) and Gutenberg and Richter (1965). The application of more recent travel-time tables and analysis techniques can cause significant revision of contemporary locations. Since then there have been many special studies that include Central American earthquakes, mostly involving computer relocation. Some of these are listed in the notes on our catalogue, and we adopt their results if appropriate.

These studies often use more stations than are available to larger agencies, and may use specially selected regional travel times.

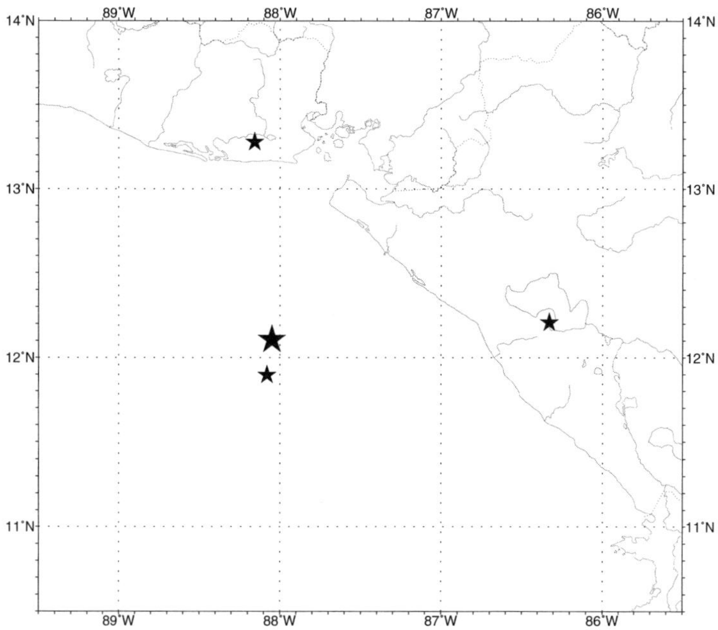

Fig. 2.1 Small stars show locations given by three national agencies for a small earthquake on 3 September 1992, using their own stations only. Large star shows position obtained by ISC by combining readings from all agencies. (From Adams and Richardson, 1996)

2.2.4 *Our Relocations*

We re-investigated instrumental information for about 60 earthquakes between 1900 and 1947. If there were enough readings, we carried out a computer relocation using the standard location program of ISC (Adams, Hughes and McGregor, 1982). This uses the Jeffreys-Bullen (1940) tables and locates using *P* phases only, but will give residuals for other main phases, including core phases such as *PKP*. The depth is determined geometrically in a four-parameter solution, but if this cannot be done, the program fixes the depth on that of the initially assumed origin. If enough reports of depth phases *pP* or *sP* are available, the program will use these to determine a depth.

For these relocations we chose *P* readings from all available sources, particularly from station bulletins, but readings reported from Milne instruments were not precise enough to be used in *P*-phase locations. Sometimes these relocations showed substantial shifts compared with contemporary estimates, and on other occasions they could only confirm the earlier position within the given errors. Often the data were not adequate to give a formal solution, but were consistent with the position given. Even if an instrumental solution were obtained, a macroseismic position would be preferred if it were better determined. Some of our instrumental re-evaluations gave good determinations of depth, often supported by depth phases. In our catalogue in Appendix B instrumental relocations carried out by us are identified as agency *AR*.

Although the Milne stations did not record body waves well enough to be used for location, nor even the onset time of surface waves, they did report consistently the arrival time of the maximum amplitude of the surface waves. For some early events we could use these times to carry out a graphical location on a globe, assuming this arrival to be an Airy phase travelling at a velocity of about 3 km/s (Ambraseys and Adams, 1986). The fact that this velocity is so low helps to reduce errors arising from uncertain timing, and such relocations could sometimes be established with an accuracy of a degree or so, which compares favourably with errors found by other instrumental techniques early in the 20th century.

2.3 Location in Different Time Periods

2.3.1 *Up to 1920*

The earliest instrumental locations on a global scale were made from 1898, using the network of medium-period Milne instruments. Such locations were necessarily inexact, but the station readings provide a control as to the general area of the shock. Milne seismographs remained recording until the 1920s, and were the predominant instruments until the more general use of shorter-period seismographs from the 1910s. Contemporary locations were hampered by a lack of accurate travel-time tables, but in many cases the readings have been re-interpreted in the light of later knowledge. The Milne instruments were not suited for the detection of short-period body waves, but did record well the longer-period surface waves. Many of the larger shocks of the early period have been relocated in the publications of Gutenberg, and solutions are given also by BAAS, based on the Milne readings.

The earliest reported shocks are restricted to those of significant magnitude, say 6 and above, and for these we have found that macroseismic locations prove most reliable, with instrumental readings, at best, providing confirmation of the general area of occurrence. In the period up to the mid 1920s, therefore, we have adopted our own relocations for most events. These are mainly from macroseismic information (given in our catalogue as *AM* or *Am*), but instrumental information may be used to improve (*AR*) or confirm (*Ar*) these positions.

In the early instrumental period, there was a preponderance of instruments in Europe, which resulted in an east-west uncertainty in location. This was improved as readings from more regional and American stations became available. There often arises a real ambiguity about the choice of stations to use; for example, acceptable solutions may be found using either of two particular regional groups of stations, but the two groups will not give a consistent combined solution.

The notes on individual earthquakes in Appendix A give instances of discrepancies between various solutions. Note for example that for the Costa Rican earthquake of 21 June 1900, our macroseismic position is about 1,300 km southwest of Gutenberg's (1958) estimate. From 1903 onwards instrumental readings start to play an increasing part in location, and the earthquake on 1 July 1907, in Honduras, is the earliest for which we can obtain a reliable instrumental position, making use of readings from regional stations at Tacubaya in Mexico, Puerto Rico and eastern America, in addition to many in Europe. Instrumental readings also suggest a sub-crustal depth. Again, this event was grossly mislocated by BAAS, which place it at 0°, 105°W. Because of this, it has not been included in some previous Central American catalogues. Figure 2.2 shows those events up to 1930 for which our redeterminations were at least 500 km from positions originally given by BAAS, ISS or Gutenberg.

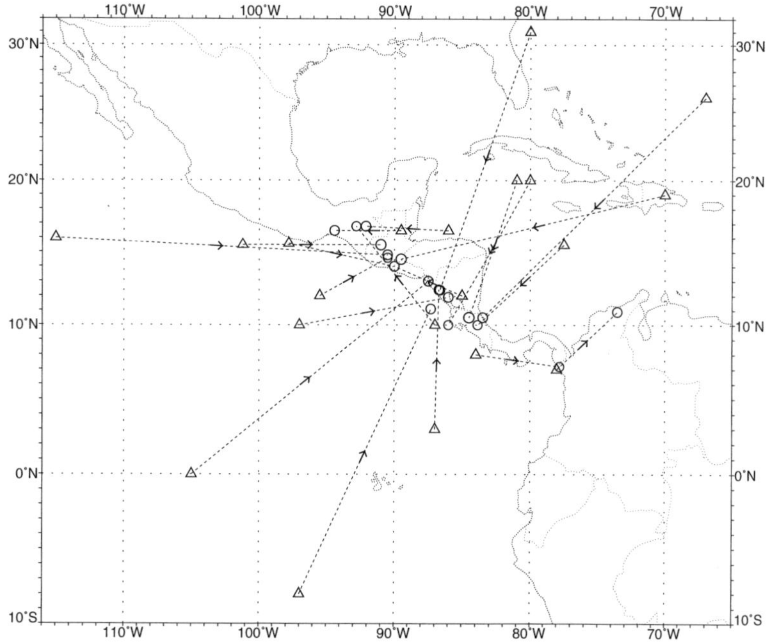

Fig. 2.2 Events from 1898 to 1930 for which our redeterminations (circles) are at least 500 km from positions originally given by BAAS, ISS or Gutenberg (triangles).

In the late 1910s instrumental solutions had become more reliable. An example is the earthquake on 23 September 1916, located by us in Nicaragua from nine P phases, with a standard error of about 20 km, with supporting felt reports. Our position is about 1,000 km to the east of that given by BAAS.

The determination of depth during the early period is very poor, and most solutions are given as *normal* by default. The first earthquake in our catalogue to be positively identified as intermediate is in 1921, and of nearly 200 events in our catalogue up to 1930, only six are allocated depths of more than 60 km, these all resulting from relocations by either us or Gutenberg. Another 13 earthquakes are designated $n+$, suggesting a depth of between 45 and 60 km.

2.3.2 *From 1920 to 1950*

For earthquakes from about 1918 onwards we adopt more of the solutions found by the ISS. Some of the larger earthquakes, however, were still subjected to special studies by workers such as Gutenberg, but well constrained macroseismic studies still give the most reliable locations. Depth, however, remains difficult to establish, and in this period most events are simply classified as *normal*, although a few are given intermediate depth.

An early example of a reliable instrumental relocation and depth estimation is that for the earthquake of 28 March 1921 in Nicaragua. We used P readings from 43 of the stations listed in ISS, ranging from Balboa Heights in Panama (8.1°) to Wellington in New Zealand (104°), and noted consistent readings of PKP phases out to Jakarta in Indonesia (165°). This analysis gave a position about 150 km to the northeast of that found by Gutenberg, with formal standard errors of about 10 km, but established a depth of 170 ± 16 km which compensated for the shift in position. There is additional evidence for this depth in some reported later phases, and in the wide spread of felt reports.

With improving analysis techniques and increasing number of reporting stations, contemporary instrumental solutions improved through the 1930s, and we adopt a larger proportion of determinations from ISS, although the best solutions still remain those found macroseismically. Depth remains predominantly given as *normal*. For events from the early 1940s, we have available a growing number of computer relocations made as part of special studies. Throughout the war years of the 1940s there is less reporting of felt effects in local sources.

An example of a well-determined solution with agreement between macroseismic and instrumental information is that of the earthquake in Nicaragua on 26 January 1947. Nearly one hundred station readings were used to determine an epicentre with a standard error of less than ten kilometres. Although the nearest station (Balboa Heights in Panama) was more than 7° distant, the spread of station distances was such that the depth could be well established as 151 ± 14 km, with the value from depth phases being 166 ± 2 km. (These have been rounded in our Catalogue to 160 km).

2.3.3 *From 1950 to 1963*

During this time the information available for instrumental solutions continued to increase, with most being computer determined. Some later earthquakes are reported by several hundred stations world-wide. From the mid 1950s we have available a growing number of results from special studies, but major earthquakes, and especially those of magnitude 7 or more continue to be re-examined by us. One result of computer location is that the program will often specify a depth, even if this is not well controlled. We list these values in our Catalogue, many giving subcrustal depths in the range 33–100 km. Such values are not necessarily well established, and should be treated with caution.

In the 1950s and 1960s there was some fall off in felt reporting, but more information is available during the more recent decades.

2.3.4 *From 1964 to 1995*

1964 marked the beginning of analysis by the International Seismological Centre (ISC), with increasingly more stations becoming available and a more carefully monitored computer location program. Although the earthquakes' positions improved in reliability, there still remained much uncertainty in depth, particularly for events in the crust and uppermost mantle.

A significant improvement was achieved by Engdahl, van der Hilst and Buland (1998), who relocated the best controlled events in the ISC lists using an improved global travel time model (aka135; Kennett *et al.*, 1995). For each event they retained the data set used by ISC, so errors due to mis-association will remain, unless obvious from arrival-time residuals. In addition to the first arrival they took account of later arriving phases such as *PcP*, *PKP* and *S* and paid particular attention to correct interpretation of near-source reflections, where there can be confusion in oceanic areas between the reflections from the sea bed (*pP*, *sP*) and from the water surface (*PwP*). Their catalogue may be considered as a refinement of that of ISC, and in particular their determinations of depth more reliable. In this period we adopt the solutions of Engdahl *et al.* when these are available, unless there is an exceptionally well-determined macroseismic position. For these macroseismically located events we adopt the depths given by Engdahl *et al.* For some events of interest not well enough recorded to be included in the catalogue of Engdahl *et al.*, we adopt the ISC position.

As a consequence of the better determination of depth in the catalogue of Engdahl *et al.*, many more events are allocated subcrustal depths in the later years. During the period covered by this catalogue, 1964 to 1995, 36% of all earthquakes over M_S 5 have been allocated depths of more than 60 km. By contrast, of 197 events in the period up to 1930 only six are deeper than 60 km. This apparent increase in the number of intermediate and deep earthquakes must not be given any geophysical significance, but the later figures should probably be considered as giving the true proportion for the entire period of study.

Chapter 3

Determination of Earthquake Size

3.1 General Principles

The most widely accepted measure of earthquake size is magnitude, derived from instrumental measurements. Many different types of magnitude have been developed, depending on the type of instrument used and the parameters measured. The main magnitude used by us is surface-wave magnitude M_S, which we discuss in detail below. To a lesser extent we make use of body-wave magnitude m_b, which we also describe. In recent decades measurements of the more meaningful parameter seismic moment M_O have become available, and these are now the preferred measure of earthquake size. These estimates of size depend on instrumental information, which is often not available during early times or for small events. In such instances size has to be estimated in other ways, such as from felt effects, or from the number of seismic stations recording the event. A common feature of magnitude scales is that the determination generally depends on the logarithm of the maximum recorded amplitude of the phase being considered, with an adjustment for the epicentral distance and sometimes for focal depth. Thus a change in one unit of magnitude corresponds to a factor of ten in recorded amplitude. The relationship between magnitude scales derived from different phases is not simple, however, particularly as different scales may be derived from phases of different period, reflecting energy levels from different parts of the earthquake's source spectrum.

3.2 Surface-Wave Magnitude (M_S)

There is considerable discussion in the literature as to the best practice for the determination of surface-wave magnitude.

Gutenberg and Richter (1936) were the first to use a magnitude based on surface waves from shallow events, M_{GR}. It was devised to extend to teleseismic distances the local magnitude M_L which had been defined in the previous year (Richter, 1935), and was more thoroughly developed in a subsequent paper by Gutenberg (1945).

According to Abe (1981), it is equivalent to present surface-wave magnitude.

The original M_{GR} scale was based on the maximum horizontal ground displacement A_{max}, but it was specified that measurements were to be made at periods near 20s. Gutenberg's worksheets, which are kept at the Millikan Memorial Library of the California Institute of Technology in Los Angeles, show, however, that Gutenberg himself frequently did not observe this rule. He used periods ranging from 10 to 25s, and quite often values of amplitude and period which differ from those published in station bulletins. There are different interpretations of the structure of Gutenberg's magnitude, M_{GR}, of its changes over the time of its development and of the method he used to choose maximum phase amplitudes (Båth, 1969; Abe, 1981; Ambraseys and Melville, 1982; Lienkaemper, 1984).

3.2.1 *Prague Formula for Surface-Wave Magnitude*

An improvement to Gutenberg's scale was made by Soloviev (1955) who proposed a surface-wave magnitude based on measurements of the maximum ground particle velocity $(A/T)_{max}$. This physical quantity describes the seismic energy flux better than the ground displacement A_{max} at a fixed period of 20s. Soloviev's scale is not restricted to a given period, and the magnitude determined at a particular station, $M_{s,i}$ can be calculated within the broad range of distances from 4° to 80°. He defined the general formula for the surface-wave magnitude at a particular station as

$$M_{s,i} = log(A/T)_{max} + s(\Delta, h) + C(M) \tag{3.1}$$

where A and T are respectively the ground displacement in micrometres, and the period in seconds associated with the maximum particle velocity $(A/T)_{max}$, $s(\Delta, h)$ is an empirical calibration function which expresses the change of ground particle velocity with epicentral distance Δ and focal depth h, and $C(M)$ is a correction term which allows for the effects at the recording site, wave path and focal mechanism (Soloviev and Shebalin, 1957).

Kárník *et al.* (1962) and Vaněk *et al.* (1962), following Soloviev, proposed the following calibration relation

$$s(\Delta) = 1.66log(\Delta) + 3.3 \tag{3.2}$$

which they derived originally from the weighted average of 14 attenuation functions existing at that time for epicentral distances between 20° and 160°, and for a wide range of surface-wave periods, to determine the value of $(A/T)_{max}$. These 14 attenuation functions, and subsequent functions used to control equation (3.2) are given in Soloviev (1961,p.115) and Kárník (1968, pp.56-60). See also Lienkaemper, (1984). Later, the validity of equation (3.2) was confirmed additionally for smaller distances down to a few degrees (Kárník and Christoskov, 1977; Kárník, 1977).

Use of the calibration relation (3.2) was proposed by IASPEI in 1967, specifically to avoid the limitations imposed by the restriction to waves of period near 20s in Gutenberg's method. Equation (3.1), commonly referred to as the original "Prague formula", was then defined as

$$M_{s,i} = log(A/T)_{max} + 1.66log(\Delta) + 3.3 + C_i \quad (3.3)$$

where C_i is a station correction term which allows for the effects at the recording site and wave path. Recommended period ranges corresponding to maximum amplitudes of surface waves at different epicentral distances were also given by IASPEI(1967), Kárník (1962) and Willmore (1979). The Prague formula was devised to be used with events shallower than 40 to 50 km, and to have a depth adjustment for deeper events. We will not discuss here the derivation of depth correction.

With few exceptions, however, workers and agencies do not use the Prague formula according to its full original definition. Since the mid-to-late 1970s, surface-wave magnitudes reported by both the National Earthquake Information Service (NEIS) and by the International Seismological Centre (ISC) have been computed and published using the Prague formula, but these agencies select data using different criteria, neither of which is consistent with the original definition of the formula.

Up to 1975 NEIS published estimates of M_S from readings on horizontal components at individual stations, but from May 1975 the assessment has been made only from the vertical component of the surface wave within the period range 18 to 22 s and for distances between 20° and 160° (see Lienkaemper, 1984 for details). It is theoretically more correct to use the vertical component rather than the horizontal ones because the vertical component records only waves of Rayleigh type, while the horizontal components record both Love and Rayleigh waves, with resulting complication in attenuation characteristics (Willmore, 1979). No depth or station corrections are applied by NEIS and M_S magnitudes are not generally computed for events with focal depths greater than 50 km.

Before 1971 ISC neither reported long-period amplitudes and periods nor calculated M_S. Between 1971 and 1976 it reported amplitudes and periods for all components, and M_S was calculated by combining vectorially the maximum reported amplitudes of the two horizontal components at periods near 20s for stations in the distance range 20° to 160°, and using the attenuation relationship from the Prague formula. These determinations were given with the station readings only and were not included with the epicentres. Between 1976 and 1978 magnitude determinations from vertical components were included for the small number of events for which data were available, the distance range was extended to between 5° and 160° and the range of allowable periods to between 10 and 60s. Since 1978 event magnitudes determined using these criteria have been given with epicentres, but only those stations at distances between 20° and 160° have ever been used in the averaging, and M_S is given only for events at depths of 60 km or less. Thus M_S magnitudes reported for events by ISC are calculated without determinations from stations at distances less than 20°. This sometimes results in ISC giving magnitudes for individual stations, without any magnitude being assigned to the event as a whole (Ambraseys and Free, 1997).

NEIS and ISC thus use somewhat different selection criteria in choosing stations for the calculation of M_S, and for a particular event the number of stations used and their distribution in azimuth may be different. In addition, because of its larger data set ISC usually uses more station readings in determining event magnitudes than NEIS, but neither agency reports standard deviations of their estimates.

In some instances readings of surface-wave amplitudes and periods from particular networks do not reach the main agencies. For instance, amplitude and period data from Uppsala (UPP) and Kiruna (KIR), stations that have a long history of full reporting of these values in their bulletins, are not reported to the ISC. Additionally, only a fraction of the relatively large amount of such data that is available from the Chinese, and particularly from the former USSR networks, is usually reported by ISC and used in its bulletins. The absence of these data from the calculation of surface-wave magnitude can introduce further azimuthal and distance bias in the assessment of mean M_S values, particularly for earthquakes observed by few stations.

3.2.2 *Station Corrections for Magnitude*

The station correction C_i in equation (3.3) for a particular station i is defined as the mean of the residual $M_S - M_{s,i}$ over a period of time

$$C_i = \Sigma_j \left(M_{s,j} - M_{s,i,j}\right)/N \qquad (3.4)$$

where N is the number of events observed by the station.

To the best of our knowledge, the first systematic estimation of station corrections from earthquakes in Europe was made by Kárník (1968), who used the original Prague formula to assess corrections for 170 stations, chiefly European. He used earthquakes in Europe and adjacent regions, with no restriction on magnitude or depth, during a 48-year period between 1904 and 1951. In this European data set almost all observing stations are inside the study area, and the paths propagate across the Eurasian continent. Kárník's station corrections are from horizontal amplitudes.

Various other authors have calculated magnitude station corrections for different regions of the Earth. These include Christoskov *et al.* (1983) for the European area, Ambraseys and Melville (1982) for Iran, and Rezapour and Pearce (1998) for a global data set.

Using Kárník's procedure, Ambraseys (1995) calculated station corrections from Central American shallow earthquakes of $M_S > 5.0$ for a 27-year period from 1904 to 1930. The study area is that taken for our main study, and is very small, bounded by 7°N, 17°N, and 80°W, 93°W. All observing stations are outside the study area and the propagation paths are oceanic and mixed.

Details of these studies and the values of the station corrections are given in Ambraseys and Douglas (1999, 2000) and their values vary for different areas. The corrections found by Ambraseys (1995) for Central America have been applied in M_S determinations made by us in our catalogue.

3.2.3 *Distance Correction for the Original Prague Formula*

Since its adoption by IASPEI in 1967 there has been much debate about the adequacy of the amplitude-distance term of $M_{s,i}$ in equation (3.2) (Evernden, 1971; Marshall and Basham, 1973; Nuttli, 1973; Von Seggern, 1977; Christoskov *et al.*, 1983; Panza *et al.*, 1989; Herak and Herak, 1993; Vaněk, 1995; Rezapour and Pearce, 1998). It must be pointed out, however, that most of these authors examined the Prague formula at periods near 20s in the distance range 6° or 20° to 160°, using M_S estimates made by NEIS or ISC, and not from station readings made according to the original definition of the scale.

Von Seggern (1977) found that the amplitudes of Rayleigh waves near a period of 20s attenuate slower, at a rate of $1.08 log(\Delta)$, as compared with $1.66 log(\Delta)$ in the Prague formula. Similar results were obtained by Christoskov *et al.* (1983) who derived a homogeneous M_S magnitude system for the region of the Eurasian continent, using 32 reference seismic stations. They derived calibration functions for surface-wave magnitudes from 4,862 long period horizontal observations from 286 shallow events for the five years from 1966 to 1970, in the distance range from 20° to 100°, and with a successive optimisation of station adjustments with respect to the Russian station Obninsk (OBN). These authors noted that a simple linear function cannot express accurately the value of the calibration function over the whole range of distances.

We found evidence for the need of an adjustment to the distance term of the Prague formula in a recent unpublished study of a dataset which was for the determination of station corrections for magnitudes associated with earthquakes that triggered strong-motion instruments in the European area (Ambraseys, 1990). The data set consists of 4,800 station magnitudes $M_{s,i}$, and shows that the constraint of the selection of data to periods in the range 18 to 24s resulted in a noticeable M_S distance dependence for $\Delta < 20°$, causing an increase of the distance coefficient to 1.66 beyond that distance, a situation similar to that described by Evernden (1971). However, the inclusion of data from events recorded over short distances, with surface-wave (Lg) periods down to 3s reduced the need to adjust the distance term in the Prague formula, mainly by reducing the bias implicit in choosing only wave amplitudes that have associated periods near 20s, which also reduces the number of data points at close distances.

Herak and Herak (1993) used the data selection criteria employed by NEIS and a data set of 250 earthquakes reported by ISC and NEIS world-wide, and vertical amplitudes recorded in the 18 to 22s period range for an undefined distance range. They found that the rate of attenuation of the Rayleigh-wave amplitude with distance is slower than defined in the Prague formula. They used a procedure in which the residuals between individual $M_{s,i}$ and M_S values extrapolated for each event to a distance of 100°, termed the "representative" M_S when plotted against $log(\Delta)$, shows that the Prague formula needs modification. They proposed that it should be replaced by

$$M_{s,i} = log(A/T)_{max} + 1.094 log(\Delta) + 4.429 \quad (3.5)$$

A similar observation can be made with data exclusively from the European region, by using the data selection criteria and regression procedure adopted by Herak and Herak (1993), that is, the use of vertical component and periods in the 18 to 22s range (Ambraseys and Free, 1997). They find that

$$M_{s,i} = log(A/T)_{max} + 0.947 log(\Delta) + 4.77 \quad (3.6)$$

From 700 earthquakes in the larger European region of magnitude in the range $3.5 < M_S < 7.4$, between 1977 and 1995, these authors re-examined the question of whether the original Prague formula needs an adjustment to the distance term. They examined the distance dependency of the residuals from station magnitudes and "representative" M_S values at distances 60° and 100° degrees, and separately they examined the distance dependence of residuals from station magnitude and seismic moment. They found that the restriction of the data to the 18 to 22s period range causes the original Prague formula to require a correction of its distance term for both global and regional data, and also that the selection of data over the much broader range implicit in the original version of the Prague formula reduces this requirement. They concluded that the use of the optimum "representative" M_S is at 60° and that the correlation of M_S with seismic moment M_O confirms that distance dependence $dM = M_{i,3} - M_{i,1}$, where $M_{i,1}$ and $M_{i,3}$ are the station magnitudes calculated without and with distance correction, remains statistically significant but small

$$dM = 0.518 - 0.282 log(\Delta) \quad (3.7)$$

when they adhere to the original definition of the Prague formula. The methods they used for calculating distance effects and separately using moment M_O were based on different principles and data, and it is reassuring that the results are so similar.

3.2.4 *Averaging Station Magnitudes to Give Event Magnitude M_S*

Station magnitudes $M_{s,i}$ estimated for the same event by different stations often diverge. Some such divergences represent real irregularities in wave propagation while systematic station magnitude errors may arise from other factors, and these can be corrected using station corrections. Some authors reject from the summation uncorrected station magnitudes $M_{s,i}$ which fall outside a given range from the mean. In their various works Gutenberg and Kárník used *ad hoc* ranges of exclusion of a whole and half magnitude unit respectively, and Lienkaemper (1984) used probabilistic criteria to exclude $M_{s,i}$ values from averaging, without prior scrutiny of the quality of data.

The method of averaging the station magnitudes $M_{s,i}$ to calculate event magnitude M_S is an important procedure, the effects of which are often overlooked. It is usual for the magnitude of an earthquake M_S to be calculated as the arithmetic mean of station magnitudes $M_{s,i}$, that is, by taking the mean of the $log(A/T)$ terms from different stations. However, the average seismic energy density at a station is proportional to (A/T) and not to $log(A/T)$ and therefore when the $M_{s,i}$ values are averaged, the M_S value is underestimated. This would not be a problem if the underestimation was constant but it varies depending on the distribution of station magnitudes.

Consider the general station magnitude equation

$$M_{s,i} = log(A/T) + blog(\Delta) + c \tag{3.8}$$

where b and c are constants. This can be transformed to

$$10^{M_{s,i}} = 10^c \Delta^b (A/T) \tag{3.9}$$

Therefore $10^{M_{s,i}}$ is proportional to the energy density at the station. Hence the new event magnitude, $M_{s,n}$, could be more correctly defined as

$$M_{s,n} = log[\Sigma 10^{M_{s,i}}/N] \tag{3.10}$$

Note that if there is only one station magnitude then $M_S = M_{s,n}$ therefore the original station magnitude definition is correct.

It can be shown, by means of an expansion, that

$$M_S \leq M_{s,n} \tag{3.11}$$

as long as the differences between $10^{M_{s,i}}$ and $10^{M_{s,n}}$ are not too great. The difference between $M_{s,n}$ and M_S can be estimated by R, where

$$R = \frac{\sum(10^{M_{s,i}} - 10^{M_{s,n}})^2}{2Nln(10)(10^{M_{s,n}})^2} \tag{3.12}$$

which depends on the variance of $10^{M_{s,i}}$. Although there is no exact relationship between R and the variance of $M_{s,i}$, they are roughly proportional. Thus the underestimation in magnitude by using M_S rather than $M_{s,n}$ is greatest when the station magnitudes are widely distributed about the event magnitude. R is a good estimator of the error when the underestimation is less than about 0.25 but a poor estimate for larger errors. This is due to the higher order terms in the expansion becoming more important.

3.2.5 *Errors in M_S Determinations*

Errors in M_S magnitudes published by major agencies for more recent events are rare, but they do occur. Some gross errors exist in the routine calculation of M_S values by ISC. These are chiefly due to confusion between nanometres and micrometres in the reported amplitude, which results in individual station magnitudes being incorrect by three units. Such occurrences are found, for example, in the events on 6 July 1984 at 04h 08m and on 25 October 1984 at 09h 49m; the former was in Iran and allocated an M_S of 5.8 by ISC based on three station determinations of 4.6, 5.0 and 7.9. The largest value results from interpreting an amplitude in nanometres as micrometres. Reducing it by three gives an event magnitude of 4.8, in agreement with that given by Moscow. The second event was in Greece and given an M_S of 5.8 based on five station magnitudes of 4.2, 8.0, 7.3, 4.7 and 4.7. Removing the effect of the highest two values reduces the magnitude to 4.6, similar to a value of 4.8 found by Moscow. Such errors have less effect when the number of station magnitudes is larger, and fortunately do not occur frequently. ISC has now modified it analysis procedure to detect suspiciously large ranges of station magnitudes in any given event.

3.3 Body-Wave Magnitude (m_b)

This magnitude was introduced by Gutenberg and Richter (1956) to be able to make use of the growing number of readings of body-wave phases that were becoming available, and in particular to be able to estimate the size of deep earthquakes, which did not generate well-developed surface waves. Its form was

$$m_b = log(A/T)_{max} + Q(\Delta, h) \tag{3.13}$$

where $(A/T)_{max}$ is the maximum value of the ratio of ground amplitude in micrometres and the period in seconds, and $Q(\Delta, h)$ is a calibrating function of distance Δ and depth h. Tables were produced for both the vertical and horizontal components of the P phase and for PP, and S. The original calibrating function was developed when most instruments were still of intermediate period, but it is now widely used for short-period instruments, which can lead to inappropriate results. Various improved calibrating functions have now been produced, such as that of Veith and Clawson (1972), but major agencies such as ISC tend to still use the original Gutenberg and Richter values. ISC restricts the distance range to between 21° and 100° and the period to less than 3s.

We use m_B to denote body-wave magnitudes from intermediate-period or broadband instruments, and m_b for those derived from short-periods. Most of the early determinations made by us will be m_B, while modern body-wave magnitudes, particularly those since the start of operation of ISC in 1964, will tend to be m_b.

As magnitudes derived from surface waves and body waves sample different parts of the earthquake's radiation spectrum, differences in spectral characteristics of earthquakes make it inappropriate to seek globally applicable relationships between these magnitudes.

3.4 Seismic Moment (M_O)

Since the 1970s the concept of seismic moment has become the favoured measure of earthquake size. In its original concept it is proportional to the area of fault break multiplied by the amount of slip, and is thus theoretically capable of being determined by field observations. It can be shown to be related to total energy release, and is now routinely determined from analysis of seismograph records, most easily from those recorded digitally. For earthquakes of 1981 and later, a group based at Harvard University led by Dziewonski (Dziewonski and Woodhouse, 1983) has routinely published centroid-moment tensor mechanism solutions, including values of seismic moment, for all earthquakes of magnitude above about $5\frac{1}{2}$. NEIS also carries out moment determinations; results from both these sources are reproduced by ISC. Originally determinations of moment were given in units of dyne·cm, and this is still widely used in the United States, but we use the SI unit of newton·metres, which is now generally preferred.

3.5 Other Measures of Earthquake Size

When appropriate instrumental information for the above methods is not available, an earthquake's size may be estimated in other ways. The duration of an instrumental record, either the whole or a specific part of it, is often used to determine a duration magnitude M_D. The earthquake's felt effects and the number of stations recording it may also be used to give an estimate of size.

3.6 Existing Magnitude Estimates for Our Study Area

In general, there are two kind of parametric catalogues for surface-wave magnitudes: (*a*) catalogues in which M_S was calculated homogeneously with a standard method from station magnitudes using amplitude and period data with or without station corrections, and (*b*) less reliable catalogues in which event magnitudes, originally of the first kind or derived from other scales, have been "adjusted" empirically to fulfil some criteria of homogeneity and completeness of the catalogue, without resorting again to instrumental data. When applied to datasets of the last two to three decades of relatively homogeneous information, these empirical techniques may show up detection and reporting changes of a global network, and occasionally magnitude shifts that can help "adjust" the data (Habermann, 1987; Perez and Scholtz, 1984; Pacheco and Sykes, 1992). However, such techniques fail to identify regional magnitude biases and they are not applicable to datasets such as ours, which span a period of 95 years of changing instrumentation, network density and distribution.

There are relatively few M_S estimates of the first kind for events before the mid-1960s in our area. No magnitudes were calculated by the International Seismological Summary (ISS) in the period of its operation up to 1963, and it is not until January 1978 that ISC started to calculate mean surface-wave event magnitudes using the Prague formula. For earthquakes after the mid-1960s estimates of non-homogeneous surface-wave magnitudes roughly equivalent to M_S have been made by different national centres using a variety of different formulae.

For the period before 1949 the main source of magnitudes of the larger earthquakes in our area are the calculations of Gutenberg and Richter (1954), while for the period 1950 to 1958 these are the estimates in Gutenberg's unpublished work-sheets.

The most recent catalogue of the second kind is by Perez (1999). Assuming that the rate at which earthquakes of $M_S > 6.0$ occur throughout the world is constant and typical of all periods for shallow ($h < 70$ km) events, he adjusted M_S event magnitudes reported by ISS/ISC and NEIC so that they satisfy this postulate. It is hardly to be expected, however, that a natural system would be conservative in terms of M_S, which is not a physical parameter (Pacheco and Sykes, 1992). Perez's dataset includes magnitudes from different scales, a number of spurious events and magnitudes determined by the Large Aperture Seismic Array (LAO) which were included in international listings, such as that of ISC. However, LAO determinations are sometimes completely spurious, and some of their high assigned magnitudes are reported by only a few high-gain stations in the USA. Some of these reported events resulted from automatic analysis which misinterpreted core phases such as $PKKP$ from large events as P from a reverse azimuth, with their high apparent velocity resulting in their being ascribed to spurious events in the far range of P-wave distance (Ambraseys and Adams, 1986). Another source of contamination of this dataset comes from the use of M estimates reported in the ISC Historical File in which magnitudes before 1977 have been adopted from other sources without calculation. ISC did not calculate surface-wave magnitudes before 1978.

With the advent of computer relocation several workers published catalogues of revised earthquake locations for this region, among them being Molnar and Sykes (1969), Dewey and Algermissen (1974) and Güendel (1986). Some of these reassessments include the determination of magnitude, and we have used these if no other data are available.

For m_b, if we are unable to redetermine magnitudes ourselves, we have included short-period mean body-wave magnitudes m_b published by ISC from its inception in 1964, or later revisions by Engdahl *et al.* (1998).

3.7 Our Redeterminations

3.7.1 *Choice of Magnitude Scale*

Our preferred measure of earthquake size is seismic moment, but traditionally for shallow events in our area the seismicity has been evaluated in terms of surface-wave magnitude M_S, and almost all earthquake catalogues for the region employ this scale. The main reason for our preference is that the global, and in particular the European, seismological network is more than capable of providing reliable data for the uniform determination of M_S, not only for recent events, but also for earlier earthquakes, allowing calibration by comparison with determinations made with modern instruments. In addition, there are more stations available, and with better distribution of azimuth, for the calculation of M_S than there are for m_b, particularly in the earlier period. In the period up to 1930, we could use 1244 amplitude-period readings from bulletins to determine surface-wave magnitude, while only 182 were available for body-wave magnitudes. M_S is also the magnitude scale that correlates best with seismic moment M_O.

Furthermore, for engineering purposes, existing magnitude-frequency relationships that describe earthquake exposure in various parts of our area are most often in terms of M_S, and this is the magnitude scale that is generally employed in the derivation of attenuation relationships of strong ground motion for hazard analysis. In contrast with Europe and the former Soviet Union, U.S. practice is to use local magnitude M_L instead of M_S, perhaps one of the reasons being that the seismographic network in the Americas before the early 1960s was not suitable to provide sufficient data to estimate M_S reliably.

Occasionally, the local magnitude M_L is used also in our area, but it is found to be far less robust due to saturation at larger magnitudes, lack of proper calibration to local conditions, and bias that can be introduced because of the small number of stations generally available for its determination and their poor distribution in azimuth. M_S will thus be the principal magnitude used in this study.

We sought to re-assess surface-wave magnitudes for all events in our area from 1898 to 1995, irrespective of depth, which meet at least one of the following criteria,

(1) M_S greater than 5.6
(2) reliable estimates of seismic moment, to enable investigation of relationships between M_S and M_O
(3) assigned M_S greater than 5.6 by other workers or agencies
(4) special interest because of association with surface faulting, triggering of strong-motion networks or exceptionally high damage.

A detailed study of magnitudes of Central American earthquakes for the period 1898 to 1930 has already been published by Ambraseys (1995).

3.7.2 *Basic Amplitude and Period Data*

For the period 1900 to 1971 amplitude and period data for the calculation of station magnitudes $M_{s,i}$ were taken only from station bulletins. After that more stations started reporting to ISC, and data also became available through its bulletins, but the amount of available data became very variable. Some stations reported to ISC regularly, others discontinuously and some only occasionally or too late to be included in its Bulletin.

First we discarded, on specific grounds, suspect amplitude and period data and then calculated $M_{s,i}$ with station corrections using the original Prague formula. We find that magnitudes from many stations, particularly in the early and middle periods, required corrections by as much as ± 0.3 to ± 0.5 magnitude units, the application of which considerably reduced large differences from the mean with the result that no *ad hoc* exclusion criteria are needed. These differences are far too great to be regarded as observational error and represent real time variations.

3.7.3 *Focal Depth*

The Prague formula was devised to be used with shallow events and should have a depth adjustment for deeper events, (e.g. Ambraseys and Free, 1997; Herak *et al.*, 2000), but the uncertainty of depth determination makes the evaluation of such corrections extremely unreliable. In this study we re-appraised M_S only for events with $h \leq 40$ km, and use the formula without correction. We find, however, that the standard errors of the mean surface-wave magnitude for subcrustal and deeper events are generally greater than for shallow events. For some earlier events this is the result of wrong phase identification, with other long-period phases being incorrectly reported in bulletins as surface waves.

3.7.4 *Variation with Time*

For practical purposes our re-evaluation is divided into three broad, overlapping periods of observation, dictated chiefly by the type of instruments available; an early period up to 1918 in which the majority of instruments world-wide were undamped or lightly damped pendulums; a middle period from 1903 to 1971, of predominantly medium-period damped analogue instruments; and a modern period after 1971, with an increasing number of digital seismographs.

For events in the early period 1898 to 1918 one method to calculate an equivalent $M_{s,i}$, when no standard damped seismographs were available, is to use the maximum amplitude from the Milne pendulums, culled from the *Shide Circulars* (1899–1918), using the formula

$$M^* = log(2A) + 1.25log(\Delta) + 4.06 \quad (3.14)$$

where $2A$ is the peak-to-peak trace amplitude in millimetres and Δ the epicentral distance in degrees. This formula was originally derived from 23 earthquakes in Iran for which both M_S and M^* trace amplitudes were available (Ambraseys and Melville, 1982). Additional data from another 95 shallow earthquakes in Eastern Europe, in the Mediterranean region, western Asia and Africa showed that the constant term in equation (3.14) may be magnitude dependent, with its value increasing from 4.0 to 4.4 as the magnitude increases from 6.0 to 7.0. Abe (1981) proposed a similar method, but for earthquakes of M_S less than about 6.7 in the European area and Central America, we find a constant term of 4.04 gives more uniform residuals than his method. For our estimates in Central America we use the value given in the formula above, as the mode of magnitudes used is less than 6.5. Our values of M^* are included in our parametric table. The reliability of this formula is shown by the fact that for events for which we have both M_S and M^* the mean values differ by only 0.02 ± 0.15 units of magnitude, with M^* being greater.

For the middle period, which starts in 1903 with the operation of analogue seismographs at Potsdam, Göttingen, Uppsala, Leipzig and La Paz, we used amplitude and period readings of long-period phases culled from station bulletins. We used the original Prague formula, with modifications by Ambraseys and Free (1997), to adjust M_S estimates from close-in stations for station corrections and distance.

We investigated station corrections for 37 stations operating in the first part of the century, but found they had little effect on average magnitudes except when the number of stations was small; less than five in our case (Ambraseys, 1995). For some stations that reported regularly we could examine the variation of station correction with time. For De Bilt (DBN) in the Netherlands, which gives a continuous source of readings, the correction remains almost constant with a mean value of -0.24. For La Paz (LPZ) in Bolivia, the correction is zero up to 1917, and thereafter remains constant at +0.30. Some stations, particularly those in Mexico and North America, had corrections that varied erratically with time.

In the calculation of M_S for the middle period we used only ground amplitudes as reported in station bulletins. Where trace amplitudes were given, because of the uncertainties in the calibration constants of analogue seismographs, particularly before the late 1940s, no attempt was made to convert these values into ground amplitudes, and they were not used. We made most estimates of M_S from amplitudes recorded on long-period horizontal components of medium-period seismographs, which were the most common during the middle period. When two components were available, the amplitudes were combined vectorially; when only one horizontal component was available, 0.1 was added to the station magnitude. If long-period vertical amplitudes were available, a separate estimate of M_S was made. The difference between determinations from vertical and horizontal components was small, with that from the horizontals being systematically higher by 0.1(±0.1) unit of magnitude. Station magnitudes were obtained by averaging values from vertical and horizontal components. Surprisingly, we found that before the 1960s instruments at near stations were often overloaded, resulting in some larger events having magnitudes determined from fewer stations than smaller ones.

From stations which reported amplitude and period data from more than one seismograph, we preferred readings from standard instruments (e.g. Wiechert, Milne–Shaw, Galitzin). We did not use readings of maximum amplitude if no corresponding period was given, and we discarded readings that were obviously grossly in error, and could not be confirmed by referring to preliminary bulletins, or to seismograms. These rules were generally used in the selection of data for the middle period and the early part of the recent period.

The same procedure was used for the calculation of M_S for the modern period after 1971, but about one third of the amplitude and period readings were extracted from the *Bulletin of ISC* and two thirds from bulletins of seismographic stations that did not contribute readings to the ISC.

3.7.5 *Practical Difficulties*

Some difficulties were encountered in extracting amplitudes from station bulletins. Before the mid-1920s, amplitudes in some station bulletins were often given with no indication of whether they were measured from peak-to-peak or from the base line. For some stations, such as Potsdam (POT), assuming that measurements were from the base line resulted in station magnitudes about 0.3 units high, indicating that this bulletin was, in fact, reporting double amplitudes. In some instances bulletins would change their reporting convention between double and single amplitudes without notice, a change which in some cases could be confirmed only by comparing preliminary with final bulletins or even by referring to seismograms.

Sometimes there were also errors in attributing phases to the correct event when earthquakes occurred close together in time. We find also that the standard errors of the mean M_S for subcrustal and deeper events of the middle period are generally greater than for shallow events. For some earlier events this is the result of the wrong phase identification with other long-period phases being incorrectly reported in bulletins as surface waves.

During the middle and modern periods there is sometimes confusion in the convention for summing amplitudes and periods for horizontal components. Some bulletins publish a single maximum amplitude and period without specifying whether this comes from a single component, or from the geometric sum of the horizontal amplitudes of two components and the arithmetic mean of the corresponding periods.

Between 1945 and the 1970s there was a drastic decrease in the reporting in bulletins of the amplitude and period data needed to assess M_S. Many bulletins stopped reporting maximum amplitudes of long-period phases, and listed only their periods. Other stations stopped reporting amplitudes and periods of maximum phases altogether, listing instead this information for body waves and giving an unspecified station magnitude. These estimates are of little use as it is not known how they were calculated; they are "virus" magnitudes that propagate from publication to publication.

3.7.6 *Summary*

Summarising the rules we applied in our own re-appraisals;

(1) Estimates from the vertical and horizontal components were averaged to give station magnitudes.
(2) From stations which reported amplitude and period data from more than one seismograph, we preferred readings from the most modern standard instrument.
(3) We did not use readings of maximum amplitude if no corresponding period was given.
(4) We discarded readings that were obviously grossly in error, and could not be rectified by referring to preliminary bulletins, or to seismograms.

Chapter 4

Macroseismic Information and Location

Our main sources of macroseismic information for this study are the local and the international press, supplemented by a variety of published and unpublished local and foreign reports such as government documents, consular correspondence, shipping and weather reports and technical papers.

In assessing intensities we use the Medvedev-Sponheuer-Kárník (MSK) scale. In some cases this has been done with a bare minimum of information and some of our intensities must be regarded with caution. Macroseismic information for Central American earthquakes is rather poor and therefore subject to misinterpretation. Our methods of intensity assessment have been given elsewhere, e.g. Ambraseys and Melville (1982), Ambraseys and Adams (1998). In particular, we assign intensity as the mode of individual observed values within a given area, not the maximum.

Later events in our Catalogue are mainly well located instrumentally, with only small earthquakes which attracted attention through being strongly felt having to be located macroseismically, because of lack of instrumental data. This procedure is self regulating, because if an event is small and strongly felt, its epicentre cannot be far from the area of strongest shaking. For early events, the macroseismic data often give more reliable positions than the uncertain instrumental readings then available. For the first 150 events in our Catalogue, up to 1925, about two thirds are located macroseismically.

Shallow events and occasionally aftershocks with well defined macroseismic epicentral regions on land, are assigned a macroseismic epicentre, which is defined as the centre of gravity of the highest isoseismal. For recent large-magnitude, shallow earthquakes, macroseismic epicentres are given only when no reliable instrumental location is available. In these cases, the macroseismic position is defined either as for smaller events or, in the presence of a coseismic surface fault rupture, by the mid-point of the rupture. Macroseismic maps for many events are shown with our Case Histories. In many cases we comment on a difference of a few tens of kilometres between our macroseismic epicentre and an instrumental one, both apparently well determined. This is not unexpected, for the instrumentally determined focus is the point of initiation of rupture, which is not necessarily the centre of energy release. One would expect macroseismic epicentres to be closer to those found by centroid-moment-tensor analysis, as both are related to energy release. The surface expression of a dipping fault can also be displaced from an epicentre above a deep-seated focus, giving rise to an apparent discrepancy.

For the location of subcrustal events on land during the early period up to 1963, we give preference to the macroseismic location, unless estimates from special studies are available. After 1964 adopted locations and depths of subcrustal and intermediate depth events are always instrumental.

Our Case Histories in Appendix A document many cases, particularly for early events, where our macroseismic position differs significantly from an instrumental one. An early example is the earthquake located by us in Costa Rica on 21 June 1900, about 1,300 km southwest of Gutenberg and Richter's (1954) position. Macroseismic information for the earthquake on 28 February 1904 confirms the general area given by BAAS, in the Pacific Ocean south of western Panama, rather than that of Gutenberg, which is about 330 km to the east. In other cases we prefer good macroseismic determinations to our own instrumental locations. For the event near Panama on 28 May 1914 these are about 400 km apart. There are several instances where our macroseismic positions can improve on poorly controlled instrumental solutions given by ISS; for an earthquake in Guatemala on 4 January 1918, the ISS position is 450 km to the south of ours. Our preferred macroseismic position for the earthquake in the same area three weeks later on 25 January is 650 km from the ISS position, and 170 km from our best instrumental relocation.

Macroseismic information can improve on poorly controlled instrumental locations even in more recent periods. An earthquake in Mexico on 3 September 1936 was located by ISS more than 300 km from its preferred macroseismic position, and one in Guatemala on 3 April 1937 was more than 400 km away.

Chapter 5

Description of Seismicity

It is not our intention to undertake a detailed analysis of the seismicity of Central America, but rather to provide information to help in further studies. Nevertheless, we give a brief discussion of the main features of the seismicity as shown by our work.

Figure 5.1 plots all 1,829 events given locations in our catalogue in Appendix B. This shows the general concentration of earthquakes along the Pacific coast of Central America, with some activity also extending into the Caribbean Sea north of Honduras along the boundary of the Caribbean Plate, and to the east of Nicaragua. This and later plots extend for one degree in each direction beyond our main study area of 7°N to 17°N, and 93°W to 80°W, so any apparent fall-off of activity towards the edge of our plots is not necessarily real. In this plot, and in those following, positions have been subjected to random perturbation, related to the precision of their location, to prevent spurious lineations appearing along lines of latitude and longitude.

In order to examine the seismicity more precisely, the variation in detection capability and accuracy with time has to be allowed for. We first examined the variation of detectability with time. Figure 5.2 shows the number of events by magnitude range per decade for earthquakes given depths of 60 km or less. This suggests that from the 1920s the list is complete for M_S greater than 5, from the 1910s for M_S greater than 6, and that for the first decade of the century not all earthquakes of M_S 6 have been detected.

In Figure 5.3 we show the allocation of depth with time for earthquakes with M_S greater than 5. Up to the 1950s most events were assigned *normal* depth as a default, which is here taken as 33 km. In the first two decades, no events were allocated depths greater than 50 km, but from 1920 onwards some were recognised as of intermediate depth, down to a maximum of about 220 km. From 1964 most depths in our catalogue are those of Engdahl *et al.* (1998), which are given as the best determination in kilometres, with no default values. From this time on the plot shows a very broad distribution of depth in the upper 100 km, with no fall off in numbers at depths less than 80 km. In view of this, we arbitrarily take a

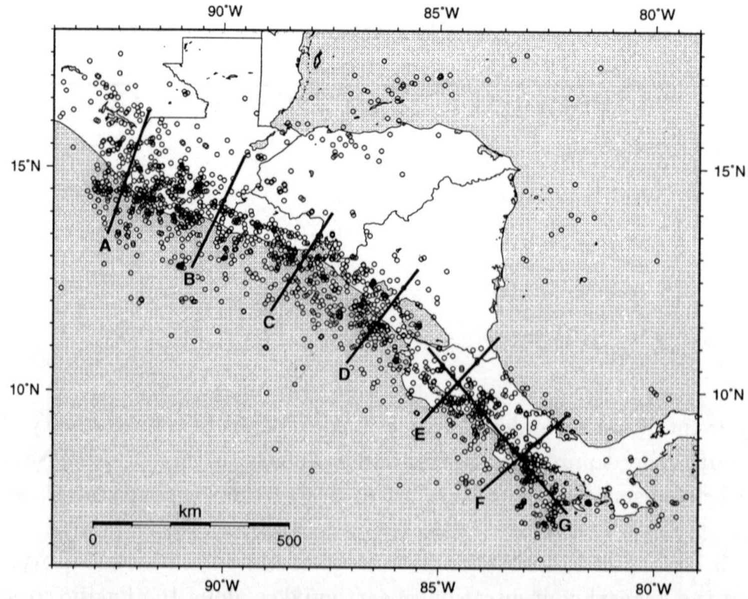

Fig. 5.1 All events regardless of magnitude or depth that are given locations in our catalogue (1829 events). Also shown are lines along which later cross-sections are plotted. In this and other maps, positions are plotted with small random displacements related to the precision of their location in order to avoid artificial lineations along lines of latitude and longitude.

depth of 80 km as the boundary between normal and intermediate earthquakes in our discussion. The slight peak near 100 km for the 1950s probably arises from the default depth 0.01 earth radius frequently used by ISS.

Figure 5.4 plots all normal depth earthquakes, 1,529 events, giving a very similar distribution to that shown in Figure 5.1. The 222 normal earthquakes with M_S greater than 6 are shown in Figure 5.5; these larger events are in general likely to be better located than smaller ones. The best located earthquakes, however, are likely to be those from 1964, and Figure 5.6 shows normal depth earthquakes for that period. Comparing Figures 5.4 and 5.6 shows that the later earthquakes form a better-defined zone of activity, with less scatter extending into the Pacific Ocean. The zone of normal depth activity is generally about 200 km wide, but extends further inland through Guatemala towards the boundary of the Caribbean Plate, and also to the Caribbean coast of Costa Rica and Panama.

All 300 intermediate depth earthquakes (depths greater than 80 km) are plotted in Figure 5.7, and those of m_b greater than 6 in Figure 5.8. The more precisely located events, since 1964, are shown in Figure 5.9. These deeper earthquakes are in general displaced shoreward compared with the shallower events in Figure 5.6, with only a few located offshore. Only a few intermediate events occur further south east than central Costa Rica. The relationship between shallow and deep events is best shown by the six sections, **A** to **F**, across the line of activity at the positions

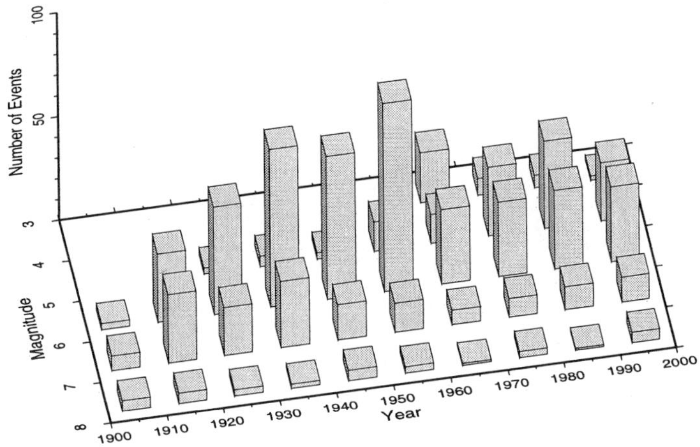

Fig. 5.2 Distribution by magnitude and decade of earthquakes of depth 60 km or less. No allowance is made for the fact that the last time interval plots events for five years only.

shown in Figure 5.1. These sections are given in Figures 5.10, 5.11 and 5.12. There appears to be no clear break in seismicity with depth, and most events occur at depths less than 100 km. Clearly defined zones of dipping activity are apparent in sections **A** to **D**, from Guatemala to central Nicaragua, with earthquakes extending to depths of 200 km or more. Section **E** in northwestern Costa Rica shows two earthquakes at depths near 200 km as the only ones deeper than 100 km, while activity across the section near the Costa Rica-Panama border (section **F**) extends only to depths of about 60 km. A roughly northwest-southeast section along the line of activity through Costa Rica shows that the intermediate activity is greatly reduced southeast of central Costa Rica (section **G**, Figure 5.13).

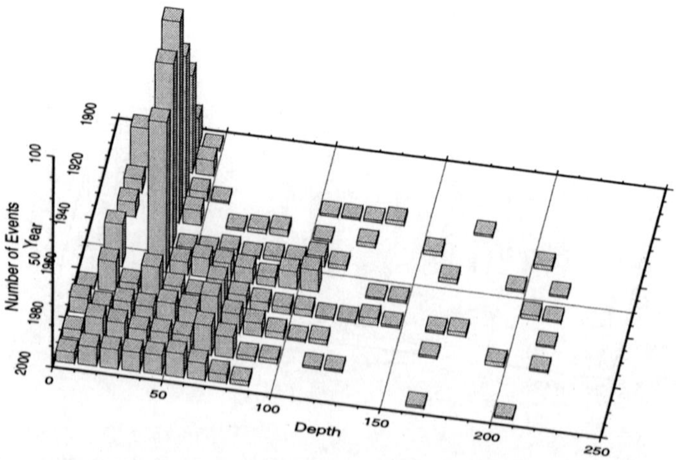

Fig. 5.3 Distribution of earthquakes of $Ms \geq 5$ by depth, plotted by decade. No allowance is made for the fact that the last time interval plots events for five years only.

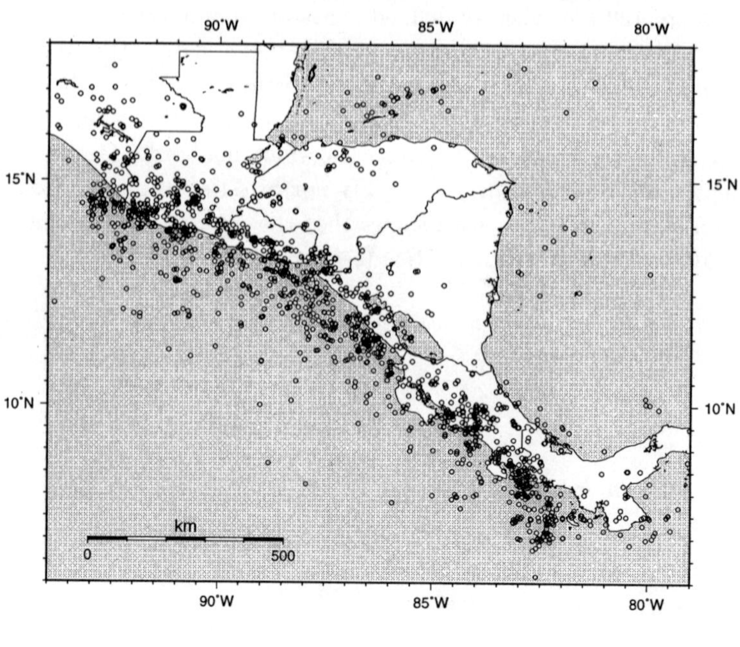

Fig. 5.4 All earthquakes allocated normal depths (s, n, $n+$ or $h \leq 80$ km). (1529 events)

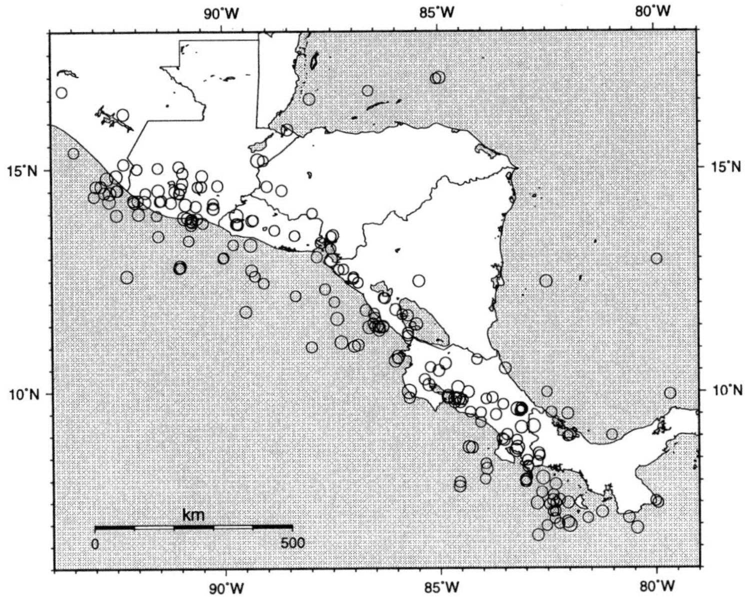

Fig. 5.5 All normal depth earthquakes with $M_S \geq 6$. (222 events)

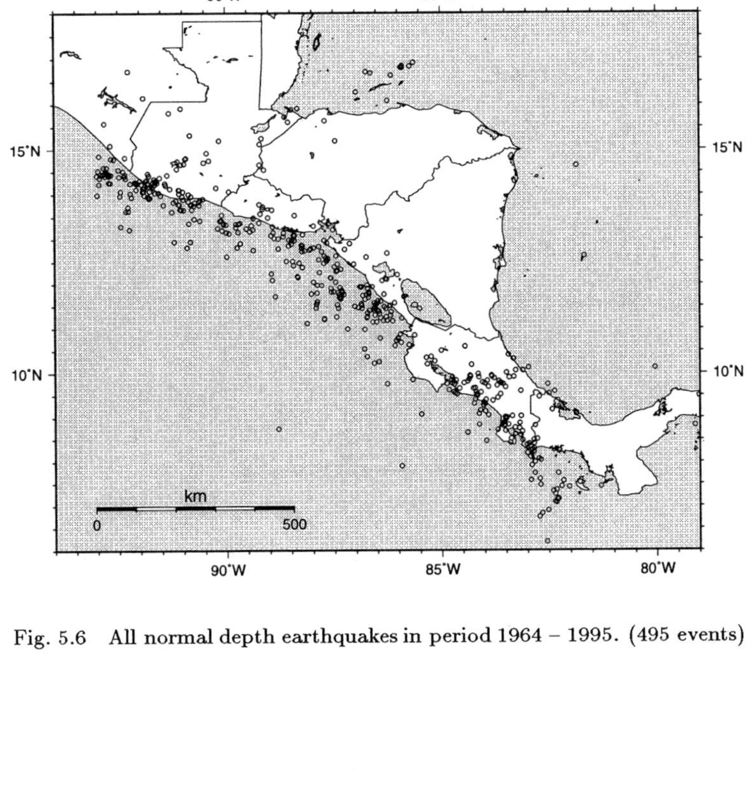

Fig. 5.6 All normal depth earthquakes in period 1964 – 1995. (495 events)

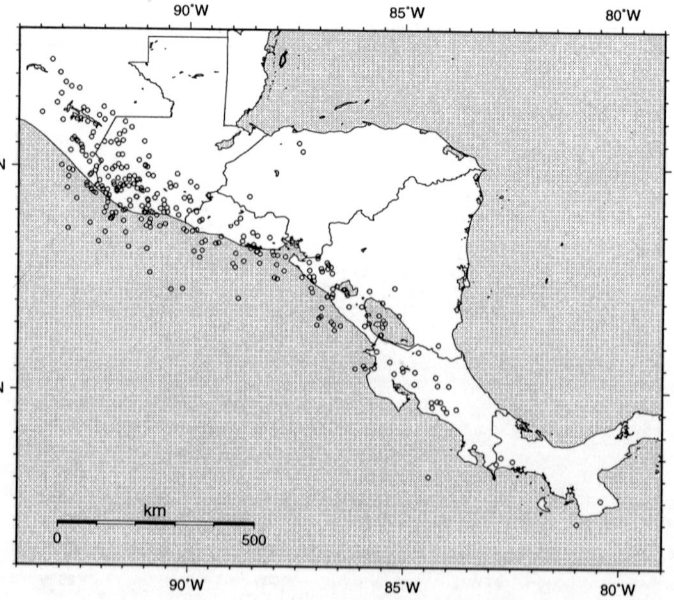

Fig. 5.7 All earthquakes allocated intermediate depths (i or $h > 80$ km). (300 events)

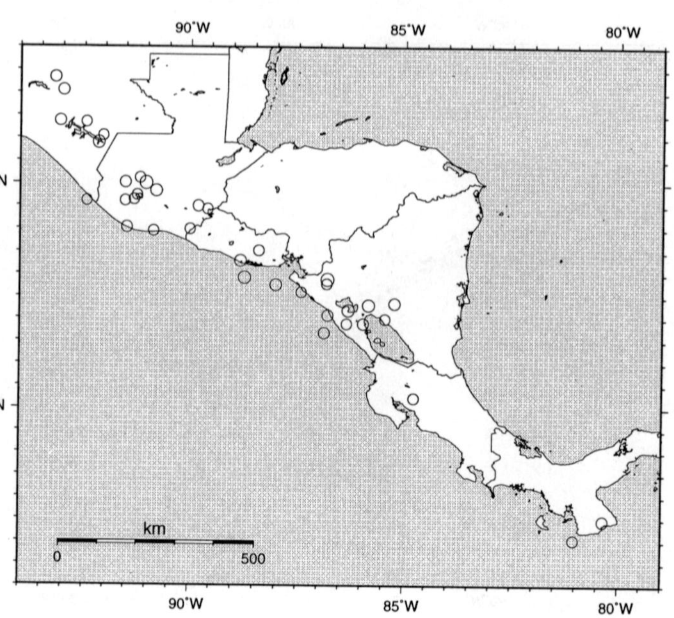

Fig. 5.8 All intermediate depth earthquakes with $m_b \geq 6$. (37 events)

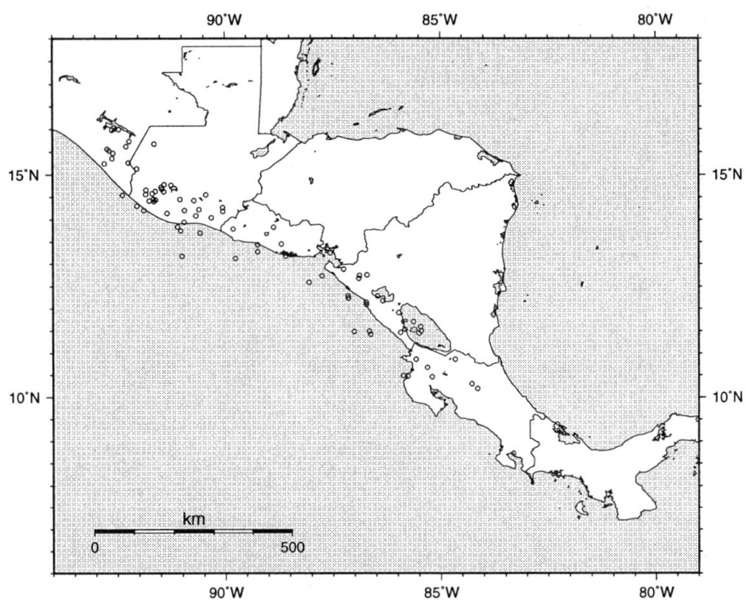

Fig. 5.9 All intermediate depth earthquakes in period 1964 – 1995. (82 events)

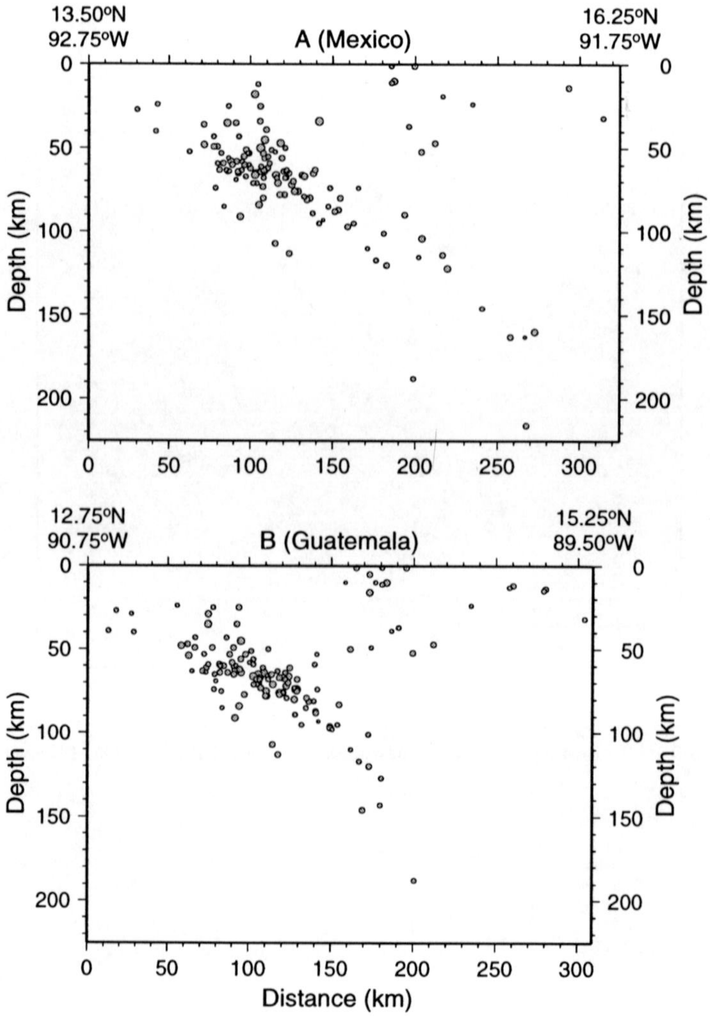

Fig. 5.10 Depths of earthquakes in period 1964 – 1995 along cross sections **A** (Mexico) and **B** (Guatemala) shown in Figure 5.1.

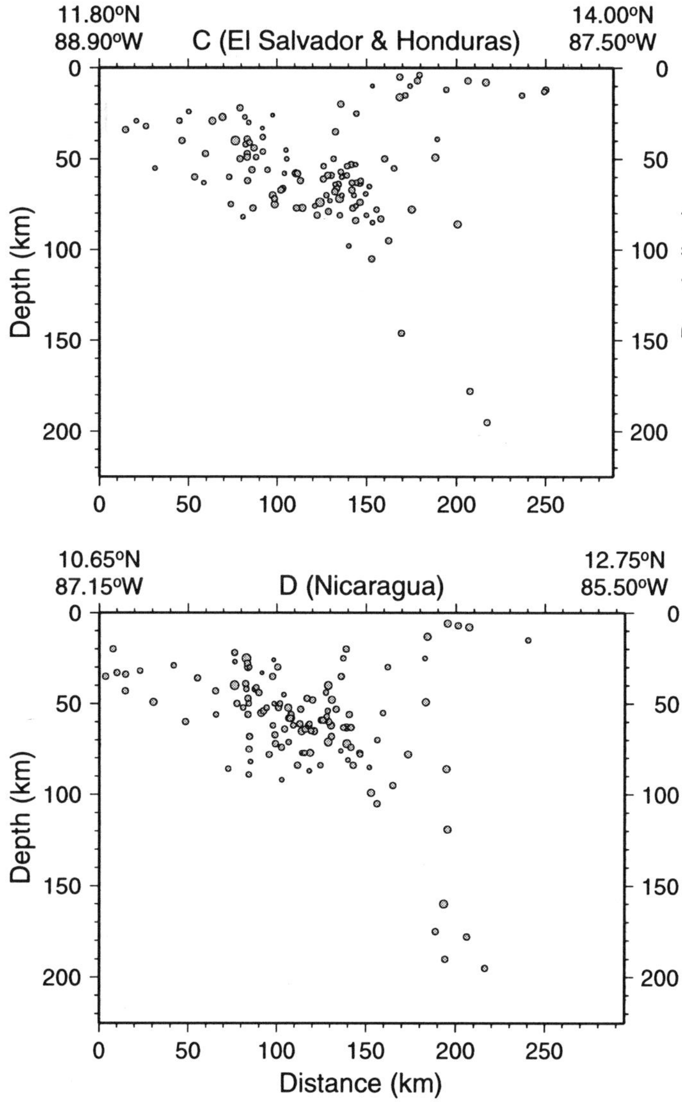

Fig. 5.11 Depths of earthquakes in period 1964 – 1995 along cross sections **C** (El Salvador and Honduras) and **D** (Nicaragua) shown in Figure 5.1.

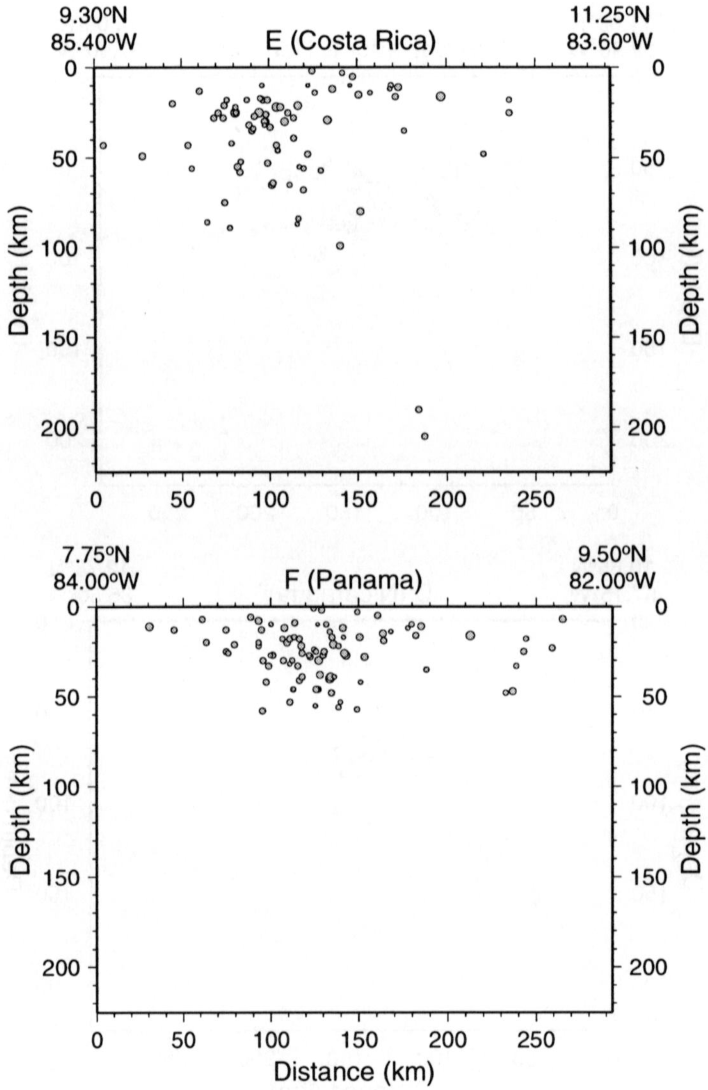

Fig. 5.12 Depths of earthquakes in period 1964 – 1995 along cross sections **E** (Costa Rica) and **F** (Panama) shown in Figure 5.1.

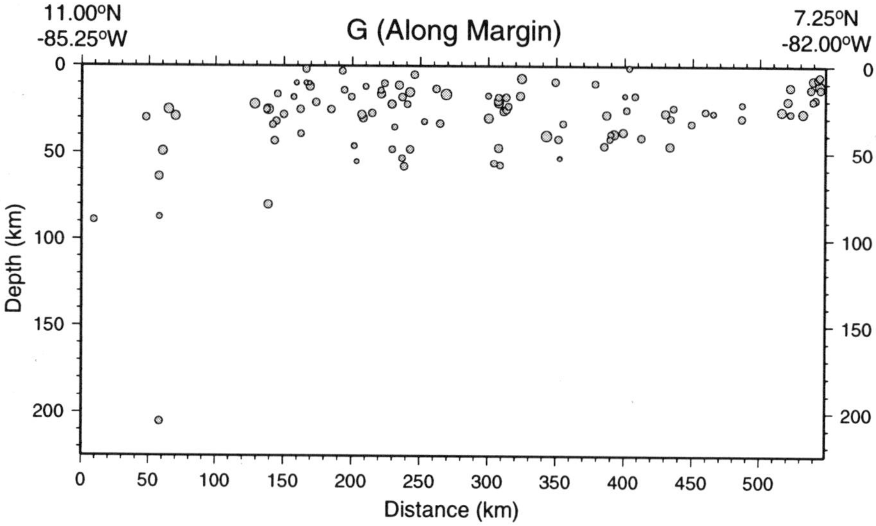

Fig. 5.13 Depths of earthquakes in period 1964 – 1995 along the arc in Costa Rica and Panama. Section **G** in Figure 5.1.

Chapter 6

Conclusions

We have produced a catalogue of more than 1,800 earthquakes in the Central American region, by combining wide ranging information and a variety of techniques. The early events, from the beginning of the 20th Century, are located largely by assessing their reported felt effects, but from the beginning of instrumental earthquake recording there is a gradual incorporation of this information. Because of early instrumental difficulties, particularly with timing and phase identification, at first instrumental readings could only confirm what was already known from macroseismic reports.

The earliest instrumental readings we used were not good enough to be used with modern computer-based techniques, but leant themselves to graphical solutions. This technique had the advantage of showing visually the degree of discrepancy among the readings, and could often locate events within a certain area without establishing an accurate position.

Gradually, as the precision of recording improved, the instrumental information became more reliable, until recent events can be located with an accuracy of a few kilometres. We have made no formal attempt to give the precision of our locations, but this is reflected in the number of significant figures given in the table in Appendix B. Similarly, the determination of focal depth remains difficult, but becomes better established with time. This is the poorest established of the focal parameters.

We have taken particular care to try to determine the magnitudes of listed events in a uniform manner, and especially to derive surface-wave magnitude M_S in a consistent way, resorting to original information in station bulletins whenever this is available. We have not attempted to convert one type of magnitude to another, nor to compare them directly, as we believe that differences in magnitudes determined from phases that arise from different parts of the radiation spectrum reflect different spectral characteristics of the earthquake, and that to seek relationships among them is not meaningful.

We believe that our catalogue is the most uniform and complete yet produced for this area, and trust that it will be used as a basis for further study, including problems of tectonics and seismic hazard.

Appendix A
Case Histories of Earthquakes

A.1 Explanation

Descriptive notes are given for all events which we consider to be of particular interest.

We aim to discuss all earthquakes with recalculated magnitudes of 5.6 or more, but also discuss smaller events that are noteworthy because of their individual destructiveness, or because they give a general illustration of the effects of earthquakes in the study area. We have also included earthquakes which have been assigned magnitude greater than 5.6 in other works, but which on re-examination prove to be smaller.

Our accounts are based on primary information retrieved from the various sources quoted. These accounts contain the essential data available and an assessment of this material in the light of the relevant factors, thus illustrating some of the macroseismic and instrumental problems associated with evaluating the seismicity of the region under study. Earlier events tend to be discussed in more detail, reflecting the greater uncertainty of their location.

Case histories of some events are very brief, either for lack of details, or in cases because the sources quoted provide all the necessary information without complication or need for further comment. In most cases, discussion of the earthquakes is as full as possible, particularly when there is a wealth of data or when it is necessary to demonstrate characteristic problems of handling the source material. Our accounts of these events are written with seismologists, earth scientists, engineers and researchers in related fields in mind.

For the later period, from the mid 1950s, events are generally better reported elsewhere, and pruning has perforce been more ruthless. Omissions are either large, well-documented events for which a complete list of references is given, or less well-known events, the accounts of which perhaps do not give rise to any particular points of interest. Details of these omitted earthquakes can be found in other publications. We have not ourselves studied the effects of earthquakes after 1993, but for the sake of completeness have extracted some information reported from NEIC and other agencies quoted in the *ISC Bulletins*.

Each entry starts with a line giving the date (and time if there is more than one event on a particular day), a magnitude if one is available, mention of associated tsunami activity, a reference to any figures describing the event, and the country where the event occurs, or is close to if it is offshore. Our general references to macroseismic information are given in square brackets following each event, while references to particular points in the discussion are given in the text in the normal way. In our descriptions we generally use the term *felt* to indicate an intensity of *III* or more. *Perceptible* implies that there were isolated reports of intensity less than *III*.

We include isoseismal maps if there is enough of interest to warrant this. In these maps filled circles indicate an earthquake's epicentre, an open circle a locality where the earthquake was reported felt, with MSK intensity if available, and a dot an unspecified place where the earthquake was felt.

A.2 Individual Case Histories

1898 Apr 29 M_S 6.8 Figure A.1 Nicaragua

This strong earthquake caused concentrated damage in northwestern Nicaragua, where many houses were ruined, but with no loss of life.

In Chinandega 60 adobe houses in the centre of the town were totally destroyed or damaged beyond repair. The church of Guadelupe was shattered and the church of San Antonio was ruined. Timber framed houses, however, suffered only some damage. Near the railway station the shock caused ground cracks.

More ground fractures were reported from near El Viejo and from the vicinity of the railway station of Amaya. Both settlements were almost totally destroyed.

Chichigalpa and the nearby sugar estates of San Antonio were also damaged but details are lacking. An old masonry dam and a narrow and comparatively high arched aqueduct did not sustain any serious damage.

In this area, with the exception of the foundations of the bridge over Rio Chiquito near Leon, which were cracked, the earthquake did no damage whatever to the railroad bridges of iron or masonry, nor to well-built structures, in spite of the fact that ground motions were severe. At Siman, a locality not identified, large trees were uprooted and in other places rockfalls blocked the railway line.

In Leon one house was totally destroyed and about 340, mainly one-storey adobe houses, were damaged beyond repair, but none of them fell. The cathedral suffered no serious damage but the small church of San Sebastian partly collapsed. No solidly built wooden or masonry buildings suffered greater damage than the formation of cracks over the doors and between windows.

In the nearby village of Telica the church, townhall and a few houses were damaged.

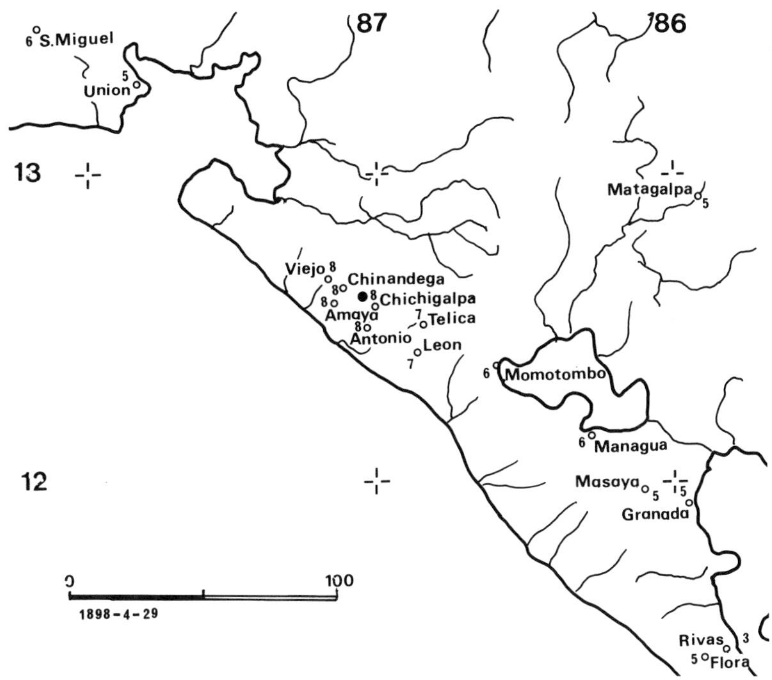

Fig. A.1 Felt effects of earthquake of 29 April 1898. In this and subsequent figures filled circles indicate an earthquake's epicentre, an open circle a locality where the earthquake was reported felt, with MSK intensity if available, and a dot an unspecified place where the earthquake was felt.

In Momotombo timber houses suffered little or no damage.

In Managua the earthquake caused sporadic damage. No house collapsed or was damaged beyond repair. The parish church, the hospital, prison, the hotels Italia and Gran, the chain stores Washington, and the American consulate suffered different degrees of reparable damage.

In the Gulf of Fonseca at San Miguel and La Unión in El Salvador, as well as at certain localities in southwest Honduras, which are not specified, the earthquake caused some slight damage, but details are lacking.

The earthquake was rather strong at Matagalpa and La Flora, was widely felt at Masaya, Granada and Rivas, and was perceptible in places further east, along the route of the proposed Nicaragua Canal, as far as Greytown (San Juan del Norte) and San José in Costa Rica.

Some contemporary writers grossly exaggerate the effects of this event while others, who were associated with the feasibility study of the Nicaragua Canal, tend to play down the damage caused. Material losses in Nicaragua exceeded three million pesos.

Aftershocks continued until 12 May.

The earthquake is well located macroseismically, (Figure A.1).

It was recorded by a Ewing pendulum at San José de Costa Rica, by undamped seismographs in Canada and Europe, and by four Milne instruments (BAAS 1899). From these Gutenberg (1958) gives M_S 7.5, and a focus somewhat deeper than normal at 12.0°N, 86.0°W, which is 150 km away from its macroseismic position. We can find no macroseismic or instrumental evidence that the earthquake was deeper than normal, or that its magnitude was 7.5, let alone 7.9 as given by Duda (1965). The distribution of damage and felt effects, as well as the long aftershock sequence suggest a crustal event of about M_S 6.8.

[Crawford 1898,1902a, Feldman 1993; Hayes 1899; Jones 1903; Jorgensen 1933; PHRL; PPCF; Sapper 1899, 1902b; Severit 1933; Sieberg 1932; Viquez 1910; Walker 1899]

1899 Mar 25 14h M_S 6.2 Mexico

This was a damaging earthquake in Chiapas in Mexico.

In Tuxtla Guetierrez, San Bartolome and Zinacantlan many private houses and public buildings were damaged, but there was no loss of life.

The shock, the time of which is not given in our sources of macroseismic information, lasted 50 seconds and was followed by a long series of aftershocks.

This is probably the earthquake recorded by a number of Canadian and European seismograph stations, but not by the Ewing pendulum at San José. BAAS (1911.58) assigns this event an origin time 14h 27m (UT), and mislocates it in the Pacific Ocean off the shore of Costa Rica.

[Feldman 1993]

1899 Mar 25 17h M_S <5.3 El Salvador

This earthquake in El Salvador was preceded by many small foreshocks. It did some damage at San Vicente, particularly to the parish church and to a number of houses, and caused injuries to a few people. Damage extended to the nearby villages of Tepetitan, Santa Caterina, Iztepeque, Apastepeque and Guadelupe, which are all located within a radius of 5 km.

The earthquake was felt at Zacatecoluca, but there is no evidence that it was perceptible at San Salvador, about 50 km from the epicentre.

Press reports time the earthquake either at 04h 09m (22h 13m, 24 March UT) or at 23h 00m (17h 04m, 25 March UT), but we can find no instrumental data or records from the Ewing pendulum at San José at these times. The only event recorded on 25 March was at about 20h (UT) and if it occurred in Central America its surface wave magnitude would have been less than 5.4.

[BAAS 1899; Feldman 1993; Martinez 1978; Milne 1911b; ODC: Viquez 1910]

1899 Jun 10 Honduras

An earthquake in Honduras, which lasted four seconds, caused considerable panic and some slight damage at Tegucigalpa, as well in other places, which are not mentioned by name. The cathedral at Tegucigalpa was seriously damaged.

[BAAS 1911.21; Feldman 1993, ODC]

1900 Jun 21 | M_S 7.2 | Figure A.2 | Costa Rica

This was a major earthquake which was strongly felt in northwestern Costa Rica where strong ground movements lasted intermittently for more than one minute.

The earthquake, which was preceded by many small shocks, caused widespread damage in the district of Guanacaste. At Filadelfia the church was seriously damaged and the roofs of many houses, mostly of timber construction, fell in. Minor damage extended to Nicoya, La Cruz and to Puntarenas, where material losses were considerable. The shock damaged the principal church at Rivas and there is some evidence for damage at Santa Rita.

The earthquake was very strong at Sarchi, Orotina, and less so at Alajuela, Heredía, San José, and Cartago. It was felt all along the Pacific coast of Costa Rica, in southern Nicaragua and probably at Limón.

Aftershocks continued to be felt intermittently for almost two months.

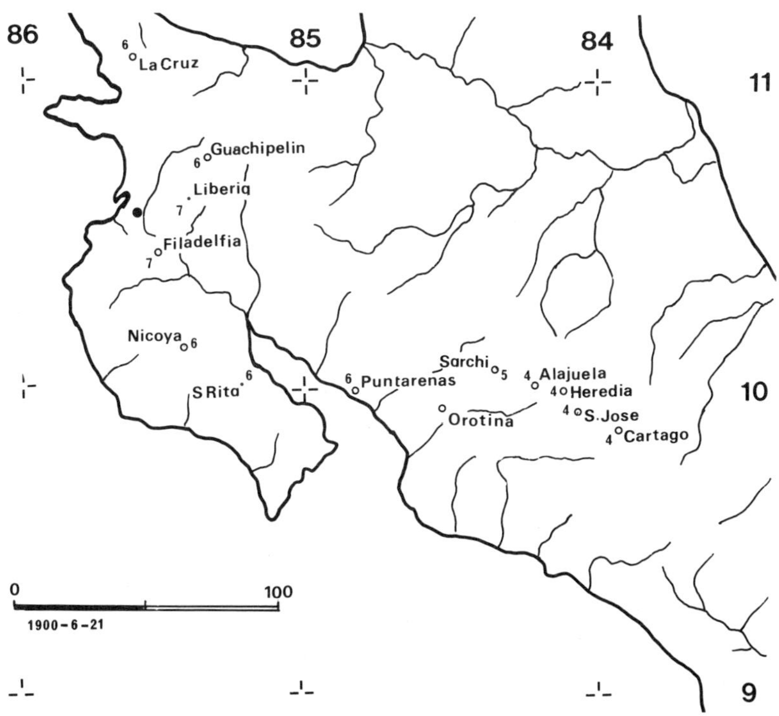

Fig. A.2 Felt effects of earthquake of 21 June 1900.

This event is fairly well located macroseismically (Figure A.2). It was well recorded by the primitive world network but Milne readings are very confused and cannot be used to locate the earthquake, other than to confirm that it is in the Central American region and that it occurred about 20h 58m UT.

The macroseismic location of this event is about 1,300 km southwest of

Gutenberg's (1956, 1958) estimate at 20.0°N, 80.0°W, which is between Jamaica and Cuba, and from where there is no corroborating evidence that the shock was felt. Because of its mislocation this event is not included in existing lists of Central American earthquakes.

The macroseismic information and the long aftershock sequence suggest an event in the lower crust of magnitude M_S 7.2, well below M_S 7.9 given by Duda (1965).

[Jorgensen 1933; Miyamura 1980; PPLB; PRPB; Tristan 1912c; Viquez 1910]

1900 Nov 9 $\mathbf{M_S}$ 6.9 El Salvador

No macroseismic information has been found for this strong earthquake which is located off the coast of El Salvador. It was recorded instrumentally in Canada and Europe but not by the Ewing pendulum in San José. Milne readings are not good, but phases can be found consistent with the position allocated by Gutenberg (1958), which we adopt, but the error could be several hundred kilometres. It is possible that this event is outside our study area.

Gutenberg (1958) gives a very rough instrumental position for this event in the Pacific off the coast of El Salvador and an origin time about 16h 10m, which is about two hours before a magnitude 6.6 earthquake in Japan, the location of which is corroborated by macroseismic information. Some of the amplitude readings of long waves reported for this earthquake by Gutenberg belong to the second event in Japan. The readings from Batavia given in the *Shide Circulars* are also very late and probably refer to a more local event. These readings have been omitted from our magnitude calculations.

[BAAS 1911; Gutenberg 1958; SHC; Utsu 1982; Viquez 1910]

1901 Oct 8 $\mathbf{M_S}$ 7.0 Tsunami Nicaragua

We have little information about this earthquake, which had an epicentre in the Pacific Ocean off the coast of Nicaragua. It caused some small damage at a number of unspecified places along the west coast of Nicaragua.

At San Juan del Sur a seismic sea-wave, about 1.5m above tide level, washed away loose lumber and other articles, but the earthquake did no other damage.

The shock was strongly felt in Managua as well as in the southwest part of Nicaragua and was reported from as far south as San José in Costa Rica.

The rather poor macroseismic information places this earthquake in the Pacific off the coast of southern Nicaragua. Gutenberg (1958) locates this event east of the Bay of Fonseca on the border of Honduras and Nicaragua at 13.0°N, 87.0°W, but there is no information that the shock was felt there. The earthquake, which was recorded by a Ewing pendulum at San José in Costa Rica, was quite well recorded by Milne stations, and its position, west of Rivas off the Pacific coast of Nicaragua, is consistent with readings of a variety of phases. The readings alone are not good enough, however, to be able to choose between the two alternative positions without the help of macroseismic information.

[Crawford 1902b; Jones 1903; STR; Viquez 1910]

1902 Jan 18 Ms 6.3 Figure A.3 Guatemala

A severe earthquake, probably a forerunner of the major earthquake of 19 April 1902, was felt over much of the southwestern part of Guatemala.

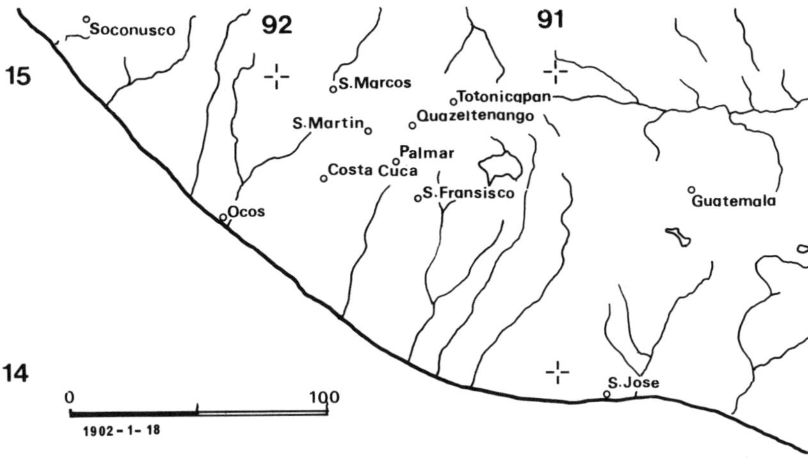

Fig. A.3 Felt effects of earthquake of 18 January 1902.

Much of San Fransisco Zapotitlan was shaken down, the shock also destroying buildings and walls in several large plantations, including those of Palmar and others near San Martin.

At Quezaltenango, Totonicapan, San Marcos and in the plantations of Costa Cuca almost all the houses suffered some damage and a few were ruined.

The shock was strongly felt at Guatemala City and along the Pacific coast from Soconusco in Mexico to San José, where it caused some damage due to the liquefaction of beach sand deposits. This was particularly noticeable in the deltas of the Suchiate and Naranjo rivers, where at Ocós, foundations of structures sunk 60cm into the ground and three ridges, sloping gently towards the sea, but steeper towards the land, were formed in the sand, running for a distance of about 1.6 km parallel to the coast-line. Similar ground deformations were observed on the Mexican coast near San Benito.

It is not known how far the earthquake was felt.

[Anderson 1908; Rockstroh 1903; Sapper 1902a; Sieberg 1932]

1902 Feb 27 Tsunami Figure A.4 El Salvador

On 26 February at 19h local time (27 February 00h 36m UT) a tidal wave flooded the coast of Ahuachapan in El Salvador from Garita Palmera to Barra del Paz and beyond, a distance of about 120 km. The height of the wave is not mentioned, but about 100 persons were drowned at San Diego and 85 at Barra del Paz. There was some property damage on the coast of Acajutla
indexAcajutla and Cara Sucia.

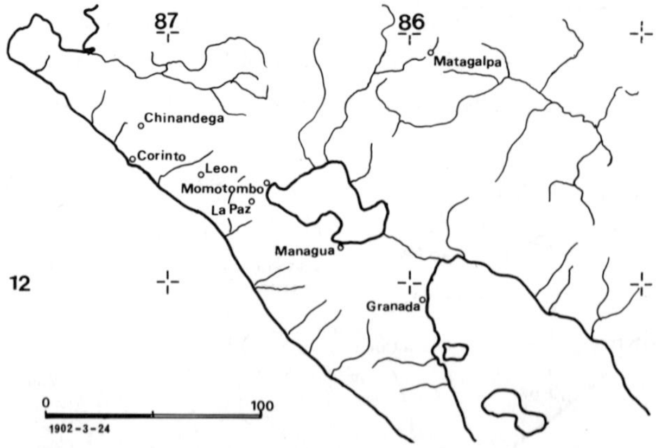

Fig. A.4 Felt effects of earthquake of 27 February 1902.

Modern writers associate this event with a spurious earthquake on 26 February 1902 (Cruz and Wyss 1983) or with the earthquake of 19 April 1903.

There is, however, no evidence that this tidal wave was triggered by an earthquake.

[Anderson 1908; Cruz and Wyss 1983; Jordan and Martinez 1980; List 1903; Sieberg 1932; Schultz 1963]

1902 Mar 24 M_S 6.6 Tsunami Figure A.5 Nicaragua

This earthquake in western Nicaragua was reported as lasting 17 seconds.

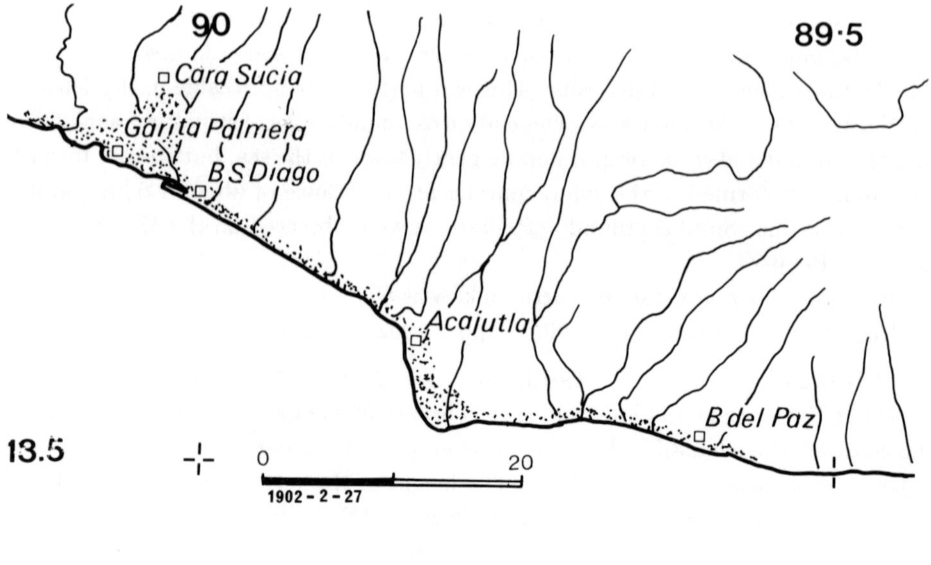

Fig. A.5 Felt effects of earthquake of 24 March 1902.

The damage caused by the main shock, and by a strong aftershock a day later, was mainly to houses of local type at Puerto Momotombo, where several houses were destroyed and many badly damaged. A part of the old rotten dock, condemned long before as unsafe, collapsed with a locomotive and two loaded cars.

At La Paz, Leon, Nagarote, and Managua many adobe houses were badly cracked, but no damage was done to solid masonry and timber buildings.

The shock was strongly felt at Corinto, Chinandega, Matagalpa and Granada.

It was felt in northeastern Nicaragua beyond Matagalpa as far as the Caribbean coast, and also in northern Costa Rica.

Following the earthquake, high, short waves were set up in Lake Managua. These continued to break against its southwestern shore for some time, but caused no damage.

Instrumental reading confirm the longitude of the epicentre, but north-south control is very poor.

[Crawford 1902a; Jones 1903]

1902 Mar 25 M_S 6.3 Nicaragua

This was an aftershock of the earthquake in Nicaragua nine hours earlier. It was felt over the same area with comparable intensity and was particularly strong in the region between Momotombo, Chinandega and Corinto.

The shock was recorded at a number of seismograph stations, but the data are not enough to establish the epicentral region with certainty, so we have adopted a position close to that of the main shock.

[Crawford 1902a; Jones 1903]

1902 Apr 19 M_S 7.5 Figure A.6 Guatemala

This major earthquake in Guatemala extended the destruction caused by the shock of 18 January (M_S 6.3) and affected a much larger area.

The towns of Acatenango, Nahuala, Quezaltenango, San Antonio Suchitepequez, San Carlos Sija, Santa Caterina, San Juan de la Laguna, San Juan del Ostuncalco, San Marcos, San Mateo, San Pedro, Sacatepeque, San Vicente, Pacaya, Tejutla, and Tuncanche were almost totally destroyed. The shock was accompanied by fires, which added to the destruction.

The total loss of life in the epicentral area is not known, but it was large; at Quezaltenango 129 people lost their lives, about 500 died in the Quezaltengo District and 49 at Pedro Sacatepeque. It is probable that at least 1,500 people were killed in Guatemala.

The shock lasted about 40 seconds and caused extensive damage in the region between Tacapa, Mazatenango, Patzizia and Santa Cruz Quiche where almost all masonry houses were ruined, while in many places native timber dwellings suffered scarcely at all. Sporadic damage extended to Guatemala City and further west in western El Salvador as well as to the north as far as Nenton. Along the Pacific coast the earthquake caused much damage, but not as great destruction as on the high lands.

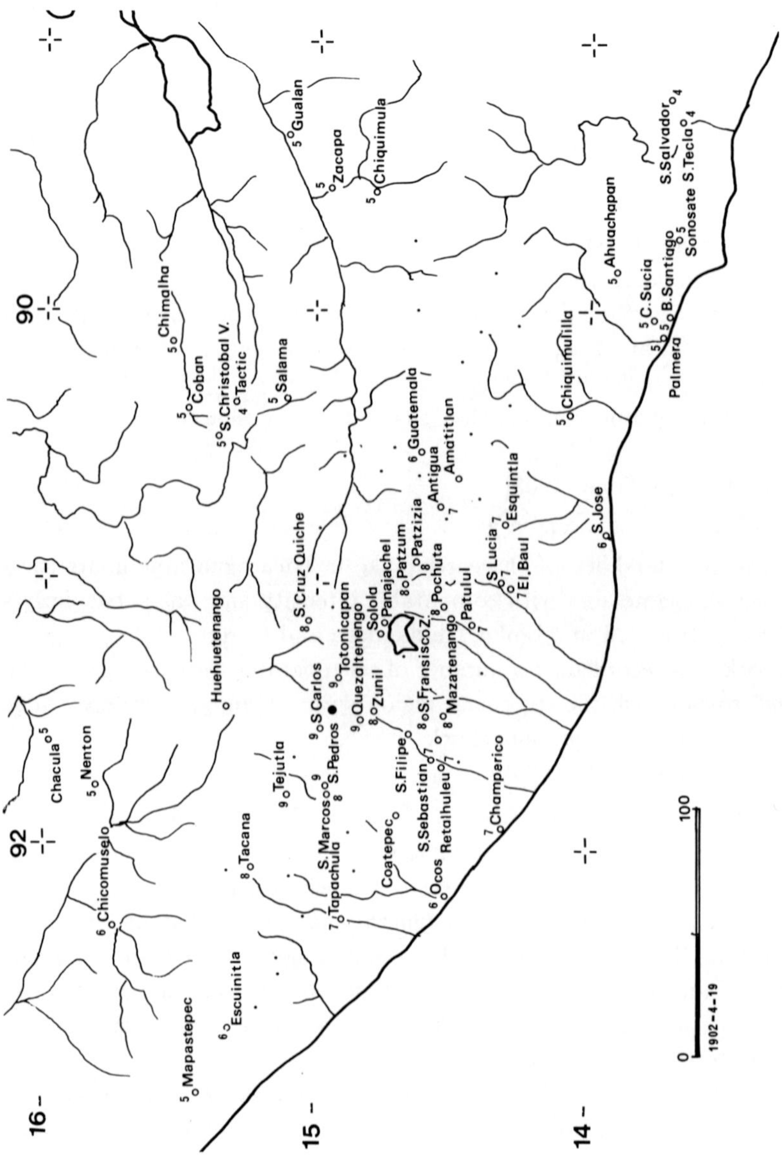

Fig. A.6 Felt effects of earthquake of 19 April 1902.

Landslides and ground fractures of unknown origin were reported from along the axis of the epicentral zone. Enormous landslips dammed the rivers Naranjo and Ixtacapa, and destroyed hundreds of thousands of coffee trees. The railway lines between Retalhuleu and the port of Champerico, and between Ocós and Coatepec were interrupted by the falling of bridges and damage to the road. The railway between Guatemala City and the port of San José was not affected.

At the port of Ocós large areas of saturated beach sands liquefied again, and the settlement of the beach extended further inland than on 18 January, causing damage to the railway line. At Ocós only three houses remained standing after the second shock and the harbour facilities were totally destroyed.

The earthquake set up standing waves in the lake of San Cristobal in Alta Verapaz and caused water to slosh from a reservoir at Chilalha, 170 km from the epicentral area.

The shock was widely felt throughout Guatemala and all along the coastal area of Central America from Chiapas in Mexico to Comayagua in Honduras (380 km), in El Salvador and as far southeast as Managua in Nicaragua (580 km), but not in Costa Rica. In the north the shock was reported felt from Mexico City, Livingston, Puerto Barrios and further east from the north coast of British and Spanish Honduras.

The earthquake was followed by a large number of relatively small aftershocks most of them reported from the Costa Cuca area of Coatepec.

There is no evidence that the earthquake was associated with a seismic sea-wave and modern writers wrongly associate this earthquake with the tidal wave of 27 February in El Salvador.

A special study of the earthquake using instrumental data was undertaken shortly after the event in connection with an insurance policy. Blocks of buildings in Guatemala City and elsewhere were set on fire by the overturning of lamps, and burnt down at this time. Some of these buildings were insured against fire, but not, as stipulated in the policy, against fire caused by earthquake, and it became a matter of importance to decide whether the lamps were upset by the earthquake, as claimed by the insurance company, or before the earthquake, as claimed by the owners. If it could be shown that the fires at Quezaltenengo, Amatitlan and Guatemala City broke out before the earthquake certain fire insurance companies were liable. The result of these investigations was that 14h 21m (UT) was established as the time of the earthquake, one of the earliest practical applications of observational seismology.

This earthquake is very well located macroseismically (Figure A.6).

Instrumental readings of maximum phases on Milne seismographs and body phases from European stations confirm the position of the event, provided it is recognised that readings at stations at epicentral distances beyond about 80° belong to two other, separate, events which occurred about the time of the Guatemala earthquake; one in the Indian Ocean at 14h 34m and the other in China at 14h 36m. Gutenberg (1958) uses Milne amplitudes from all three earthquakes to calculate M, the value of which he overestimates.

Macroseismic effects suggest a shallow focus.

[Anderson 1908; Carden 1911; Eisen 1903; Figueroa 1973; Jordan and Martinez 1908; List 1903; Martinez 1976; Milne 1911; Nishenko and Singh 1987; ODC; Oldham 1905; PDCA; Rockstroh 1902, 1903; Sapper 1902a; Sieberg 1932; STR; Viquez 1910]

1902 Sep 23 **M_S 7.6** **Mexico**

This was a major earthquake in the district of Chiapas in Mexico. It caused extensive damage in the region between Tuxla and Comitan and sporadic damage over a large area, which extended into Guatemata and Honduras from where there is little or no information. The earthquake was felt in south Mexico, throughout Guatemala including the district of Peten, in British and Spanish Honduras and as far as San Salvador. It was perceptible in Mexico City.

Macroseismic information for this earthquake is good, and instrumental readings from stations world-wide agree well with its macroseismic position in Chiapas in Mexico, provided it is recognised that some of the readings from distant stations belong to an earthquake in the south of the Indian Ocean, which occurred 13 minutes later at 20h 29m. There is no evidence that this was a subcrustal event. The magnitude of 8.2 to 8.4 assigned to this event by some authors is grossly overestimated.

[Anderson 1908; ERD; Rockstroh 1903; Figueroa 1973; Milne 1911a; Nishenko and Singh 1987; STR]

1902 Oct 24 **M_S 5.2** **Guatemala**

This strong earthquake in Guatemala occurred just before the eruption of the volcano of Santa Maria. The shock was strongly felt in the region of Quezaltenango, and was followed by many aftershocks.

[Anderson 1908; ERD; STR]

1903 Jan 14 **M_S 7.8** **Mexico**

This major earthquake had an epicentre off the south coast of Mexico somewhere in the Gulf of Tehuantepec.

The earthquake caused no damage but it was strong along the coast of the Gulf of Tehuantepec, between Tapachula and Salina Cruz, and further inland at Tuxtla Gutierrez, Comitan and San Cristobal.

The shock was felt over a large area and in most places from which it was reported it lasted more than one minute. In Mexico the shock was generally felt as far as Chilpancingo, throughout the state of Tabasco, Comitan, and in Guatemala as far as Chiacam, Cobán, Las Mercedes and San Martin Jilotepeque, and it was perceptible in Veracruz and Mexico City. Allegedly, the shock was perceptible at San José in Costa Rica. Macroseismic data are fairly good and confirm that this was an intraplate earthquake with an epicentre off the coast of Chiapas. The shock was very well recorded, and unusually for the Milne era, many arrivals could be identified as *P*. Although timings were not good enough to allow a formal solution to be found, it appears that the fit favours a location in the Gulf of Tehuantepec at 15°N, 95°W, outside our area, compared with Gutenberg's (1958) which is even further west at 15°N, 98°W. The location of 15°N, 93°W suggested by Nishenko and Singh (1987) is not supported by macroseismic data. Instrumental data also confirm that a magnitude greater than 8.0, often assigned to this event, is an overestimation.

[Figueroa 1970; Nishenko and Singh 1987; Rudolph 1905; STR; Viquez 1910]

1903 Feb 28 M_S 5.8 Costa Rica

This earthquake occurred somewhere in northeastern Costa Rica. It was reported felt from San Isidoro de Alajuela and from the region of Los Guatusos, southeast of Lake Nicaragua, where it was very strong.

The shock was recorded at 04h 09m local time (09h 45m UT) by a Ewing pendulum at San José, where it was generally felt.

BAAS (1911) place this shock between Cuba and Jamaica, about 1,100 km from north Costa Rica, but there is no evidence that the shock was felt there.

This event was widely recorded, but times and amplitudes are difficult to separate from another event that occurred 13 minutes later. The most likely position is southeast of Lake Nicaragua.

[Rudolph 1905; Viquez 1910]

1904 Jan 20 M_S $7\frac{3}{4}$ Panama

This earthquake most probably had an epicentre in the Pacific, south of western Panama.

The shock was very strong in the Chiriqui District and it was felt throughout Panama and Costa Rica, but caused no damage.

The earthquake was strong at Panama City, and at San José, where it was recorded by a Ewing pendulum.

On 21 January a ship sailing near the Revilla Griedo Islands (20°N, 110°W) reported floating tree-trunks and bodies of dead animals, but this is too far away from Chiriqui to be associated with this earthquake.

Macroseismic information for this shock is just sufficient to suggest a location off the coast of Chiriqui, near the position adopted by BAAS. The earthquake was recorded worldwide and Gutenberg places its epicentre in the Gulf of Panama, about 330 km east of Chiriqui. From four undamped or lightly damped pendulums at Cordova, Hamburg, Porto Rico and Tiflis, he calculates a magnitude of $7frac34$.

[BAAS; ERD; Gutenberg; Oddone 1907; Viquez 1910]

1904 Dec 2 M_S 6.9 Panama

This earthquake originated in the Pacific, probably in the same area as the shock of January 20, south of western Panama.

The shock was perceptible at Quito and allegedly at Las Mercedes in Guatemala. Felt reports from other parts of Central America are lacking.

The location of this event is very uncertain. There is no macroseismic information that could help fix its general position. BAAS place it seaward of the Middle American Trench but we have assumed that it probably originated from the same area as the earthquake of 20 January 1904.

[BAAS; Oddone 1907; STR]

1904 Dec 20 M_S 7.5 Figure A.7 Panama

This was a major earthquake with an epicentral area extending along the Atlantic shore of Costa Rica and Panama from Limón and Coto to Bocas del Toro in

Panama.

The shock was severe at Bocas del Toro, where there was damage to most of the few houses that were built after the fire of March 1904. In the Gulf of Bocas del Toro the shock was very strong on board a man-of-war at anchor. It is also reported that as a result of the earthquake a part of the islet of Sapodilla subsided.

There is some evidence that damage was more serious in the mountain region of the Cordillera Madre and in northwestern Panama between Coto, Boqueron, David and Bocas del Toro where houses and native wattle and daub huts were thrown down and landslides blocked streams, but details are lacking, except that the earthquake caused the collapse of houses at Bugaba and David.

Fig. A.7 Felt effects of earthquake of 20 December 1904.

At Limón the shock caused great panic and did some damage to a number of houses and shops but not those built of renforced concrete. The earthquake was of such a long duration that water in tanks was thrown out.

It is said that submerged flat reefs near Limón were uplifted and became visible after the earthquake, but this information is in need of authentication.

Sporadic minor damage extended over a large area. In the region of San José, from Siquirres to Alajuela and San Marcos and along the railway line to Limón, the shock was strong enough to set church bells ringing, to crack walls of buildings and to cause people to flee from their houses.

At the Caldera saltpans, near Puntarenas, the earthquake caused water to slosh out of ponds. The shock was felt in El Salvador, at Santiago and Colón in Panama and it was perceptible as far as Las Mercedes in Guatemala and allegedly at Quito in Equador.

The data are insufficient to determine the macroseismic position of the event, which is tentatively placed between Coto and the Gulf of Bocas del Toro.

The earthquake was followed by strong aftershocks, which lasted up to the end of the month, some of them causing concern in San José.

Instrumenental readings from American and European stations agree well with the general macroseismic position of this event (Figure A.7). Damage and felt reports confine the epicentral area to the southwest side of the Cordillera Madre, a region of very low population density.

[Barquero and Rojas 1994b; Boschini 1989; Camacho and Viquez 1992; Feldman 1984; Miyamura 1980; ODC; Oddone 1907; PDRC; PMTC; PPLB; PRPB; Sieberg 1932; Viquez 1910]

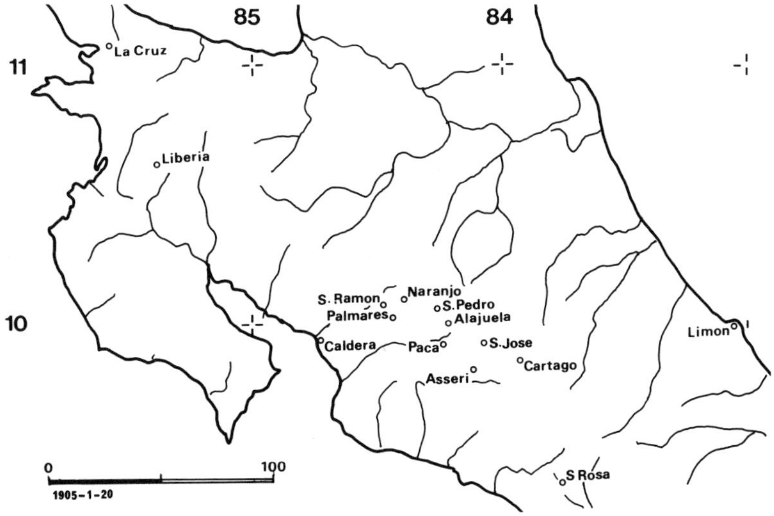

Fig. A.8 Felt effects of earthquake of 20 January 1905.

1905 Jan 20 **M_S 6.5** **Figure A.8** **Costa Rica**

A strong earthquake in Costa Rica caused sporadic damage over a relatively large area between San Ramón, Aseri, Cartago and San Pedro. In this area the shock, which lasted about 20 seconds, damaged churches at Desamperados, Heredía, Naranjo and San José, and cracked the walls of many houses and of a number of public buildings.

The shock was strong at Limón, Santa Rosa and La Cruz.

Viquez (1910) confuses the effects of this earthquake with those produced by the earthquake of 20 December 1904.

The earthquake was recorded by a Ewing pendulum at San José and by a considerable number of seismographs in the Americas and Europe.

The relatively large epicentral area of low intensity suggests the focus was not very shallow.

[Christiansen and Ziemendorff 1909; Miyamura 1980; ODC; PDRC; PPLB; PRPB; Szirtes 1909; Viquez 1910]

1905 Dec 28 **Costa Rica**

A local earthquake caused minor damage to a few houses in Cartago and San José in Costa Rica.

No instrumental data have been found for this event.

[Jorgensen 1933; Miyamura 1960; PRPB]

1906 Jan 3 **Nicaragua**

This earthquake in Nicaragua occurred during a period of re-activation of the volcano of Masaya.

The earthquake shook down several old houses in the town of Masaya, and volcanic dust fell lightly on all the country close to the volcano. No one was killed, nor was railway traffic interrupted.

The area over which the shock was felt was limited to between Managua and the Lake of Nicaragua.

[ERD; NEN; Palmer 1945.128; Sieberg 1932]

1906 Jun 20 **M_S 6.1** **Figure A.9** **El Salvador**

This was a strong earthquake in El Salvador.

Damage was reported from the region between Santa Ana and Suchitoto but no details are available.

The shock was strong at Comasagua, Guayabal, San Martin, and San Salvador, and was felt at San Miguel, Santa Rosa, El Baul and in the District of Cobán in Guatemala.

At San José the shock was recorded by a Ewing pendulum but it was not felt in the city, although it was reported as being so in the local press.

The earthquake was recorded at a number of seismograph stations but the data are not enough to establish its epicentral region which must be between Santa Ana and Guayabal.

[ERD; Jordan and Martinez 1980; Martinez 1978; ODC; Scheu 1911; STR; Szirtes 1910; Viquez 1910]

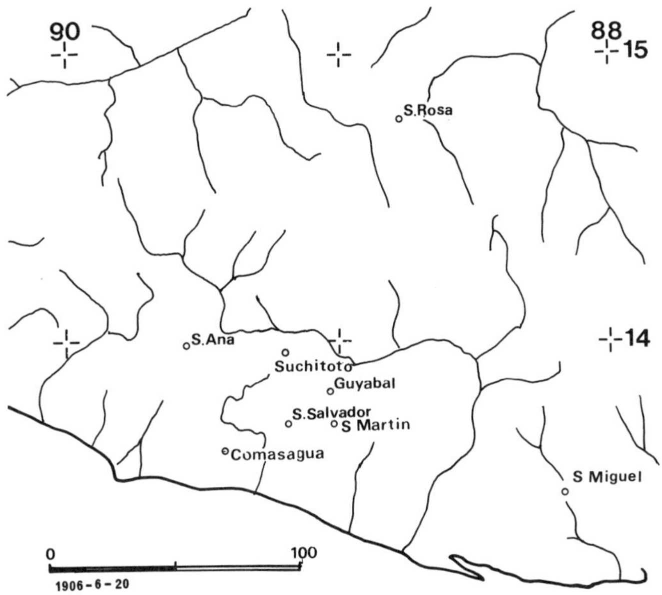

Fig. A.9 Felt effects of earthquake of 20 June 1906.

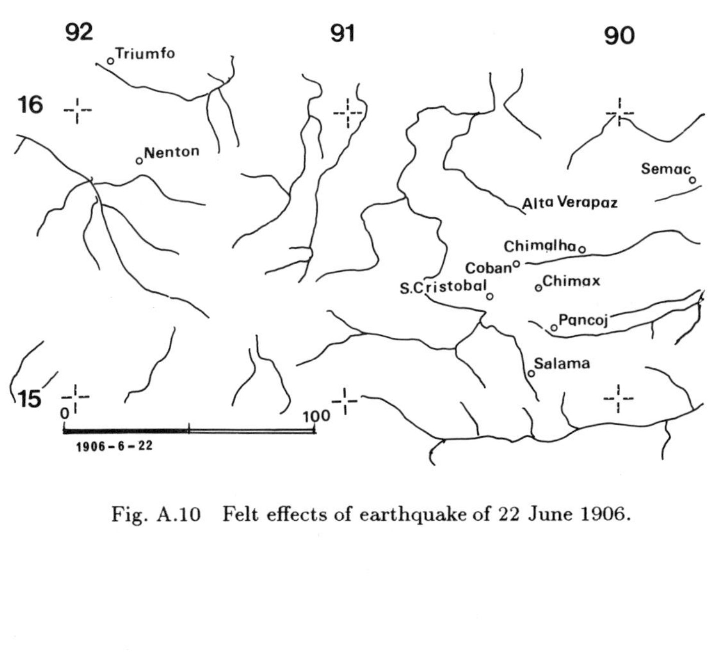

Fig. A.10 Felt effects of earthquake of 22 June 1906.

1906 Jun 22 **M_S 5.3** **Figure A.10** **Guatemala**

This was a small earthquake in northeastern Guatemala in the sparsely populated region of Alta Verapaz. The shock was rather strongly felt at Chimalha, Chimax, Cobán, Panjoc, San Cristobal and Semac, but caused no damage.

The earthquake was reported from Triumfo in Mexico, where allegedly it caused some damage, most probably the result of a different event. It was followed by many small aftershocks which were reported from farms around San Cristobal.

The earthquake was recorded instrumentally but the data are insufficient to locate the epicentral region, even approximately.

[ERD; STR; Szirtes 1910]

1907 May 31 **Guatemala**

This locally damaging earthquake in Alta Verapaz in Guatemala was accompanied by many foreshocks and aftershocks. The earthquake caused the collapse of houses at San Cristobal and did some damage at Cobán. The shock was strong at a number of farms at Mercedes, Setal, Sertiquiche, Chijolm, Soledad and in other farmsteads. The Guatemalan press does not mention this event.

An aftershock at 20h 30m on June 7 added to the damage.

These shocks were recorded only at the nearest seismograph station of Tacubaya.

[ERD; Scheu and Lais 1912; Szirtes 1912b]

1907 Jul 1 **M_S 7.4** **Honduras**

This was a relatively large magnitude shock in the Gulf of Fonseca in Honduras, which was felt strongly at Amapala and Tigre Island. No other macroseismic data have been found for this earthquake from mainland Honduras, El Salvador or Nicaragua.

Reports that the shock was also felt at Mercedes and Soledad about 300 km away may refer to the continuing aftershock activity in Guatemala of the earthquakes in June and July, or they may indicate a subcrustal depth. The earthquake was followed by many aftershocks during July and August.

This event, which had a surface-wave magnitude of 7.4, is grossly mislocated by BAAS(1912.11) at 0°, 105°W and is not listed in standard catalogues. It was well recorded instrumentally, particularly in Europe, and our recalculated solution is offshore and close to the macroseismic position given by Szirtes (1912a), which is near the island of Tigre. There is some seismological evidence for a subcrustal focus, perhaps as deep as 100 km.

[Lais 1913; Scheu and Lais 1912; Szirtes 1912a; Sieberg 1932; Szirtes 1912b]

1907 Sep 23 **M_S 6.5** **Guatemala**

This very strong earthquake in Guatemala did considerable damage to farms and estates in the upper reaches of the Motagua River, but details are lacking. The shock was strong in Guatemala City, where contrary to foreign press reports, it caused no damage. The shock was felt as far as Chinax, Cobán, and Finca Setal in Alta Verapaz.

The earthquake was well recorded by seismographs. BAAS(1912) place the earthquake in the Pacific west of Mexico. Szirtes (1912a) reports readings from 30 stations, which he interprets as giving a solution near Guatemala City, which is closer to the preferred macroseismic position.

[Lais 1913; PDCA; Scheu and Lais 1912; Sieberg 1932; Szirtes 1912b; Van de Putte 1924]

1907 Oct 6 M_S 5.8 Costa Rica

This earthquake in Costa Rica was preceded and followed by many smaller shocks, and caused considerable concern. The shock was strongly felt at San José.

We can find no entry for this event in the bulletin of the seismological station of San José and the few readings from other stations add nothing to the macroseismic position.

[Lais 1913; Scheu and Lais 1912; Szirtes 1912a; Szirtes 1912b]

1907 Dec 30 M_S 7.4 Nicaragua

Felt reports suggest that this major earthquake occurred in the Pacific off the coast of Nicaragua. The shock was widely felt from Corinto and Managua in Nicaragua, to San José in Costa Rica. There is no evidence that it caused damage.

Without giving details of his calculations, Szirtes (1912a) gives an epicentre at the city of Managua, a location adopted by later writers; Duda (1965), Abe and Noguchi (1983a), Abe and Noguchi (1983b). BAAS(1912.13) also grossly mislocate the event at 8.0°S, 97.0°W. The earthquake was reasonably well recorded by seismographs world-wide and by a Ewing pendulum at San José in Costa Rica. Our recalculated position is in the Pacific Ocean.

[ERD; Viquez 1910]

1908 Feb 1 M_S 6.6 Costa Rica

Very little is known about this earthquake in Costa Rica. The shock was recorded by a Ewing pendulum at about 17h 27m (23h 27m UT) at San José, where it was strongly felt. Information from other parts of the country is lacking, except from Greytown (San Juan del Norte) where it is said that in 1908 the town had suffered some damage from an earthquake and from high winds.

From a small number of instrumental readings, BAAS(1912) identifies two events at 23h; one in the Pacific at 23h 17m and another northeast of the Bahamas at 23h 22m. We can find no macroseismic evidence that earthquakes were felt in these localities. The instrumental readings are not particularly consistent but they are in fair agreement with the allocated position in the Guatusos region in northern Costa Rica.

[Miyamura 1980; PPLB; Sieberg 1917a; 1917b; Viquez 1910]

1908 May 19 Honduras

This earthquake in the District of Copán in Honduras ruined the villages of Cucuyagua and La Unión and caused some damage at Santa Rosa and to nearby settlements on the east flanks of the Cordillera El Marendon. The earthquake was followed by many aftershocks, the strongest of which occurred on 28 August.

The event was not reported in the Guatemalan press, and no instrumental readings were found.

[Chacon 1909; Sieberg 1932; STR]

1908 Jul 19 **Guatemala**

A strong earthquake in Guatemala caused some damage at San Luis Malacatan. The shock, which lasted about 40 seconds, was felt at Quezaltenango and in other parts of the district of San Marcos, which are not specified; it was followed by aftershocks.

There was no instrumental information from which to assess the magnitude of this event.

[STR]

1909 Aug 16 **M_S 6.9** **Nicaragua**

This earthquake was widely felt in Nicaragua south of Matagalpa, and in northern Costa Rica, but the available information is insufficient to define the macroseismic epicentral area of the event unambiguously. We know that in Managua the shock caused some concern, and that it was strong at Rivas and along the coast of Lake Nicaragua. The earthquake was perceptible at Bluefields and San José.

The distribution of the few felt reports available confines the epicentral region to Nicaragua, more than 200 km from the location determined by BAAS (1912.19), which is in Costa Rica. Instrumental readings confirm a location in Lake Managua.

[STR]

1909 Aug 30 **M_S 6.1** **Panama**

This earthquake was felt throughout Panama. It was particularly strong in the district of Chiriqui, but it is not known whether it caused any damage. The shock was felt along the north coast of Panama, from the Golfo de los Mosquitos to Nombre de Dios. The earthquake was felt in the Canal Zone, where it was sufficiently strong to stop a pendulum clock at Gatun and to be felt by many people.

The shock was widely recorded by European and American stations.

[Camacho and Viquez 1992; CNR: 1909.iii.5; PINF; PNTC]

1910 Jan 1 **M_S 6.9** **Honduras**

This strong earthquake had an epicentre off the north coast of Honduras. The shock was very strong at Swan Island, and was felt in Yukatan at Vigia Chico, Iturbide and Tekax, as well as in Jamaica at Moy Hall, Kingston and Mt. Holstein. Unconfirmed reports add that the shock was also felt as far as Mexico City.

The event was well recorded instrumentally and readings confirm its general position near the Swan Islands, but east-west control is poor. Gutenberg (1958) assigns to this event a depth of about 60 km, for which we can find no evidence.

[BAAS 1913; TAC]

1910 Mar 10 **Costa Rica**

This strong earthquake at Santa Maria de Dota in Costa Rica was reported from San Pablo, Corralillo and San José; it was also felt in the south of the country.

The shock was recorded only at San José.

[Tristan 1925; Viquez 1910]

1910 Apr 13 **M_S 5.6** **Figure A.11** **Costa Rica**

These were the first of a series of damaging earthquakes in central Costa Rica.

The first shock occurred at 06h 13m (UT) but the greatest damage was done 28 minutes later by a second shock at 06h 41m (UT), which caused damage to villages a few kilometres southwest of Cartago. A third shock the following day, at 14h 06m on 14 April, added to the damage.

The effects of these shocks are difficult to separate. The earthquake at 06h 41m was felt throughout Costa Rica, but serious damage was confined within a small area of about 6 km radius that included the villages of San Miguel, Jerico and Tobosi. There were no casualties. Sporadic damage extended over a larger area that included Aseri, Cartago, San José, Tres Rios and Tejar.

The shock was widely felt by most people at Limón and Filadelfia, more than 100 km from the epicentral area, and caused some concern in the region of Bocas del Toro, 220 km away. In the south the shock was perceptible in the region of the Golfo del Dulce and in the north at La Cruz.

Of the three shocks of the series only the second and largest was recorded at the nearest station of Tacubaya, at a distance of 17.6°, by a long-period (36 s) small magnification Bosch-Omori undamped instrument, and at larger distances by some more sensitive instruments of the world network.

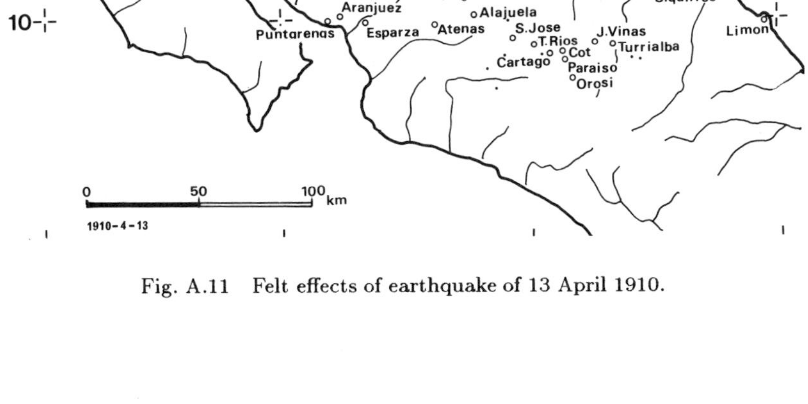

Fig. A.11 Felt effects of earthquake of 13 April 1910.

[Barquero and Rojas 1944b; Calvert and Calvert 1917; CNR: 1910.no.34; Güendel 1986; Michaud 1910; Michaud 1912; Montero and Miyamura 1981; ODC; Pittier 1910; PRPN; Sieberg 1932; Tristan 1925; USNA; Viquez 1910]

1910 May 5 M_S 6.1 Figure A.12 Costa Rica

This was the largest shock of the sequence in Costa Rica that began in April.

It affected the Department of Cartago, an area aleady damaged a few weeks earlier, and almost totally destroyed Cartago, Paraiso and Tierra Blanca, where about 500 people were killed and thousands made homeless. Other villages within a radius of 20 km were seriously damaged.

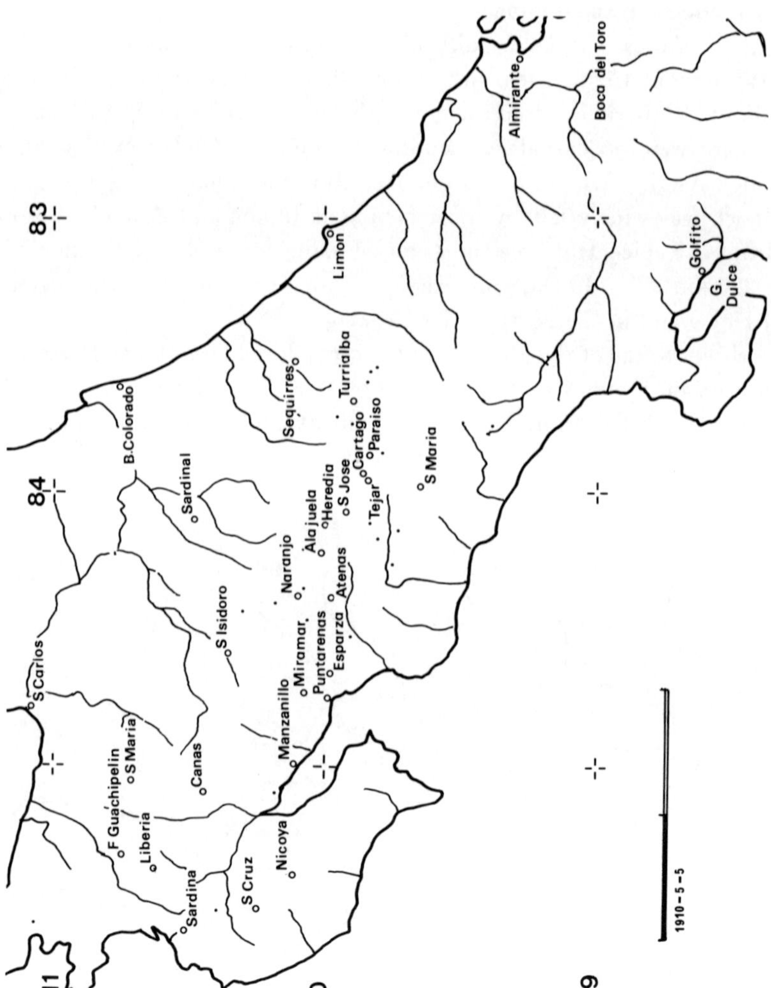

Fig. A.12 Felt effects of earthquake of 5 May 1910.

San José, though not in ruins, was left with badly shattered and cracked houses, with nearly every building in the city in need of considerable repair. Agua Caliente, Heredía, Orosi, Paraiso, Tejar, and Tierra Blanca sustained serious losses.

The shock was felt throughout Costa Rica, from Rivas in Nicaragua, La Cruz and Boca di Colorado in the north to the Golfo del Dulce and Bocas del Toro in Panama in the southeast.

The shock was not felt at Managua and in the Canal Zone but it was reported from other places in southern Nicaragua and western Panama, which are not named in the reports.

The earthquake was widely recorded; the origin time assessed by the Seismological Investigation Committee is 00h 26m (BAAS: 1913a.4).

[Anonymous 1911; Barquero and Rojas 1994b; Calvert and Calvert 1917; CNR: 1910.290,299; Feldman 1984; Guardia and Marin 1910; Güendel 1986; Kirkpatrick 1920; Michaud 1910; Miyamura 1980; Montero and Miyamura 1981, 1982; Pittier 1910; PRPB; Sieberg 1932; Tristan 1925; USNA; Viquez 1910]

1910 May 14 **Costa Rica**

This strong shock felt in the region of Santa Maria de Dota caused no damage, but was widely felt.

It was recorded only at San José.

[Viquez 1910]

1910 Dec 21 **M_S 6.2** **Panama**

This earthquake was preceded by a strong foreshock at 23h 15m on December 20. The main shock was strong in the Districts of Chiriqui and Bocas del Toro, but details of the damage it caused are lacking. At Bocas del Toro the shock aroused practically everyone and stopped a pendulum clock. The main shock was felt at Culebra 250 km away.

Instrumental readings from Tacubaya and stations to the northeast agree well with the macroseismic epicentre.

[Camacho and Viquez 1992; CNR: 1910 no.18; Kirkpatrick 1920]

1911 Jul 11 **Costa Rica**

An earthquake with an epicentre probably offshore in the Gulf of Nicoya was felt strongly along the Pacific coast of Costa Rica at Puntarenas, Las Agujas and in places along the east coast of the Peninsula of Nicoya. The earthquake was reported from other places as far as San José, where it was recorded instrumentally, and was perceptible at Cartago.

[PRS; Tristan 1912a; Tristan and Biolley 1912]

1911 Aug 25 **Costa Rica**

The facts about this earthquake are not clear. The shock was recorded by a Ewing pendulum at San José at 06h 40m local time (12h 16m UT) and was felt along the Pacific coast of northern Costa Rica. It was reported as strong from many places in the Province of Guanacaste and from Puntarenas. The shock was felt as far east as Orosi, Cartago, Curridabat and San José. However, no instrumental data have been found for this earthquake, which was felt relatively widely, and which probably had an offshore epicentre.

[Tristan 1912a; Tristan and Biolley 1912]

1911 Aug 29 M_S 5.8 Figure A.13 Costa Rica

This earthquake caused some damage in the sparsely populated area between the volcanoes Poas, Congo and Viejo in Costa Rica.

Tore Amarillo (Bajos del Toro), a settlement of 25 log cabins, and two nearby farms were completely ruined, but there were no casualties. Within a radius of about 8 km and to the north of Trojas the shock triggered rockfalls, landslides and caused cracks in the steep slopes of the volcanoes Congo and Viejo on either side of Rio Toro.

In Grecia the iron-framed church, built after the earthquake of 1910, as well as timber-framed houses, suffered insignificant damage. Minor damage extended to San Ramón, Naranjo, Sarchi, San Pedro, Alajuela and Zarcero. The shock was strongly felt at Heredía and San José.

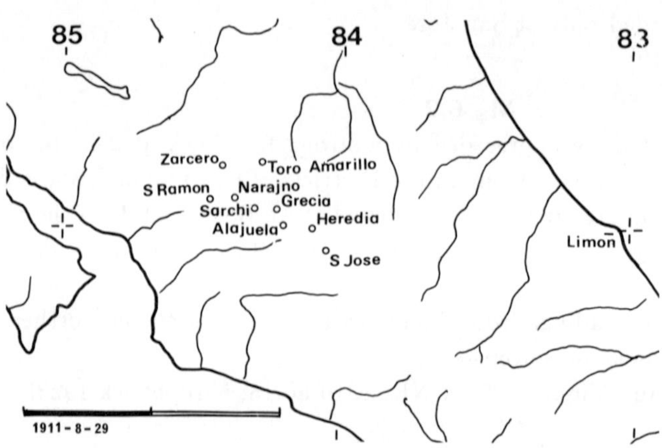

Fig. A.13 Felt effects of earthquake of 29 August 1911.

We could find no information about the far-field effects of this event.

Instrumental recording was poor. The earthquake occurred at 22h 06m local time on 28 August (03h 42m, 29 August UT) and was recorded by a Ewing pendulum at San José, as well as at a few other seismographic stations, the nearest being Tacubaya and Merida. The data are not enough to establish an epicentre.

White and Harlow (1993) date this event on 28 August at 06h 28m (UT) and use the ground amplitude of an event recorded at De Bilt in the Netherlands at this time to calculate the magnitude of the earthquake in Costa Rica. The amplitude of 8 microns they use in their calculations belongs to an unrelated earthquake in

Tunisia (Kárník 1968).

The earthquake was followed by a few weak aftershocks.

[Güendel 1986; Alfaro, Michaud and Biolley 1912; Miyamura 1980; PNTC; PPLB; Tristan 1912b; 1925; Tristan and Biolley 1912]

1911 Oct 10 — M_S 6.5 — Figure A.14 — Costa Rica

This earthquake in northern Costa Rica was felt throughout the country as well as in southern Nicaragua.

Its epicentral area seems to have been in the sparsely inhabited region of Los Guatusos where, it is said, even native huts were damaged, trees were uprooted and the ground in places opened by as much as 30 cm in long cracks.

The shock was very strong at Tenorio, Palmares, Canas, Tilaran, Arenal, Buenavista, San Carlos, Unión, Muelle and Buenavista.

In Puntarenas and La Cruz the shock caused panic and it was strongly felt in San Juan del Sur and at San José.

The earthquake was reported from all along the railway line from the Pacific to the Atlantic coast, from Rivas in the north, and from south of Santa Maria in the south, but details are insufficient to allow the assessment of intensity at these places.

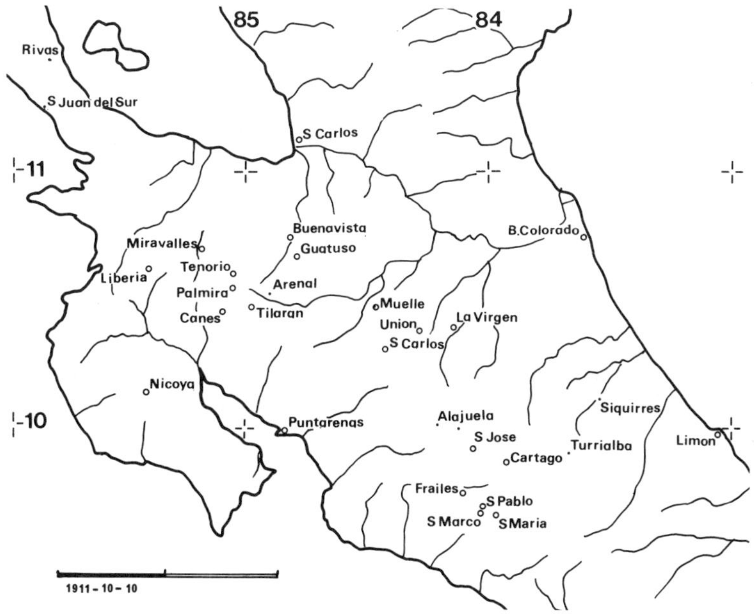

Fig. A.14 Felt effects of earthquake of 10 October 1911.

The earthquake was followed by a few weak aftershocks, and was well recorded instrumentally.

[Miyamura 1980; PNTC; Tristan 1912b; Tristan 1925; Tristan and Biolley 1912]

1911 Nov 1 **M_S 6.7** **Nicaragua**

This earthquake was felt in Nicaragua, but details are lacking.

The shock was widely recorded, but the readings can only confirm a location off the coast of Nicaragua in the Pacific.

[DBN; STR]

1912 Feb 21 **Figure A.15** **Costa Rica**

This was a locally damaging earthquake in Costa Rica.

Within a radius of about 4 km of Curridabat, and in particular at Tres Rios, all houses and public buildings not built of timber were damaged and many became uninhabitable. Slides triggered by the shock blocked Rio Chiquito which dried up temporarily.

The shock caused sporadic, minor damage to a few houses already damaged by previous shocks over an area that extended from Alajuela to Aseri, Pacayas and San Isidoro, and caused panic in nearby San José.

The earthquake was perceptible as far as Liberia and Limón. Its only instrumental recording was at Merida, at a distance of 12.3°.

[Güendel 1986; Miyamura 1980; PINF; PPLB; PRPB; USNA]

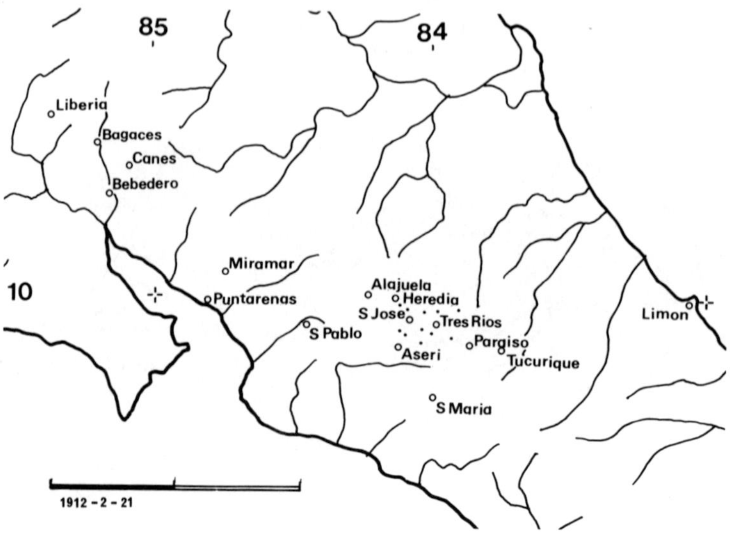

Fig. A.15 Felt effects of earthquake of 21 February 1912.

1912 Jun 6 M_S 5.1 Figure A.16 Costa Rica

This was a small, locally damaging, earthquake in the region of Sarchi near the Volcano Poas, in Costa Rica.

Its epicentral area was relatively small, confined between Bajos del Toro, Grecia, San Juan, Sarchi, Naranjo and the headwaters of the Rios Desague and Sarchi, an area of about 10 km radius. In this area the shock and the flash floods that followed destroyed a few hundred adobe houses. In all, seven people were killed by the landslides and floods.

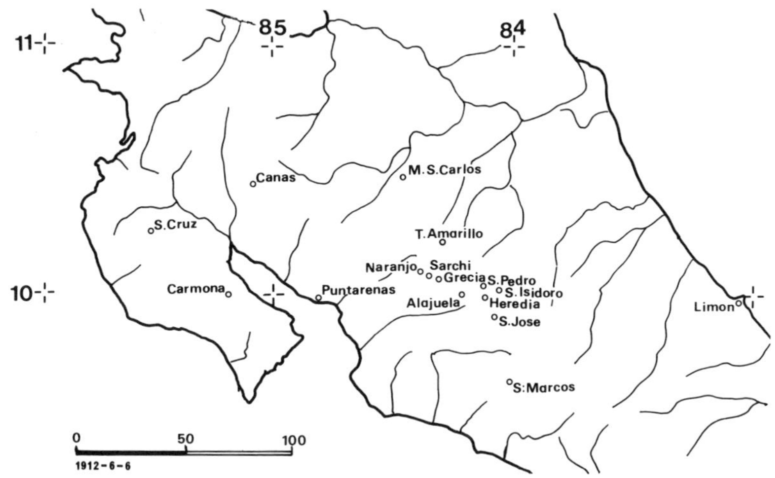

Fig. A.16 Felt effects of earthquake of 6 June 1912.

Minor damage was done at San Isidoro, San Perdor and Bajos del Toro.

The shock caused great panic but no damage at Alajuela, Barva, Heredía, and San José where hanging objects were set swinging and some clocks were stopped.

The earthquake was felt as far as Limón, to the south of San Marcos, at Carmona, Santa Cruz, and to the north of Muella San Carlos.

In the epicentral area rockfalls and massive landslides from the steep slopes of Volcano Poas dammed temporarily Rio Anonos and Rio Surchi. These dams were breached in sequence, from one to 36 hours after the earthquake, flooding the downstream areas, destroying houses and drowning a number of people.

The main shock was followed by many small aftershocks.

As far as we can tell the shock was recorded only at San José (0.5°) and Merida (12.1°) giving an approximate origin time 06h 12m (UT). The effects and size of this earthquake have been grossly exaggerated in recent reports. White and Harlow (1993) assign to this event an origin time 12h 40m (UT) and a magnitude M_S 6.5 which they calculate from a ground amplitude of 30 micrometres recorded at De Bilt in the Netherlands. However, we can find no earthquake listed in the bulletin of

De Bilt for 6 June (Van Everdingen 1915.69). The only amplitudes of 30 micrometres recorded at this station during the first two weeks of June belong to the Alaskan earthquakes of 8 June and 10 June, and to the Mexican earthquake of 12 June at 12h 43m. (Van Everdingen 1915.70).

[Güendel 1986; Miyamura 1980; PRPB; Sapper 1912; Sieberg 1932; Tristan *et al.* 1912]

1912 Jun 12 M_S 6.8 Belize

No macroseismic data have been found for this earthquake which was widely recorded instrumentally. Gutenberg and Richter (1956) give a shallow focus earthquake on the border of Belize with Honduras.

1912 Jul 18 El Salvador

A locally strong earthquake in El Salvador preceded the lava eruption of the Volcano Izalco. The shock did considerable damage to Santa Ana, Izalco and Armenia, small towns within a radius of 15 km of the volcano, where a number of people were injured. The shock was hardly perceptible at San Salvador, 45 km away.

This event was apparently too small to be recorded by near or distant seismographic stations. The magnitude of M_S 5.9 assigned to this event by White and Harlow (1993) is excessive.

[Jordan and Martinez 1980; Martinez 1978; Meyer-Abich 1952]

1912 Dec 9 M_S 7.2 Mexico

This earthquake occurred in the Pacific at some distance from the coast of Mexico.

It was stronly felt along the coast of the Gulf of Tehuantepec at Juchitan, Tapanuetepec, Tonala and further inland at Tuxtla Gutierrez, Palizada and Providencia, but caused no damage. The shock was felt in Guatemala City and it was perceptible in other parts of the Provinces of Chiapas, Oaxaca and as far as El Oro in the District of Mexico.

This large magnitude earthquake was originally mislocated by Pulkovo, in a position south of Cuba, 930 km eastnortheast of the Gulf of Tehuantepec; this false location is included in some recent earthquake catalogues, (Duda 1965).

[Gutenberg; PDCA; Schneider 1914.55; TAC]

1913 Mar 8 M_S 6.4 Figure A.17 Guatemala

This earthquake in Guatemala occurred in the district of Santa Rosa.

In Cuilapa many houses, schools, the cathedral church and the prison were destroyed by the main shock and its numerous aftershocks, with some considerable loss of life. Heavy damage and loss of life extended to Barberena and to the settlements of Cerro Redondo, Llano Grande and El Zapote.

The towns of Fraijanes, Vinas, Coatepeque and Jalpatagua were also seriously damaged.

In Guatemala City the shock caused panic and some minor non-structural damage.

In the epicentral area the earthquake triggered landslides and rock falls, which blocked roads and footpaths. A long ground crack was reported in the Cerro Los Esclavos.

The shock was strongly felt over a relatively large area, as far as Salama, 100 km to the north of Cuilapa, and was perceptible at Solola, Rio Bravo and at different parts of El Salvador.

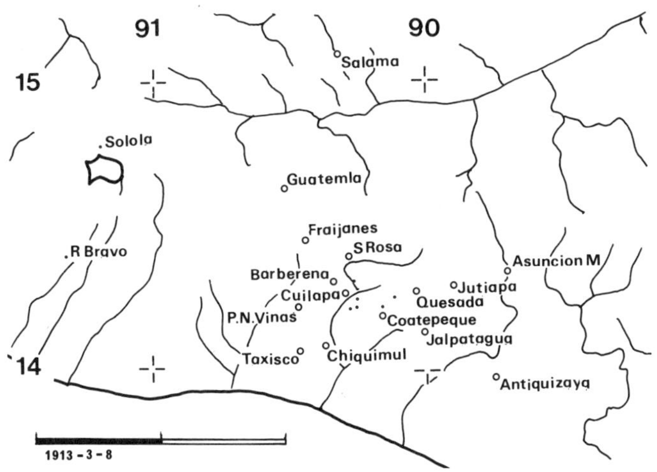

Fig. A.17 Felt effects of earthquake of 8 March 1913.

Successive aftershocks almost totally ruined Cuilapa.

The earthquake was well recorded by intermediate- and long-period instruments, as well as by 42 Milne seismograph stations worldwide (BAAS: 1913b.6), but lack of azimuthal variation prevents an accurate position from being determined, and the readings can only confirm the general area of occurrence.

White and Harlow (1993) assign a magnitude of 5.9 from De Bilt; we can find no amplitude and period data for this event listed in the station bulletin.

[PDCE; Sapper 1913.34; Van de Putte 1924; White and Sanchez 1980]

1913 Mar 13 **Honduras**

An earthquake was felt throughout the Department of La Esperanza in Honduras, causing some minor damage at Esperanza.

[PDCA]

1913 Jul 25 **M_S 6.1** **Costa Rica**

This earthquake had an epicentre in the Caribbean Sea off the south coast of Costa Rica. The shock was strongly felt in the region of the Bocas del Toro and it was reported from Limón and Cristobal.

The instrumental position agrees with the macroseismic epicentre within the errors. Gutenberg and Richter (1956) place this event 340 km to the north, off the eastern coast of Nicaragua.

[OCD; Kirkpatrick 1920]

1913 Oct 2 M_S 6.7 Figure A.18 Panama

This strong earthquake occurred in the Pacific off the coast of Los Santos in Panama.

The earthquake was preceded by a strong foreshock at 18h 48m (UT) on 1 October. Owing to the fact that very few masonry buildings exist in the province the damage was not as great as it otherwise would have been and there was no loss of life.

Much of the more serious damage was done in the south of the Azuero Peninsula.

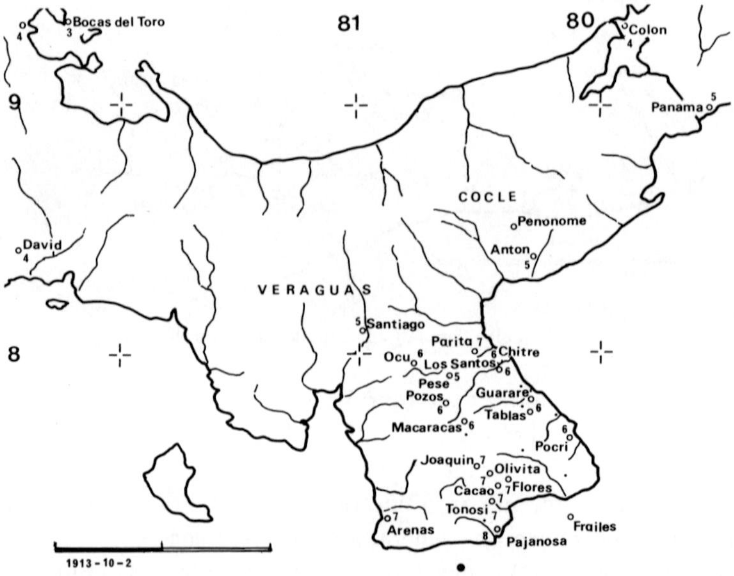

Fig. A.18 Felt effects of earthquake of 2 October 1913.

Many of the houses at Tonosi and in the valley were damaged and a few collapsed, without causing casualties. In the villages and farms of Arenas, Bebedero, Cacao, Canajagua, Flores, Joaquin, Mogolon, Olivita, and others in the Rio Viejo and Rio Joaquin valleys, all houses with tile roofs were damaged. Extensive cracking and liquefaction of the ground was observed along the banks of the river at Pajanosa, as well as on the main road to Tonosi. At Cacao and along the bank of Rio Viejo at Las Huertas the shock caused widespread liquefaction. Near Cerro Bonita it was found that the Rio Joaquin had cut a new channel, caused by the opening of fissures at a bend in its channel. The shock caused many landslides in this area, mostly of surface material from steep slopes.

Further to the north native ranches with grass roofs were nowhere seriously affected. At Guarare a few houses lost their tiles and at Macaracas the church was badly damaged and its tower, 25m high, collapsed. At Los Santos a few houses were cracked but the Government building was totally ruined. The police barracks were also badly damaged as well as the 35m high church tower. At Pesé there was no damage and at Tablas the walls of the church and its tower were cracked but private houses were little damaged.

Sporadic but minor damage extended to Santiago and to the Provinces of Veraguas and Coclé.

The shock was very strong at Anton and Penonome.

In the Panama isthmus a few houses suffered some minor damage and church bells were set ringing. The earthquake causing considerable panic but no damage to the lock structures of the canal.

The earthquake was felt in the Districts of David and Bocas del Toro but not in Costa Rica.

A few hours after the earthquake a submarine slide ruptured the Central and South American cable near the islet of Frailes del Norte, about 25 km off the south coast of Los Santos Province. It has been reported that the broken end of the cable was found at a depth of about 1,500 m, whereas, when the cable was first laid, the depth was only about 300m. Here the shock caused an abrupt change in the bathymetry that suggests submarine sliding. The tide gauge at Balboa showed no abnormal fluctuations of the sea level and except in the Gulf of San Miguel, on the east coast of the Gulf of PanamaGulf of Panama, from where it was reported that the sea behaved abnormally on the day of the earthquake, there is no evidence of a seismic sea-wave.

The earthquake was followed by a long sequence of strong aftershocks until mid-November.

The earthquake was well recorded and instrumental readings confirm a position at the southeast extremity of the peninsula of Azuero.

Macroseismic effects suggest a focus in the upper crust.

[CNR: 1913.53,73,114; Kirkpatrick 1920; Lutz 1917; MacDonald and Johnston 1913; PINF; PRPB; PTIM; Sieberg 1913]

1913 Oct 23 **M_S 6.0** **Panama**

This was a strong aftershock in the series that followed the earthquake of October 2. It caused some small additional damage in the districts of Los Santos and Coclé, particularly at Macaracas, Los Santos, Anton and Penonome; it was strongly felt in the Canal Zone.

[CNR: 1913.91; Kirkpatrick 1920; PINF; PRPB]

1914 May 28 **M_S 6.4** **Panama**

This strong earthquake near Panama affected the Province of Los Santos, at Chitré, Macaracas, Nata, Ocu, and Pedasi. It was less severe in the Districts of Coclé, Panama and Veraguas.

In the Canal Zone it caused some minor non-structural damage and set up seiches recorded by the water stage registers in different locks; the surface of the lakes in the Zone fluctuated by as much as 15 cm. The shock had no effect on the Cucaracha landslide, but an additional slide occurred a few hours after the earthquake.

On 31 May there was an aftershock which was felt at Balboa.

The macroseismic evidence places this earthquake on the west side of the Gulf of Panama, near 8°N, 80°W. BAAS report readings from 47 stations, which they interpret as giving a location derived by Pulkovo at 15.1°N, 84.8°W in northeast Honduras. Gutenberg and Richter (1965) place the event in eastern Panama more than 1,100 km from the BAAS location, at a depth of 70 km and assign it a magnitude 7.2.

The event was widely recorded in Europe, but there were few consistent readings from elsewhere. Our instrumental analysis gives a location northeast of the Isthmus at 11.1°N, 78.0°W, with little control of depth. This location is about 400 km from the macroseismic position of 8°N, 80°W. The difference between the surface-wave magnitude (M_S 6.4) and the body-wave magnitude (m_B 6.9) suggests the focus may be subcrustal.

[CNR: 1914.397-8; PRPB; Kirkpatrick 1920]

1915 Sep 7 Ms 7.7 Figure A.19 El Salvador

This was a major lower crustal earthquake with an epicentral area extending from southwestern El Salvador to southeastern Guatemala.

Maximum damage was reported from western El Salvador. At Juayua almost all the houses were ruined or damaged beyond repair, including the local church, but there were no casualties.

In the nearby village of Salcoatitan, the church, the town hall and nearly all the houses were shattered and many collapsed injuring a number of people. The road leading to Juayua was damaged by ground cracks and slides.

Also at Ahuachapan almost all houses were ruined, including public buildings, and quite a few collapsed, killing five people.

At Ataco all houses were damaged beyond repair but no one was killed or injured. The shock produced ground fractures, one of which was found running for a length of nearly one kilometre.

At Santa Catarina all houses and public buildings were ruined; many collapsed injuring a number of people. To the north of the village the earthquake triggered landslides.

The same kind of damage was sustained by Nahuizalco, where the belfry and some houses collapsed and a few people were injured. Here also the shock caused cracks in the ground and landslides.

Less severe damage extended to Santa Ana, where two people were killed, to Armenia, where one person lost his life and the shock triggered landslides and rockfalls, to Sononote, where bridges were damaged, to Tacuba, where the shock injured a number of people and to Moyuta and Jalpatagua.

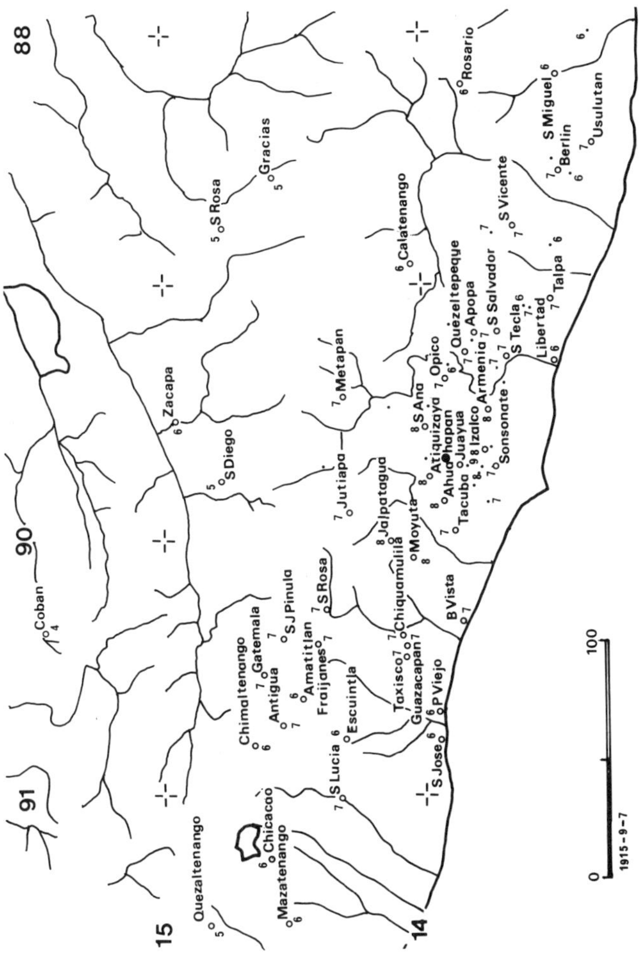

Fig. A.19 Felt effects of earthquake of 7 September 1915.

Outside this region the distribution of damage to rural houses and public buildings was irregular and extended over a very large area. In places the shock interrupted telegraph lines and triggered landslides which were enhanced by the torrential rains that affected the whole region before and after the earthquake.

Sporadic damage to houses and public buildings was reported from as far as Escuintla, Antigua, Metapan, and Usulutan.

The shock was strongly felt at Tuxtla in Mexico, Cobán and Puerto Barrios in Guatemala, at Santa Rosa and Teguchigalpa in Honduras, and all along the Pacific coast of Nicaragua at Chinandega, Leon, Managua, Masaya, Granada and San Juan del Sur. It was perceptible at San José in Costa Rica, 790 km away.

It is said that as a result of the earthquake the submarine cable on the Costa Rica coast was damaged.

The earthquake was followed by a series of relatively large magnitude aftershocks that continued to be felt for over two weeks. Details of the larger of these shocks are given in our parametric Catalogue.

This earthquake was very well recorded instrumentally. Gutenberg and Richter (1956) assign to the main event a focal depth of 80 km. We could not constrain the depth instrumentally. Our free solution gives a location at 14.0°N, 90.1°W and an excessive depth of 240 km, which does not appear to be supported by supplementary phases or from macroseismic observations. If the depth is held at 33 km the location shifts little to 14.3°N, 89.8°W and the overall standard error is slightly better, but the residuals at the closest stations are rather worse. Reports of maximum surface waves at about the times expected suggest that the earthquake is not significantly deeper than crustal. This is suported by a well-defined epicentral area of relatively high intensity, large aftershocks and the relative values of the surface- and body-wave magnitudes, M_S 7.7 and m_B 7.5. We have assumed a depth of 60 km at the macroseismic position of 13.9°N, 89.6°W.

[BSSA 1915.158; CNR: 1915.28; Jordan and Martinez 1980; Larde 1916; Martinez 1978; PDCA; Schultz 1963; Van de Putte 1924]

1915 Nov 26 M_S 5.8 Panama

This earthquake was strongly felt in the Province of Los Santos in Panama, but details are lacking. The shock was perceptible at Balboa.

[DES, CNR]

1915 Dec 1 Guatemala

A strong earthquake in Guatemala caused some damage in Guatemala City, injuring four people. The shock, which occurred in the morning, was strong enough to cause people to leave their home and camp in the open. The earthquake was felt in other Departments of the country, which are not specified.

We could find no instrumental data for this event.

[PDCA]

1915 Dec 20 Guatemala

A local earthquake in Guatemala was strongly felt at Amatitlan, Chimaltenango, Escuintla, Guatemala City and Sacatepeque. It caused no serious damage, and was followed by many aftershocks.

It was not recorded instrumentally

[BSSA.1916.48; DES; Van de Putte 1924]

1915 Dec 28 M_S 6.4 Figure A.20 Honduras

A series of damaging earthquakes affected the District of Gracias á Dios in Honduras.

Foreshocks causing some damage began on 15 December and continued intermittently until 29 December, the most damaging of the series occurring on 26 and 27 December. These are detailed in our parametric Catalogue.

The main shock took place on 28 December at about 05h 35m local time (23h

54m UT) and destroyed what was left standing in the town of Gracias and in the nearby villages of Flores, San Sebastian, Talqua and Talfua. A small area in and about Gracias slumped into the ground sufficiently to place a part of the town under water. The foreshocks and the main earthquake killed a few people and made about 4,000 homeless. The town was temporarily abandoned. Information is insufficient to separate the effects of the different events.

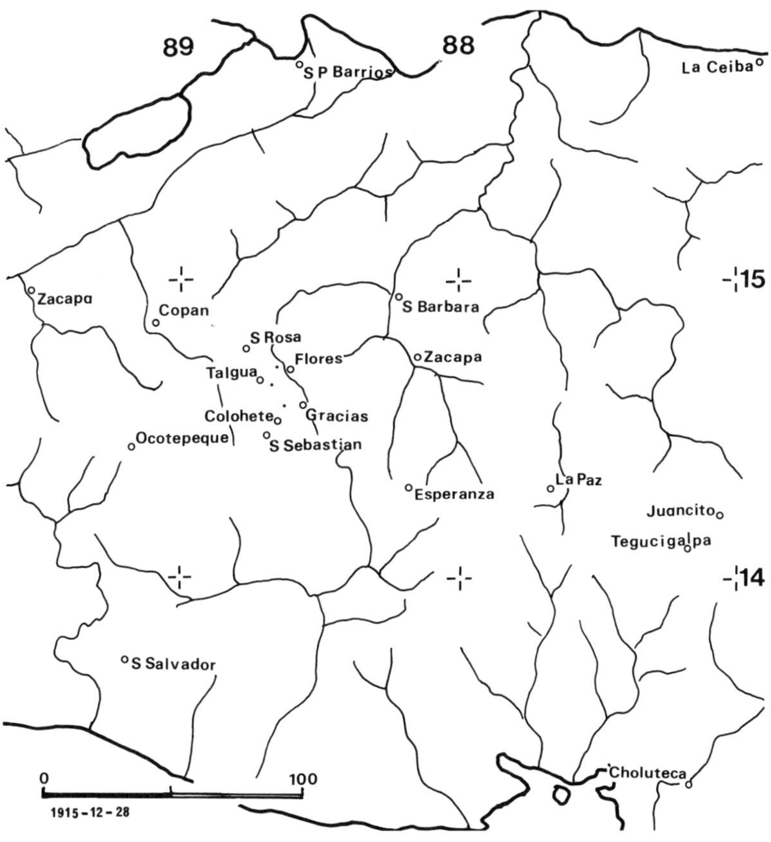

Fig. A.20 Felt effects of earthquake of 28 December 1915.

The shocks caused some damage at Santa Rosa as well as in the Departments of Copán and Santa Barbara, but there are no details about the extent of damage and sites affected. The main shock was strong at Zacapa (in Guatemala or in the Depatment of Santa Barbara), at Ocotepeque and Esperanza, and it was felt as far as Tegucigalpa, Choluteca, San Salvador and possibly at La Ceiba. Press reports allege that as a result of the earthquake a few walls collapsed in San Salvador, killing two people.

The earthquake was well recorded but the data are not enough to establish the epicentral region instrumentally.

[BSSA: 1916.48; DES; Lara 1964; Masters 1916; PDCA]

1916 Feb 8 **M_S 5.5** **Panama**

An earthquake was felt strongly in Panama City and throughout the Canal Zone, causing some concern.

The recordings from Balboa Heights suggest that the epicentre was in the vicinty of Los Santos Province, but we have no felt reports from there.

This earthquake happened about five minutes after a relatively large shock in Chile and instrumental readings are confused.

[CNR: 1916.no.25; DES; Kirkpatrick 1929]

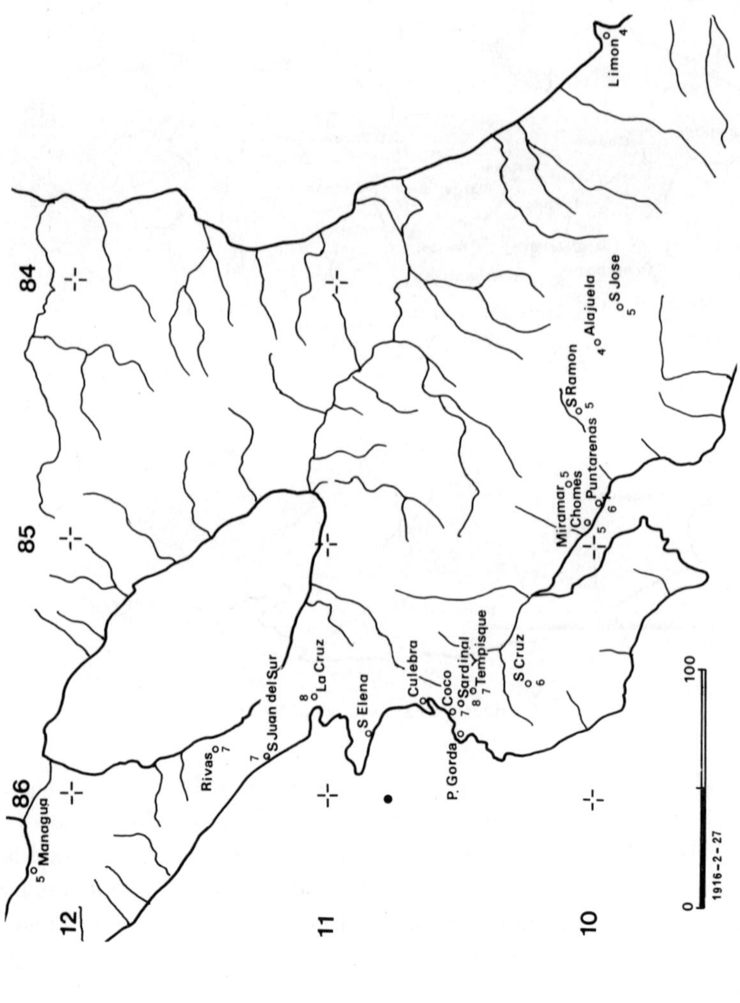

Fig. A.21 Felt effects of earthquake of 27 February 1916.

1916 Feb 27 M_S 7.3 Figure A.21 Costa Rica

This major earthquake in northwestern Costa Rica was preceded by a number of foreshocks.

The maximum intensity of this earthquake was observed in northwestern Costa Rica, southwestern Nicaragua and in Culebra Bay, Cape of Santa Elena, and Punta Gorda. In Rivas the cathedral was damaged and at Coco the shock was very severe causing cracks in the ground. There was partial or total collapse of houses at other unspecified places in the Province of Guanacaste. At Sardinal the bell towers of the church were thrown down and at Tempisque a number of houses were ruined as well as the church and the local prison. There is no evidence that the shock caused any loss of life.

The earthquake did some damage at Santa Cruz and in Puntarenas, but not at nearby Miramar and Chomes. We could find no damage reports from other places in the district of Nicoya.

Sporadic minor damage was reported from San Ramón, Alajuela and from as far as San José and Managua, where the shock was strongly felt, though no damage was done.

The shock was rather strongly felt further east in the rest of the plateau as well as at Limón and along the Caribbean coast, and probably in western Panama.

The earthquake has a relatively well-defined macroseismic area on the coast of western Costa Rica (Figure A.21). The shock was widely recorded and its instrumental relocation agrees with a position off the coast of Guanacaste. Residuals at close stations suggest a focus in the lower crust. Gutenberg and Richter (1965) place the shock 450 km away, off the coast of Guatemala.

[Barquero and Rojas 1994b; Boschini and Montero 1993; DES; Güendel 1986; Jorgensen 1933; Miyamura 1980; PINC; PINF; Tristan 1916]

1916 Mar 12 M_S 6.3 Guatemala

A strong earthquake in Guatemala allegedly caused some damage at Chichicastenango and to farms to its east, but details are lacking.

The shock, which was of long duration, was reported from Quiche, Chimaltenango, Patulul and Chicacao.

[PDCA]

1916 Mar 29 M_S 5.7 Panama

This earthquake was felt in the Canal Zone in Panama where most of the population was awakened.

The shock was recorded at Balboa Heigths at 05h 09m local time (sic) and its epicentre was estimated approximately 80 km southwest of the observatory. This is probably the shock recorded by some of the stations in the region at about 18h 50s and located roughly in Central America.

[CNR: 1916.281; DES; Kirkpatrick 1920]

1916 Apr 24 08h 02m M_S 7.3 Figure A.22 Costa Rica

This earthquake was felt throughout Costa Rica, Panama and southwestern Nicaragua.

It lasted about 30 seconds and caused sporadic, minor damage to old houses over a large area, particularly at Naranjo, Nicoya, Orotina, Puriscal, San José, and San Ramón. The shock set church bells ringing at Alajuela, Grecia, Rio Segundo, San Mateo, San Marcos, and San Ramón, and caused panic particularly in the Nicoya Peninsula and along the coast of Puntarenas.

Information from Nicaragua is lacking except that the shock caused concern in the region of San Juan del Sur, and some minor damage at Rivas.

The earthquake was rather strong at Limón, David and in the Department of Chiriqui in Panama, and it was perceptible at Balboa and in the Canal Zone. It is not known how far to the northwest the shock was felt.

The earthquake was followed by a large number of aftershocks.

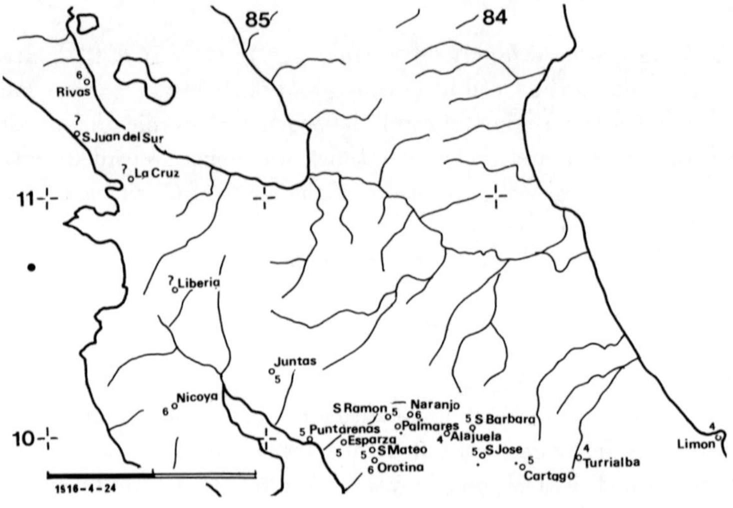

Fig. A.22 Felt effects of earthquake of 24 April 1916.

The available macroseismic information (Figure A.22), is insufficient to allow the unambiguous location of the epicentral area. The earthquake was not very well recorded. The best solution places the focus off the coast of Nicoya, near the origin of the earthquake of 27 February 1916, at a depth of 60 to 100 km, about 110 km from the location of Gutenberg and Richter, to which they assign normal depth. We prefer a depth of about 60 km.

[Camacho and Viquez 1992; CNR: 1916.314,320; Güendel 1986; Kirkpatrick 1920; Miyamura 1980; PEPO; PINC; PPLB; Reid 1917]

1916 Apr 24 08h 31m M_S 6.6 Panama

This was a strong aftershock that followed at 08h 31m, 29 minutes after the main shock. It was reported from Cartago, Esparta, Rio Segundo, San José, San Marcos and other places, but caused no damage. Because of the small time interval between the two shocks their macroseismic effects cannot be separated.

[Feldman 1984]

1916 Apr 26 M_S 6.9 Tsunami Figure A.23 Costa Rica

This strong earthquake occured in the Talamanca Mountains, a sparsely populated region of Costa Rica.

In the mountains of the region of El General and Buenos Aires ground motions were so severe that even timber houses were thrown to the ground and the ground itself opened up in places. Similar effects were also reported from the region north of David in Panama.

In the region of Bocas del Toro, the earthquake did considerable damage. At Almirante 89 houses built on stilts were thrown off their foundations and water tanks collapsed, but there was no loss of life. The shaking was so strong that it was difficult for people to stand. The sole reinforced concrete structure in the town, a three storey building, was damaged beyond repair. Water and electricity services were disrupted and the submarine telephone cable connecting Almirante with Bocas del Toro was cut.

In places the ground liquefied causing extensive damage to the quay wall at Almirante and to the railroad, the tracks of which were bent. The piers of the bridge on Cedar Creek collapsed and Careening Cay was nearly destroyed. Damage to engineering works at Almirante alone was estimated at $200,000 gold.

The shock was equally strong at Bocas del Toro, where houses were thrown off their stilts and the landing pier was damaged. The shock caused widespread liquefaction.

A seismic sea-wave set up by the earthquake carried canoes and debris 200 metres inland and destroyed strorage tanks, adding to the damage. The wave flooded Bastimentos, Isla de Carenero and other parts of the coast. The shock also caused some minor damage at Limón.

At Sixaola, Guabito and Changuinola a number of houses and water tanks were destroyed and rivers overflowed their banks, probably because they were dammed temporarily by slides.

Severe shaking causing panic and sporadic damage was reported from places along the railway line from Limón to San José, in particular at Alajuela, Atenas, Puriscal, Rio Segundo, San Joaquin, San Vicente, Santo Domingo and Remedios in Panama. At Sarapiqui the shock triggered a landslide and it set church bells ringing at Barba, Curidabat, Escasu Rio Segundo and San Joaquin. The shock was also equally strong in Panama at David and Remedios.

The earthquake was felt throughout Costa Rica from La Cruz and Barra de Colorado in the north to Las Tablas, Colón and Balboa Heights in Panama in the southeast, where the shock was strong enough to stop pendulum clocks and to be registered by water-stage gauges at Gatun, Pedro Miguel and Miraflores.

The earthquake was followed by many strong aftershocks, two of which, four and five hours after the main shock, were particularly severe.

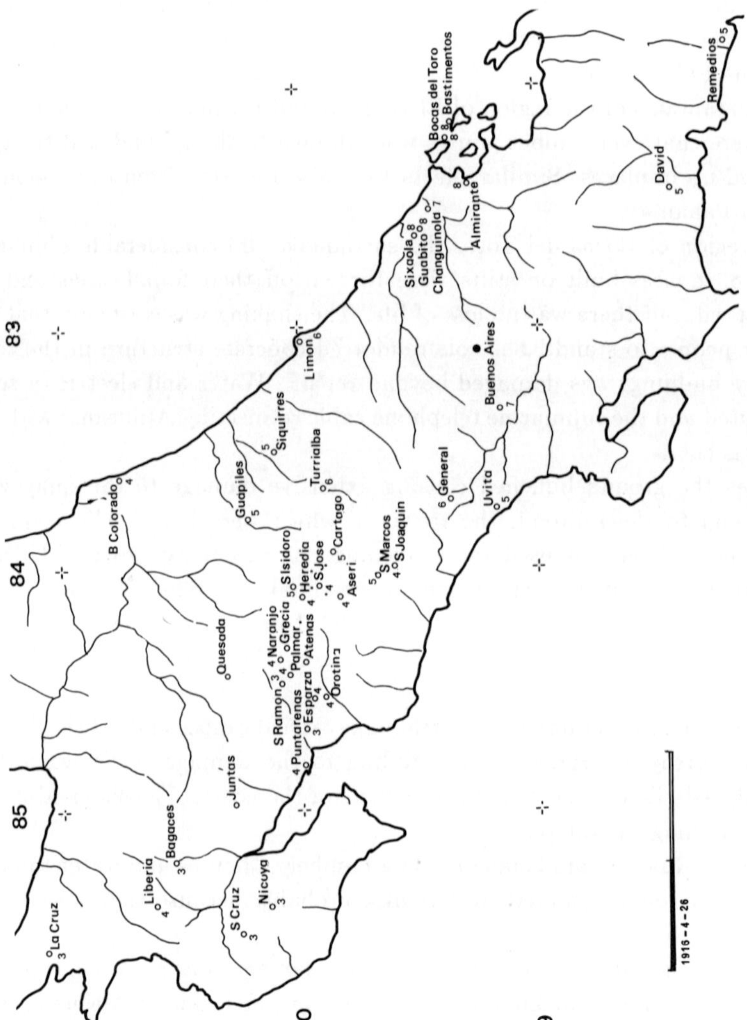

Fig. A.23 Felt effects of earthquake of 26 April 1916.

Macroseismic evidence (Figure A.23) places this earthquake in the border area between Costa Rica and Panama. It was poorly recorded instrumentally, with several inconsistencies in reported times. The best solution, which is uncertain, suggests a location in the Cordillera de Talamanca and a crustal depth, perhaps associated with a low-angle plane dipping southwest beneath the Caribbean coastal plain. Gutenberg and Richter (1956) place the shock 200 km from this position in the Gulf of Nicoya on the west coast of Costa Rica, from where we have no evidence of damage. Some of the effects in the Central Plateau and in Nicoya attributed to this earthquake in the press, belong to the earthquake of April 24.

[Barquero and Rojas 1994b; Camacho and Viquez 1992; CNR: 1916.320; DES; Güendel 1986; Kirkpatrick 1920; Miyamura 1980; PINC; PINF; PPLB; Reid 1917]

1916 May 1 Costa Rica

The earthquake of 26 April was followed by a series of shocks, which were reported from various places in Costa Rica. The earthquake of 1 May was a strong aftershock of the sequence which was particularly strong in the valley of the Rio El General and to the southwest of the Cordillera de Talamanca.

Early press reports of heavy damage to the capital and of an interruption of telegraphic communications in Costa Rica as a result of this earthquake are not founded.

The event was recorded at La Paz.

[DES; Güendel 1968; PEPO; PIMC]

1916 May 10 M_S 6.1 Costa Rica

A strong aftershock off the coast of the Golfo del Dulce in Costa Rica was perceptible in the Canal Zone in Panama.

[Kirkpatrick 1920]

1916 Sep 23 M_S 6.7 Nicaragua

An earthquake with an epicentre in Nicaragua was strong at San Juan del Sur. It was widely recorded.

[DES]

1917 Jun 8 00h 51m M_S 6.7 Figure A.24 El Salvador

This was a damaging earthquake in El Salvador, which was followed 40 minutes later by a small aftershock. The volcano of San Salvador erupted one hour and twenty minutes later.

The first shock occurred at 00h 51m (UT) and almost totally destroyed Quetzaltepeque, Opico, Nejapa, Ateos, Armenia and Tepecoyo in the district of Libertad, killing a few tens of people. Damage extended to Legartos, San Salvador, Soyapango, Santa Tecla, Zaragoza and Paisnal. The tracks of the railroad between Ateos and Nejapa and the station of Nino were destroyed and the mountain road to Ceiba was blocked by landslides.

The earthquake was very strong at Jutiapa and Nueva Vinas, and was felt at San Miguel and at Tegucigalpa in Honduras.

BAAS(1917) reports readings from 34 stations, which they interpret as giving a solution at 14.5°N, 91.0°W and an origin time of 00h 51m 28s. Our recalculated position is 205 km to the east of the BAAS location with no evidence to suggest that the depth is other than normal. This revised position is 75 km to the north

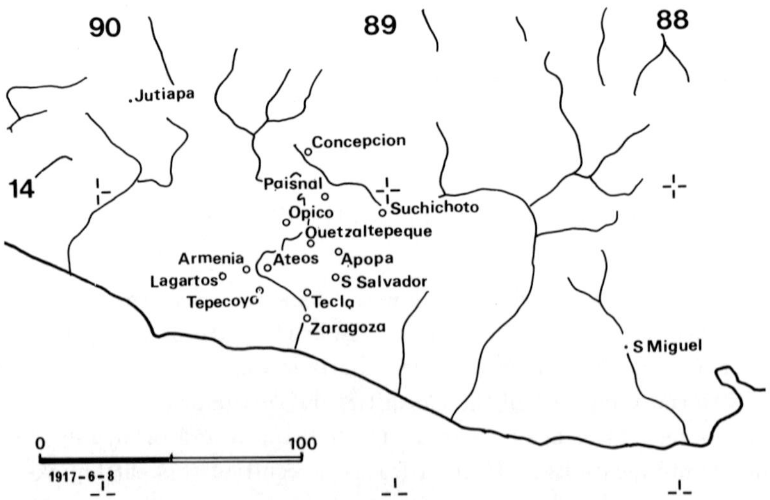

Fig. A.24 Felt effects of earthquake of 8 June 1917.

of the macroseismic epicentre. The seismological control is poor, however, and the station residuals relative to the macroseismic position are not significantly worse. In this case there is no compelling seismological reason not to adopt the macroseismic position.

La Paz recorded the two aftershocks with approximate origin times at 01h 32m and 02h 54m.

[Anonymous 1917; Arenales 1961; BSSA 1917.76,130; DES; Jordan and Martinez 1980; Martinez 1978; PDCA; PECR; PTRL; Schultz 1963; Wacher 1917]

1917 Jun 8 01h 32m **El Salvador**

About 40 minutes after the main shock, a strong aftershock did some additional damage, mainly to settlements in the eastern part of the epicentral area.

One hour and 20 minutes later, at 02h 54m, a third shock marked the beginning of the activity of the volcano of San Salvador in the vicinity; its parasitic cone Jabali erupted, throwing out ash which covered the surrounding country. The flow of lava reached the villages of Nejapa and Apopa, adding to the damage. In addition, the tracks of the Occidental Railway were covered by streams of lava.

Both events were recorded at nearby seismographic stations.

White and Harlow (1993) interpret the last group of long waves recorded at De Bilt at 02h 01m on 8 June as a major aftershock, to which they assign an origin time of 01h 30m, a position at 13.75°N, 89.27°W and a surface-wave magnitude of 6.4. We can find no evidence for such a large magnitude aftershock.

1917 Jun 13 **Guatemala**

This strong earthquake in Guatemala in the region of Mazatenango was particularly severe at Chicacao and Atitlan.

[PDCA]

1917 Jun 30 M_S 6.1 Panama

This earthquake was felt by a number of people at Balboa in the Canal Zone.

BAAS uses readings from 25 stations to place this event off the Pacific coast of Costa Rica. Our relocated position is on the Panama-Colombia border, 690 km to the east of the BAAS location. With the lack of close stations there is no instrumental control of depth, but there is no reason to assume the depth to be greater than normal.

[Kirkpatrick 1920]

1917 Oct 22 M_S 5.9 Nicaragua

We can find no macroseismic reports for this event, which BAAS place off the east coast of Honduras, close to the Mosquito Coast. The earthquake was very poorly recorded and we do not have enough data to offer any alternative to the BAAS solution.

1917 Nov 18 Guatemala

This locally damaging earthquake in Guatemala in the region of Lake Amatitlan was a forerunner of the earthquake of 26 December.

The shock ruined a number of houses, public buildings and the market place at Amatitlan, and caused serious damage to the nearby villages of Villa Nueva, San Joaquin, Villa Canales, Moran and Petapa, localities situated within a radius of about 5 km. The shock was strong at San Vicente, Antigua and Guatemala City, and it was followed by a few aftershocks.

It is alleged that the shock killed all the fish in the Lake of Amatitlan.

We could find no instrunental data for this event.

[AGCA; PDCA; Van de Putte 1924; Zamacois 1918]

1917 Dec 12 Guatemala

Another strong earthquake added to the damage at Amatitlan and Pacaya. The shock was felt in Guatemala City and it was followed by aftershocks until 18 December.

[PDCA; Van de Putte 1924]

1917 Dec 26 04h M_S 5.6 Figure A.25 Guatemala

In Guatemala three successive earthquakes of relatively small magnitude, at intervals of about one hour, caused great destruction in the capital and surrounding villages.

The first shock occurred at 04h 30m UT. In Guatemala City, Mixco and Santa Catarina Pinula many houses and public buildings were destroyed progressively by the shocks, killing about 70 people and making thousands of people homeless. Many of the better built houses in the city and in the surroundings were ruined. The Colón Theatre, the Military Academy, the hospital, most of the churches, prisons and other public buildings were shattered and many of them were destroyed by

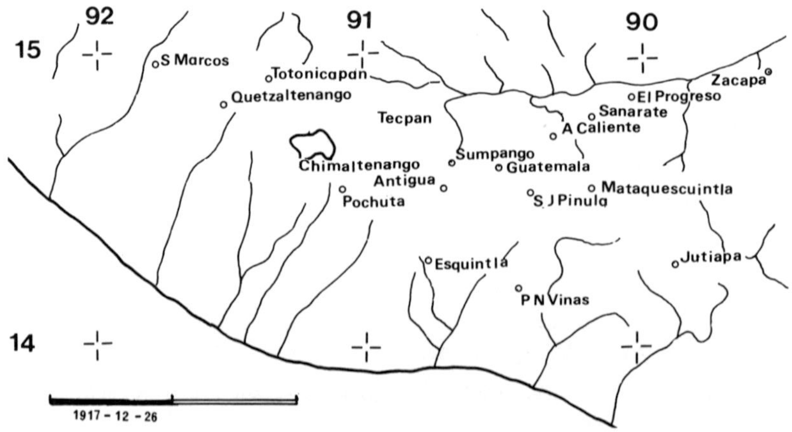

Fig. A.25 Felt effects of earthquake of 26 December 1917.

aftershocks. The dome of the cathedral church collapsed and a part of the old fort fell, killing 150 soldiers. The water mains were burst and the electric supply system was destroyed and not restored for a number of days.

At San José Pinula the barracks, school, court house and many houses were ruined. Minor damage extended to Fiscal, Cerizal, Golfo, Sanarate, and to farms northeast of Guatemala City.

Landslides disrupted the Northern Railway as well as telegraphic comunications with both the Atlantic and Pacific coasts. The long railway viaduct at Fiscal, 12 km northeast from the city, was not damaged but the railway line was cut near Sanarate and Agua Caliente.

The first shock at 04h 30m was recorded only at Tacubaya (TAC) and La Paz (LPZ).

[AEG; AGCA; Anonymous 1918; Diaz 1918; Mitchell 1918; Morley 1918; PBDN; PDCA; Roberts 1924; Saville 1918; Sieberg 1932; Van de Putte 1924; Villacorta 1926; USNA]

1917 Dec 26 05h M_S 5.6 Guatemala

The second shock on this day occurred at 05h 21m UT and its effects cannot be separated from those of the first earthquake. It was felt rather strongly at San Marcos, Totonicapan, Zacapa, Jutiapa and Pueblo Nuevo Vinas, but it was not reported from places along the Pacific coast.

This shock was also very poorly recorded. The only station giving confident *P* and *S* phases is La Paz; these are consistent with the macroseismic epicentre.

1917 Dec 26 06h M_S 5.1 Guatemala

The third shock of this sequence, at 06h 18m, was also poorly recorded, with no instrumental control. For both the second and third shocks BAAS(1917) adopts a location at 10.5°N, 91.0°W, about 460 km south of our macroseismic position.

1917 Dec 29 M_S 5.2 Guatemala

This was a strong aftershock of the earthquake of 26 December.

It caused further damage to Guatemala City and to surrounding villages. About 20 people were killed, mainly from the collapse of adobe houses already damaged. After this earthquake little remained standing in Guatemala City except outside walls, and these were demolished by the aftershock of 4 January 1918. As a result of the continuing shocks the city was evacuated.

As a result of the First World War, there is little instrumental information for this event.

[Anonymous 1918; Diaz 1918; Morley 1918; PBDN; Van de Putte 1924]

1918 Jan 4 M_S 6.0 Figure A.26 Guatemala

Another two shocks in Guatemala, two minutes apart, completed the destruction.

Much of what was left standing in Guatemala City by the previous shocks, and in the nearby villages of Los Mixcos, Villa Nueva, Santa Catarina, San José Pinula, San Joaquin Villa Canales and Villa Nueva collapsed, killing more than 100 people. After this earthquake only 20% of the houses of the city were left standing.

The several supplies of water for Guatemala City, consisting of four main supplies known as the Acatan, Minas, Pinula and Mixco, and the two private supplies of Raicare and Sance, were seriously damaged by the shock of 26 December, and this was aggrevated by the shock of 4 January. The main lines approaching the city were damaged and the distribution pipes in the streets were broken in many places. However, there were still many outlets and sources of flowing water after January 4.

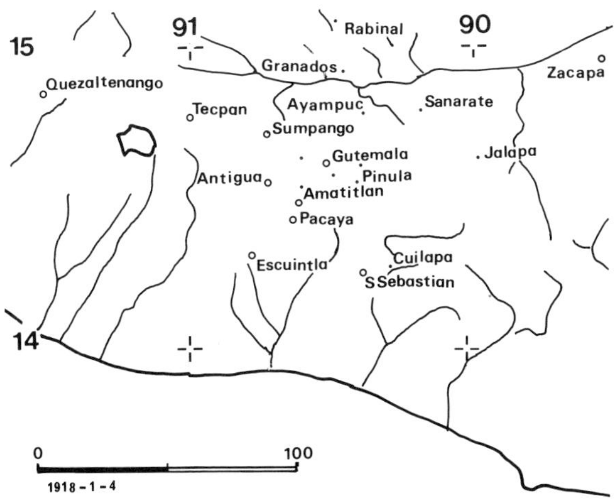

Fig. A.26 Felt effects of earthquake of 4 January 1918.

In the surroundings of the city and within a radius of 30 km, all adobe houses were totally ruined.

The shock was felt at Quezaltenango.

The ISS report readings from 16 stations for the first shock at 04h 32m and from 21 stations for the second shock, about two minutes later and give for both events a position at 10.5°N, 91.0°W, which is 450 km to the south of the epicentral area.

The first event at 04h 30m UT was very poorly recorded with only five readings of possible *P* phases.

The second shock at 04h 33m UT was somewhat better recorded giving a solution about 100 km to the north of the macroseismic epicentre with no evidence for a deep focus.

The macroseismic positions are to be preferred.

[Anonymous 1918; Diaz 1918; Mitchell 1918; Morley 1918; PBDN; PDCA; Saville 1918; Van de Putte 1924; Villacorte 1926; USNA]

1918 Jan 25 — M_S 6.2 — Figure A.27 — Guatemala

This was the last and largest shock of the series in Guatemala.

It did much additional damage and extended the area of destruction to the northeast of Guatemala City.

As the city itself had been almost entirely demolished by the previous shocks, not much new damage was done there by this shock. The statue of Christopher Colombus which was still standing, was overturned and only the Cathedral and the churches of Santa Rosa and del Carmen, both in ruins, were left standing.

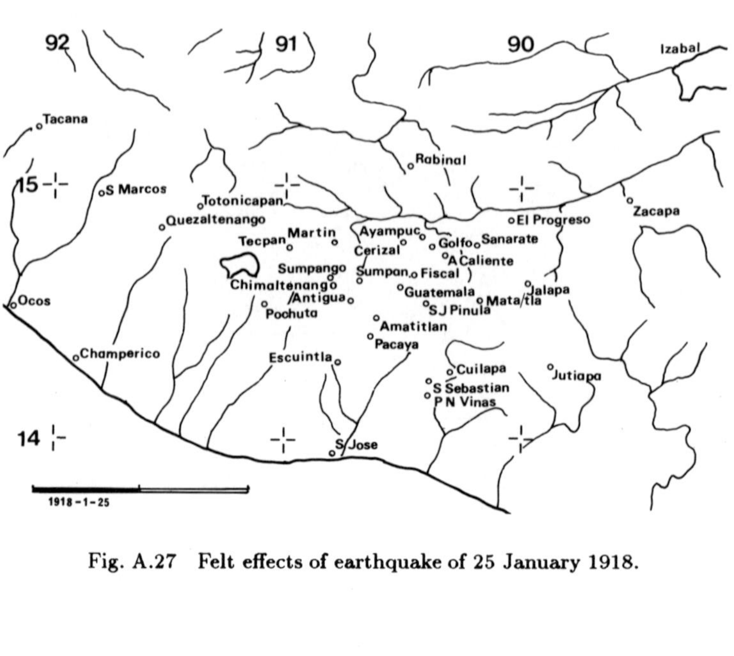

Fig. A.27 Felt effects of earthquake of 25 January 1918.

The shock damaged the water distribution system further so that nearly all the outlets and sources ceased flowing.

The villages of Mixo, San Pedro and San Juan Sacatepeque, San Pedro Ayampuc, Agua Caliente, and San José Pinula to the north and northeast of the city were ruined, with some loss of life.

In Sumpango and Antigua almost all the buildings and churches, which had survived earthquakes since the 19th century, were also ruined and many collapsed.

At Amatitlan, Villa Nueva, Petapa, and in other settlements many of the houses that had not been not destroyed by the previous shocks collapsed, injuring a number of people.

New landslides disrupted again telegraphic and rail communications, particularly along the Guatemala City - Agua Caliente section. Where it ran in deep cuts, the track was blocked by slides, and on level ground by differential settlements. The railway viaduct at Fiscal was not damaged except for its approach structures.

The shock was felt over a large area. It was strong at San Sebastian, Pueblo Nuevo Vinas, Escuintla, Sumpango, Rabinal and Jalapa.

It was felt as far as Tecpan, Ocós, Champerico and San José, along the Pacific coast, and as far as Zacapa and Izabal in the northeast of the country.

Progressive damage in this series of earthquakes makes it almost impossible to assess the distribution of intensity associated with individual events.

This earthquake seems to have been associated with surface faulting which runs in a north-south direction east of Guatemala City. The second of three unpublished field reports written on 4 February 1918 by Col. L. Leonardo, accompanied by a sketch map and addressed to the President, in free translation, reads:

> I have the honour to inform you of the results of my second visit to the region of the ground cracks in the zone of Yampuc.
>
> Crack #5, like in the zone of Santa Rosa, runs in a north-south direction.
>
> Crack #1, like #3, extends from Las Lomas de Menocal to the west of Nacahuil and on 24 of January caused the destruction of the houses of the workers and of the settlement of Nacahuil. From Las Lomas de Menocal the second line of cracks continues until it breaks the Cerro de Azapote which is located to the north of San Pedro Yampuc, breaking up the roads that it crosses on its course.
>
> Cracks #3 is nearly parallel to #1 and follows the valley of Capullin, reaching the Cerro de Azapote causing a separate rupture in the rocks from which a new spring of water issues that increased the flow of the Naranjo river. These two cracks reappear further to the north in the El Pinal de las Gracia.
>
> Crack #4 shows its largest width of 80 cm in the "ranchos de los muralles" at Lo de Reyes, goes around the tableland and crosses the Quebrada Oscura, and heads for the Golon mountain where it

> joins crack #5 that comes from the Rodriguitos, passing through the slopes of Cerro de Chile.
>
> These two lines of cracks break the Golon mountain and give rise to five streams, one of them of warm water, which together run along the course of the winter streams to the north...

There is much inconsistency among instrumental readings for this event, which ISS place in the Pacific, 610 km to the southwest of the macroseismic epicentre. Our recalculated location, which is rather poor with no control of depth, is 170 km to the southwest of the macroseismic location. The macroseismic position is to be preferred.

[AEG; AGCA; Anonymous 1918; Diaz 1918; Leonardo 1918; PDCA; PBDN; Saville 1918; Van de Putte 1924; Villacorte 1926]

1918 Feb 1 **El Salvador**

Press reports from San Salvador refer to a violent earthquake, as a result of which telegraphic communications with the interior of Guatemala were interrupted. We could find no other macroseismic or instrumental information for this event which may be confused with the earthquake of 25 January in Guatemala.

[DES; PDCA]

1918 Feb 21 **Guatemala**

A strong earthquake was felt at night at Pochuta in Guatemala.

[PDCA]

1918 Apr 5 **Guatemala**

A strong earthquake in Guatemala was felt at Huehuetenango, Quezaltenango and Sacatepeque. It is not known whether it caused any damage, but the shock was felt also in Guatemala City. The date of the event is not certain.

[PDCA]

1918 Apr 20 **Guatemala**

This was a damaging shock in southeastern Guatemala, as a result of which most of the houses at Jalpatagua became uninhabitable and quite a few collapsed, injuring a considerable number of people. The church and a number of public buildings were damaged. The shock also ruined a number of nearby farms.

The shock was felt at Santa Ana in El Salvador but it was not reported felt from Guatemala City.

We can find no evidence that the shock was recorded instrumentally.

[AGCA; PDCA]

1918 Apr 27 **M_S 5.3** **Costa Rica**

There are no macroseismic reports for this earthquake, which was very poorly recorded at 22 seismographic stations. ISS give a position in southeast Costa Rica, but the data are so poor that no redetermination can be made. There is no option but to accept the ISS solution.

1918 Jun 13 M_S **5.5** **Honduras**

There are no felt reports for this event which was not well recorded, with no station closer than 25°. The most likely recalculated solution, with estimated errors of about half a degree, is in Honduras, 120 km northwest from that of ISS.

1918 Jun 16 M_S **5.6** **Nicaragua**

This earthquake was felt strongly in Nicaragua, in Managua and at Jinotega (Matagalpa) where it caused some small damage. As a result of the main shock and its two aftershocks telegraphic communication between Managua and the rest of the country was interrupted. There is no evidence that the shock caused any destruction.

The instrumental readings were too few and too inconsistent for a reliable solution to be found. ISS suggests that European readings at this time belong to another event. It is best to adopt the macroseismic position, although there is some instrumental evidence to suggest that the epicentre might be about one degree further north.

[Jorgensen 1933]

1918 Jun 22 M_S **5.7** **Costa Rica**

This is a poorly recorded event for which we could find no macroseismic information. It could possibly be near the ISS position in eastern Costa Rica, which we adopt.

1918 Jul 31 M_S **6.2** **Nicaragua**

This earthquake has an ISS epicentre off the Pacific coast of Nicaragua. There is no macroseismic information.

1918 Oct 19 M_S **6.2** **Guatemala**

According to press reports this earthquake was destructive in Guatemala, but details are lacking.

[DBN; UPP]

1918 Oct 29 M_S **6.0** **Costa Rica**

This is a poorly recorded earthquake for which we could find no macroseismic information.

It could possibly be near the position adopted by ISS off the Pacific coast of Costa Rica, which we adopt.

1918 Nov 16 M_S **5.3** **Guatemala**

A strong earthquake in Guatemala caused some damage at Chiche, Santa Cruz and Joyabaj. The shock was felt at Guatemala City.

[BSSA; PDCA]

1919 Apr 17 M_S **7.0** **Figure A.28** **Guatemala**

This was a subcrustal earthquake in western Guatemala.

It was very strongly felt over a large area and caused sporadic damage. At Quezaltenango the shock damaged the tower of Centro America and caused considrerable concern at Auytla, Coatepeque, Mazatenango, Pajapita, Puerto de San José, Retalhuleu, San Felipe, San Joaquin, San Lucas Toliman, Santa Lucia Cotzumalhuapa, Solola, Totonicapan and Tumbador. At Tapachula, Ciudad Hidalgo and in nearby settlements of Soconusco the earthquake is said to have caused considerable damage, but details are lacking.

In Guatemala City the shock was strong enough to stop the pendulum clock at the observatory.

The shock was felt throught El Salvador where it caused some slight damage at Santa Ana.

The earthquake was followed by a few weak aftershocks. It was well recorded instrumentally and readings confirm the general area, suggesting a focal depth deeper than normal.

[Figueroa 1973; PDCA]

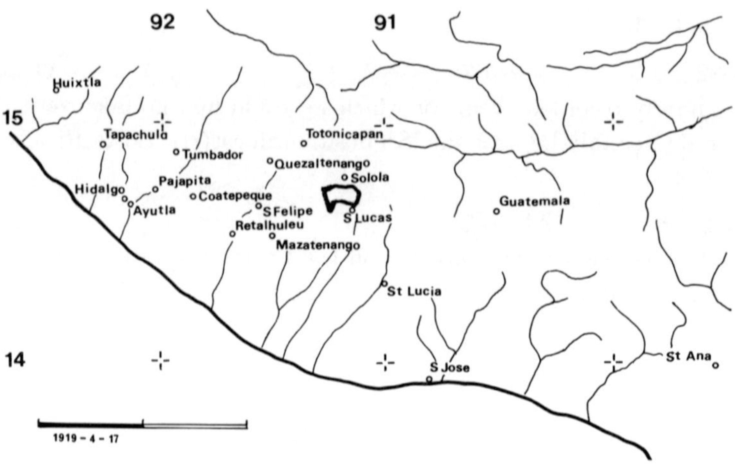

Fig. A.28 Felt effects of earthquake of 17 April 1919.

1919 Apr 27 **Nicaragua**

An earthquake in Nicaragua, probably of subcrustal origin, was felt along the entire Pacific coast from Corinto to San Juan del Sur. It did no damage, but caused considerable alarm at Masaya.

The earthquake was recorded at San Salvador and Guatemala City, but we could find no instrumental information from other, more distant, seismograph stations.

[PDCA]

1919 Apr 28 **M_S 5.9** **Figure A.29** **El Salvador**

This locally destructive shock in El Salvador occurred without foreshocks. It caused heavy damage at San Salvador and to the nearby settlements of Garita,

Soyapango, Cerro de San Jacinto, and San Marcos. In San Salvador many public buildings, including the American legation and the few reinforced concrete buildings that existed at that time, were damaged, some of them beyond repair. Much of the destruction was attributed to the weak state in which most of the houses in the region were left after the earthquakes of June 1917.

At Agua Caliente the shock triggered landslides and damaged the electric power station. Contrary to later reports this station was destroyed after the earthquake; it was washed away by the water of the lake when an unsuccessful attempt was made to open its outlet into the Jiboa river.

Less serious damage extended to Mejicanos, Ayutuxtepeque, Apopa, Tonacatepeque, Ilopango, Santo Tomas, Los Planes and Cuscatlan.

In all, about 70 people were killed and more than 500 were injured.

As a result of the earthquake a large number of fish were killed in Lake Ilopango and the temperature of the lake water rose.

The shock was strong at Moyuta, Metapan and San Miguel and was felt throughout the country, but not at Guatemala City.

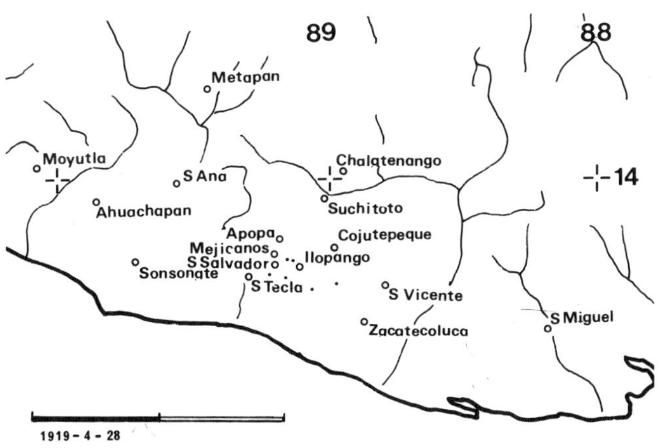

Fig. A.29 Felt effects of earthquake of 28 April 1919.

The earthquake was followed by relatively few aftershocks.

It is alleged that as a result of this shock, or of the earthquake of 27 April, a small island appeared in the Gulf of Fonseca.

This event was poorly recorded, and unusually there were no reliable readings from Europe. There is no seismological evidence for a subcrustal depth.

[DES, Jordan and Martinez 1980; Martinez 1978; Lomnitz and Schulz 1966; PDCA; PDRL; PDSS]

1919 Jun 29 **M_S 6.7** **Honduras**

This earthquake caused considerable damage in the region of the Gulf of Fonseca, but no casualties.

At La Unión, San Miguel and Choluteca many houses were damaged. The shock was strongly felt at Chinandega and it was perceptible in Managua.

Gutenberg and Richter (1956) assign to this event a focal depth of 90 km but we found no support from instrumental readings.

[Martinez 1978]

1919 Jul 22 **M_S 5.6** **Guatemala**

This relatively small magnitude earthquake was strongly felt in Guatemala City and in other parts of Guatemala.

Gutenberg and Richter (1956) place the focus of this event about 700 km away, west of Lake Managua, at a depth of 150 km.

We adopt the macroseismic position.

[DES]

1919 Oct 14 **M_S 5.7** **Guatemala**

The facts about this earthquake are not clear. The foreign press report a severe earthquake which was felt at Amatitlan, a town situated about 20 kilometres from Guatemala City, where the shock was not felt. They add that at Puerto Barrios, which is 260 km to the northeast of Amatitlan, the intensity of the shock was such as to cause general alarm. In contrast, we can find no information about these events in the Guatemalan press. It is probable that these reports refer to two different events.

ISS place the shock in the Pacific, off the coast of Nicaragua. The earthquake was too poorly recorded instrumentally to help resolve the ambiguity.

[DES]

1920 Jul 16 **M_S 5.8** **Costa Rica**

This earthquake was strongly felt throughout almost all the Central Plateau of Costa Rica and in the District of Puntarenas. There is no evidence that it caused any damage.

[Miyamura 1980; PDCR]

1920 Oct 8 **M_S 6.1** **Mexico**

This earthquake was felt along the coast of Chiapas in Mexico and was particularly strong at Mojarras.

[Figueroa 1970]

1920 Oct 28 **El Salvador**

This earthquake was felt in San Salvador and caused panic in the region between San Miguel Tepezontes, Comasagua and the coast. It was followed by a few aftershocks.

[Martinez 1978; STR]

1920 Nov 6 M_S 5.8 Guatemala

An earthquake recorded at San Salvador at 4h 19m local time, was felt in the city and throughout the central and western districts of El Salvador, but caused no damage. The shock was also reported as felt in the region of Jutiapa in Guatemala, where it caused panic.

Apparently, this is the earthquake for which ISS adopt an epicentre 2,200 km away in the Dominican Republic, from where there is no evidence that a shock was felt.

[Martinez 1978]

1920 Dec 11 M_S 6.3 Guatemala

For this this earthquake we could find no macroseismic information except that it was felt in El Salvador and in eastern Guatemala.

There were few records of this event, but they were consistent, confirming a crustal event in the region of Jalapa.

[TAC]

1921 Feb 4 M_S 7.2 Guatemala

We have no macroseismic information for this event. Gutenberg and Richter (1956) locate its focus in central Guatemala at an intermediate depth. An alternative solution for a shallow event in western Belize is given by ISS.

1921 Mar 28 M_S 7.4 Figure A.30 Nicaragua

This major intermediate-depth earthquake caused widespread but not very serious damage over a large area of eastern El Salvador, southern Honduras and northwestern Nicaragua.

At San Juan Nonualco, San Nicolas Lempa, Usulutan and Leon a few adobe houses were ruined and many suffered some damage. Zacatecoluca was badly damaged, but there were no casualties. At San Miguel, La Unión, in settlements around the Gulf of Fonseca and in Managua houses of similar construction were cracked and a few were ruined. At San Salvador the shock caused fall of plaster in a number of houses, and at San Juan del Sur there was panic among the population.

The earthquake was strongly felt all along the Pacific coast from Guatemala to Costa Rica, a distance of 750 km, and it was strong at Tegucigalpa.

The extent of the earthquake's felt area is not known, but the sparse macroseismic information available is shown in Figure A.30.

The large area over which the shock caused sporadic damage, the even larger area over which it was felt, and the absence of a well defined epicentral area all suggest an earthquake of intermediate depth. The event was well recorded worldwide. After discarding some readings that appear to be in error, arrivals from 87 stations were analysed, starting with the ISS position, and a good solution was obtained in northwestern Nicaragua. The solution is fitted well by various later phases, some of which help to confirm a depth of 170 km. Arrivals beyond the *P* range, can sometimes be identified as *PP* and *SKP*, which are phases expected at such distances.

[DES; Jordan and Martinez 1980; Martinez 1978; PDCA]

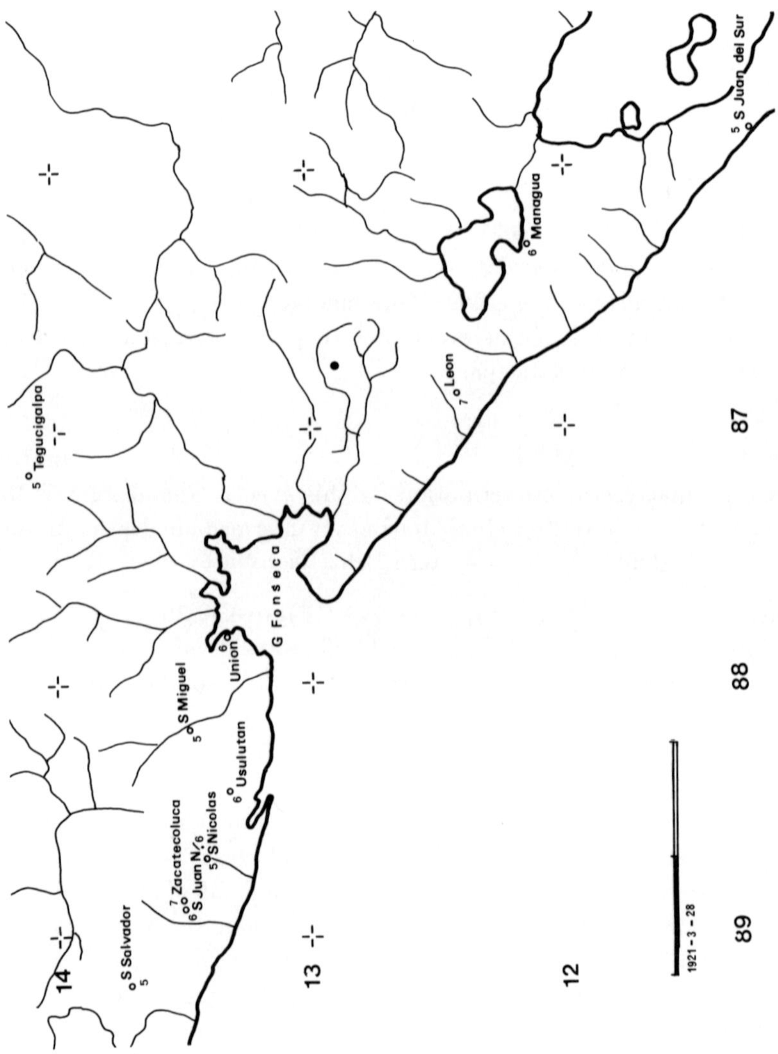

Fig. A.30 Felt effects of earthquake of 28 March 1921.

1921 May 14 El Salvador

An earthquake was felt throughout El Salvador. There have been no reports of casualties or of material losses. No instrumental data have been found for this event, and no location can be given.

[DES]

1921 Nov 13 **Nicaragua**

A strong earthquake was experienced in several towns of Nicaragua. No damage resulted, and no location can be given.

[DES]

1922 Feb 16 **M_S 6.4** **Figure A.31** **Nicaragua**

This earthquake occurred in Nicaragua.

In Granada part of the cathedral was badly damaged, as well as some houses and buildings. Also at Rivas, San Juan del Sur and in the district of Carazo some damage was done to a number of local houses. It is said that as a result of the earthquake the ground near the coast of Lake Nicaragua was fissured.

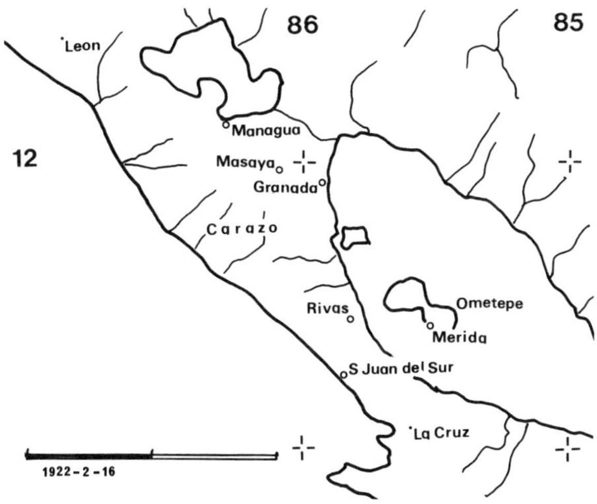

Fig. A.31 Felt effects of earthquake of 16 February 1922.

Aftershocks continued to be felt for almost a month. The strongest aftershocks occurred on 4 and 10 March.

The Ometepe Volcano, which was active at the time of the earthquake, ejected new lava and the inhabitants of Merida near the volcano fled to the mainland.

The shock was strong at Leon and in Costa Rica at La Cruz, but details are lacking.

[DES; Jorgensen 1933; STR]

1922 Apr 20 **M_S 5.8** **Guatemala**

A strong earthquake was felt in Guatemala. The available information is insufficient to establish the epicentral region, except that it must be near 15°N, 91°W.

ISS, using readings from 16 stations, place the shock 1,100 km west of Guatemala in the Pacific.

[DES]

1922 Dec 8 **M_S 5.1** **Costa Rica**

This strong earthquake in Costa Rica caused some damage to houses within a small radius around Santa Cruz di Turrialba. The earthquake was felt at San José, and was followed by many small shocks until the end of the year.

Based on readings from eight stations, ISS place the epicentre of this event 900 km northeast of Costa Rica.

[DES, Sieberg 1932]

1923 Jan 10 **El Salvador**

A strong earthquake in San Salvador followed by a long series of aftershocks caused considerable concern and the level of Lake Ilopango to be lowered. It cannot be located.

[Sieberg 1924]

1923 Feb 2 **Nicaragua**

An earthquake in Nicaragua, probably of volcanic origin, ruined a number of farms on the island of Ometepe in Lake Nicaragua. The shock was not reported felt from other parts of the country, and it was followed by a reactivation of the Volcano Ometepe. No location can be given.

[Sieberg 1924]

1923 Feb 17 **El Salvador**

This earthquake in El Salvador was strongly felt in the capital and was reported from various places in the interior, which are not specified. The shock caused no damage and was followed by an aftershock on 20 February. We cannot locate it.

[DES; Sieberg 1924]

1923 Jul 14 **Nicaragua**

A series of shocks in Nicaragua culminated in a strong earthquake at San Francisco in the District of Marazan. No locations can be given.

[DES; Sieberg 1924]

1923 Nov 29 **Nicaragua**

This earthquake was felt throughout Nicaragua.

Its epicentral area was near Jinotega where much damage was done. No instrumental data have been found.

[DES; Jorgensen 1933; Sieberg 1924]

1923 Dec 26 **M_S 4.9** **Guatemala**

A series of strong local earthquakes caused extensive cumulative damage in Guatemala City.

Many houses and public buildings were ruined, including the Cathedral and the Capitol. The shock killed a few persons and injured about 100. The aftershock sequence lasted for almost a month and affected a small area between Guatemala City and Amatitlan.

The strongest shock was recorded instrumentally only at Tacubaya (TAC) and Mazatián (MAZ).

[DES]

1924 Mar 4 M_S 7.0 Figure A.32 Costa Rica

This relatively large magnitude earthquake in Costa Rica had an epicentral area in the region of the Gulf of Nicoya. The earthquake was preceded by a small shock and was followed, one and a half hours later, by a very strong aftershock.

Maximum damage from both shocks occurred in the region between Machuca, Higuito, San Mateo, Orotina, Coyolar and the coast, including the Rio Grande Valley. In the region of Turrubares almost all rural settlements and villages were totally ruined, with some loss of life.

The banks of the Rio Grande slumped in places and large cracks opened in the ground. Many roads were blocked by slides and rockfalls.

Communications by rail were disrupted for two weeks. Segments of the Pacific Railroad between Machuca and Quebradas were destroyed, and the railway line was affected by slides and weakened bridges between San Mateo and Alajuela.

Heavy damage, in places serious with a few casualties, extended eastwards. San Pablo, Alajuela, Heredía and San José were severely damaged and nearly all the public buildings, churches, graves and barracks were ruined or damaged to the extent that they had to be rebuilt.

In San José, with the exception of a number of modern office and public buildings including wooden houses, there was not a single house which was not damaged, some of them beyond repair. The Metropolitan Cathedral, National Opera House and other public buildings were damaged. Warned by the foreshock, the people went out in the streets, so there was no loss of life and very few injuries. Because of the size of the city, the greatest financial loss occurred in San José, estimated at $750,000 gold.

Minor damage, aggravated by numerous aftershocks, extended to Puntarenas, Esparta, San Ramón, Cartago, Playón and Jaco. It is not known whether the shock did any damage in the Nicoya Peninsula.

The earthquake was felt over a large area. In the north it was reported from the district of Chontales in Nicaragua, Bluefields and San Juan del Norte on the Caribbean, and from the east coast of Lake Nicaragua. In the south it was felt strongly at Buenos Aires, in the district of Talamanca and at La Conceptión and David in Panama.

Total material losses were estimated at $1.8 million.

Damaging aftershocks continued until 28 March, adding to the damage and enlarging the area affected. The aftershock sequence terminated late in July.

The macroseismic data available for this event clearly indicate an upper crustal focus in the Gulf of Nicoya (Figure A.32). The earthquake was well recorded, giving no suspicion of depth. The unusually large area for which we have low intensity felt reports is due to the great amount of information that we have been able to retrieve, rather than the large area over which the shock was felt.

[Barquero and Rojas 1994b; Güendel 1986; Miyamura 1980; PCNT; PDCO; PDCR; PNUP; Sieberg 1932; Tristan 1925; USNA]

Fig. A.32 Felt effects of earthquake of 4 March 1924.

1924 Mar 11 **M_S 6.3** **Costa Rica**

A damaging aftershock in the District of Puriscal caused the collapse of houses in the Bajos de San Juan. The shock was strong at Esparza, San Ramon and San José.

[Miyamura 1980; PDCO; PDCR]

1924 Mar 25 **M_S 6.2** **Figure A.33** **Costa Rica**

Instrumental analysis of this earthquake of the Costa Rican sequence suggested an intermediate depth. This is consistent with the fact that it was strongly felt over a large area, but caused no material losses except in the region of Turrubares, where additional damage was done to houses already weakened by previous shocks.

The earthquake was perceptible at Rivas and Chiriqui.

[Miyamura 1980; PDCO; PTRB]

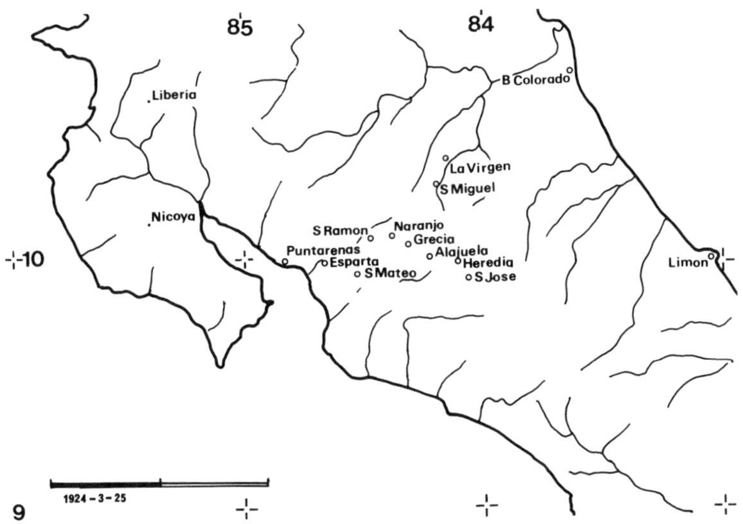

Fig. A.33 Felt effects of earthquake of 25 March 1924.

1924 Apr 23 **Costa Rica**

A small earthquake in Costa Rica was felt within a small area encompassed by La Virgen, San José, Orotina and Quesada. It was strong in the region of Naranjo, but it caused no damage there.

We could find no instrumental data for this event.

[PNUP]

1924 May 1 **M_S 6.6** **El Salvador**

No macroseismic data are available for this earthquake.

It was very well recorded instrumentally; a free solution gives a position in the Pacific south of El Salvador, and a crustal depth.

1924 Jul 12 **El Salvador**

A strong earthquake was felt at San Salvador; no damage was reported, and no location can be given.

[DES]

1924 Oct 10 **Ms 4.9** **Figure A.34** **Guatemala**

This small earthquake in Guatemala, in the region between Guatemala City, Cuilapa and Antigua, was strong, but only at Amatitlan was there damage to houses, caused by the main shock and its numerous small aftershocks.

The shock was not felt over a large area.

[DES; MEC; Sieberg 1932]

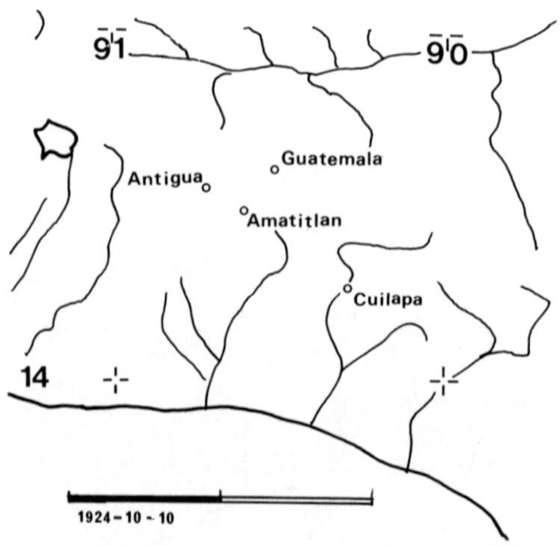

Fig. A.34 Felt effects of earthquake of 10 October 1924.

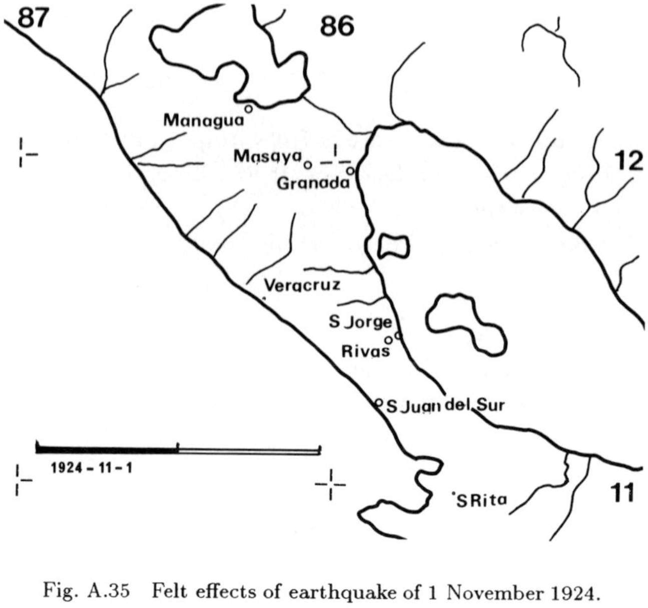

Fig. A.35 Felt effects of earthquake of 1 November 1924.

1924 Nov 1 M_S 6.0 Figure A.35 Nicaragua

This earthquake had an epicentre off the coast of Nicaragua.

Allegedly it caused some small damage in the coastal area of Veracruz in Carazo. The shock was strongly felt at Managua, Granada, Rivas, San Jorge, San Juan del Sur and at Masaya, in some of these places causing panic.

The earthquake was also reported felt from the region of Santa Rita, but details are lacking.

[DES; Jorgensen 1933]

1925 Jan 15 El Salvador

An earthquake of considerable intensity which lasted for about 30 seconds was felt in San Salvador. Its epicentral area was about 160 km from San Salvador.

[DES]

1925 Feb 10 El Salvador

This earthquake was felt in San Salvador. It is believed to have been centred on the northern coast of Nicaragua or Honduras.

[DES]

1925 Apr 4 El Salvador

An earthquake was felt at San Salvador and in the western part of the country, but details are lacking, and no location can be determined. The shock was recorded at the local seismological station.

[DES]

1925 Apr 15 Nicaragua

An earthquake, followed by two strong aftershocks, was felt in Managua. It was probably of volcanic origin; the Volcano Ometepe in Lake Nicaragua discharged lava and emitted ash, ruining the crops of nearby farms. We have no location.

[DES]

1925 Sep 30 Nicaragua

A series of local earthquakes occurred in Nicaragua. At Masaya many buildings were damaged including the cathedral. The shocks were also felt in Managua. We cannot give a location.

[DES]

1925 Oct 5 M_S 6.4 Nicaragua

This earthquake was strongly felt along the Pacific coast from Managua in Nicaragua to Alajuela in Costa Rica. No damage was reported.

Gutenberg and Richter (1965) assign this event a focal depth of 135 km.

[DES]

1925 Nov 28 M_S 5.5 Mexico

An earthquake in the District of Chiapas in Mexico caused great panic at Altamirano and Ocosingo.

[Figueroa 1973]

1925 Dec 10 **M_S 7.1** **Mexico**

This was a relatively large magnitude earthquake in the Gulf of Tehuantepec in Mexico. The shock was very strong at Zanatepec, Arriaga and other towns near the coast of Oaxaca. It is not known how far the shock was felt.

Our macroseismic evidence suggests a position two degrees further west than that given by ISS.

In the Catalogue three aftershocks are given at the ISS positions, but they too are probably further west, closer to our estimate for the position of the main shock.

[Figueroa 1973]

1925 Dec 18 **Nicaragua**

A series of strong shocks was felt at Rivas and San Juan del Sur in Nicaragua associated with the continuing eruption of the Volcano Ometepe.

[DES]

1926 Feb 8 **M_S 7.1** **El Salvador**

We could find no macroseismic information about this earthquake. The shock was well recorded, and reported in ISS. A free solution places its focus south of El Salvador in the Pacific at a depth of about 100 km. There is some scatter in residuals, and in particular many European stations seem late, perhaps being *pP*. The solution cannot be greatly in error, though, and the depth seems well established.

1926 Feb 10 **M_S 5.7** **Nicaragua**

Macroseismic information for this earthquake is lacking.

The event was very poorly recorded instrumentally and the ISS position is not well established. There is some suggestion from the readings that the location is further west, and we have adopted a position near the Gulf of Fonseca.

1926 Mar 24 **M_S 5.9** **Mexico**

A very strong earthquake was felt at Chicomuselo and Comalapa in Chiapas in Mexico. The shock was widely felt but there are no details.

[Figueroa 1973]

1926 Apr 12 **Nicaragua**

Two earthquake shocks caused some small damage in Managua and vicinity.

[DES]

1926 Jun 29 **M_S 5.4** **Guatemala**

The facts about this earthquake are not clear.

ISS adopts an epicentre in the Pacific, but a relocation by Wysession *et al.* (1991) moves it 1,850 km to the northeast to Chiapas in Mexico. This is about 200 km from the location given by Tacubaya, who place its focus at a depth of 100 km.

The shock was felt in Chiapas and probably in the Totonicapan region in Guatemala, at Quezaltenango and in Guatemala City.

[Sieberg 1932; TAC]

1926 Oct 19 M_S 5.8 Costa Rica

This earthquake in Costa Rica was very strong in the region between Turrialba and Squires, but it is not known whether it caused any damage. The shock was felt at San José.

[STR]

1926 Nov 5 M_S 7.0 Nicaragua

This was a relatively large magnitude, intermediate-depth, earthquake with a focus in Nicaragua.

It is said that the worst damage was done at Leon; allegedly, 80% of the houses sustained damage and many were totally destroyed. In Managua about 50% of the houses were damaged, the shock causing material losses of £800,000. Chinandega, Granada and Rivas also suffered some damage. At San Juan del Sur several persons were killed and severe property damage was done along the coast. Later press reports add that the overall damage was, in fact, not so serious and that the shock was perceptible in Costa Rica and El Salvador, from where details are lacking.

The shock was very violent on board two ships sailing off the coast of Costa Rica, 330 km from the instrumental epicentre.

The damage distribution and large area over which this earthquake was felt suggest an intermediate depth. We adopt, without recalculation, the solution of Gutenberg and Richter, which gives a focal depth of 135 km.

[DES; MEC; Turner 1930]

1927 May 9 M_S 6.1 Mexico

This strong earthquake in Chiapas in Mexico caused minor damage at La Providencia, Jiquipilas and Cintalapa, and was strongly felt at Tuxtla Gutierrez.

[Figuerroa 1973]

1927 Jun 22 El Salvador

A locally damaging earthquake in the Carrizal Valley in El Salvador ruined a number of houses and killed many animals. The shock was strong at Cojutepeque.

[MEC]

1927 Aug 10 M_S 6.7 Panama

This earthquake had an epicentre off the Azuero Peninsula of Panama. It was felt in the region of Herera and was reported by a ship at 7.62°N, 82.80°W. There is no evidence that the shock was felt at Balboa in Panama.

[BSSA; DBN]

1927 Dec 11 Panama

An earthquake in the Gulf of Parita, Panama, was very strong at Chitré. It was reported from Anton and Penonome, but we have no details.

The shock was strongly felt at Balboa Heights.

[DES]

1928 Mar 27 **M_S 5.4** **Guatemala**

This earthquake in northern Guatemala was felt in the district of Peten and was also reported from Pushilah in Belize. It is not known whether it caused any damage.

[DBN, Gann 1928]

1928 May 21 **M_S 5.6** **Guatemala**

An earthquake shook the entire region of Guatemala, but details are lacking.

[DES]

1928 Jul 1 **M_S 5.5** **Nicaragua**

A strong earthquake shook Managua for 30 seconds. Damage was confined to the cracking of walls.

It was very poorly recorded instrumentally. ISS adopt an epicentre 1700 km west of Nicaragua in the Pacific.

[DES; Jorgensen]

1928 Oct 25 **M_S 6.7** **Nicaragua**

A series of severe earthquakes in Nicaragua consisted of several shocks, all reported as lasting for a minute. Damage was widespread but minor.

[DES]

1929 Jan 17 **M_S 5.1** **Guatemala**

Considerable damage was done by an earthquake which shook Puerto Barrios in northeastern Guatemala. There were no casualties.

[DES]

1929 Jan 19 **M_S 5.9** **Panama**

An earthquake with an epicentre offshore from northwestern Panama was felt at Balboa.

[DBN]

1929 Jan 24 **M_S 7.1** **Guatemala**

This earthquake occurred off the coast of Guatemala. We could find no macroseismic information except that the main shock and two of its aftershocks were felt strongly on board a vessel at 12.8°N, 91.0°W.

We adopt, without further analysis, the rather uncertain position calculated by ISS, landward of the Middle American Trench.

[DES]

1929 Feb 10 **M_S 6.6** **Guatemala**

A strong earthquake was felt at Quezaltenango, Chimaltenango, and in particular at Escuintla, and in the coastal area of Guatemala. There is no evidence that the shock caused any damage.

[DES]

1929 Feb 20 El Salvador

Three shocks were felt at La Unión in El Salvador; we could find no instrumental data, and cannot locate them.

[DES]

1929 Feb 28 Nicaragua

A strong earthquake was felt at San Juan del Sur in Nicaragua; no damage was reported.

[DES]

1929 Mar 19 M_S 6.2 Mexico

This earthquake was strong at Ciudad Hidalgo and in the alluvial plain at the mouth of the river Suchiate in Guatemala.

[Figueroa 1973]

1929 Mar 21 M_S 6.4 Figure A.36 Guatemala

A large shock with an epicentre off the coast of Guatemala was strongly felt in the coastal region of San José, where it caused some panic but no damage. It was also strong at Escuintla, Ahuachapan and Sonsonate and in villages between these towns.

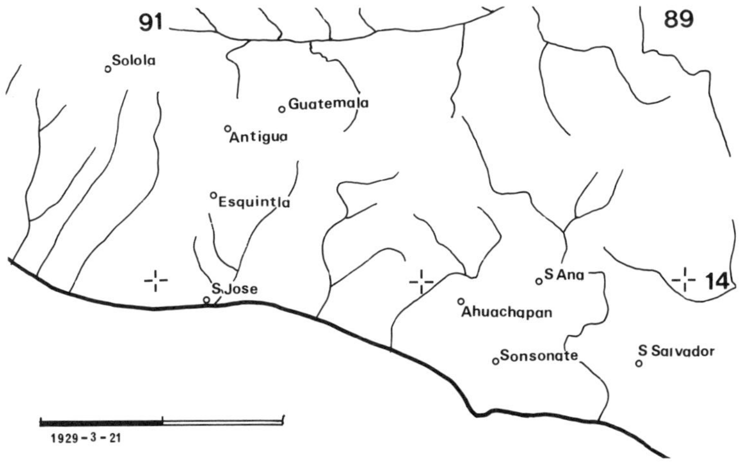

Fig. A.36 Felt effects of earthquake of 21 March 1929.

The shock was felt with varying degrees of intensity at Solola, Antigua, Guatemala City, Santa Ana, San Salvador and was perceptible at Ciudad Hidalgo in Chiapas and along the coast of El Salvador.

The shock was well recorded instrumentally, and we found a well-determined location in the Pacific off the coast of southeastern Guatemala. There is no indication that the depth is other than crustal.

[DES; Figueroa 1973; PDCA; PIMG]

1929 Nov 20 El Salvador

This strong earthquake in Chinameca in El Salvador was followed by a long series of shocks. The swarm was local and was not felt beyond Berlin, at a distance of about 20 km.

[Meyer-Abich 1952]

1929 Dec 20 M_S 5.3 Figure A.37 Guatemala

This relatively small magnitude earthquake in the region of Quezaltenango in Guatemala was preceded by many foreshocks, which began on 17 December. It caused some minor damage at Mazatenango, and was strongly felt at Palmar, Retal, Cuyotenango and Rio Bravo.

The shock was felt in the districts of Retalhuleu, Suchitepequez, Solola, San Marcos, and Totonicapan.

[PDCA]

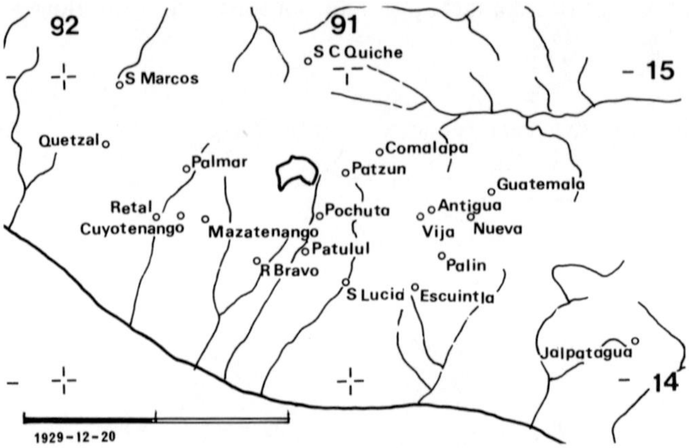

Fig. A.37 Felt effects of earthquake of 20 December 1929.

1930 Jan 20 Guatemala

This earthquake in northern Guatemala was reported as strong from the Departments of Alta and Baja Verapaz, and Huehuetenango, but it caused no damage.

The shock was recorded instrumentally only in Guatemala City and probably at Tacubaya, on low-gain Wiechert seismographs.

[White 1984]

1930 Mar 9 El Salvador

A local earthquake in the vicinity of the Volcano San Vicente was widely felt at San Salvador. It was followed by a weaker shock five hours later. They caused no damage.

The shock was recorded on the seismograph at San Salvador, but not at more distant stations.

[Martinez 1978]

1930 Apr 30 **El Salvador**

A series of earthquakes caused some alarm at San Salvador. They cannot be located.

[DES]

1930 Jun 9 **Guatemala**

This earthquake in Guatemala was followed by a long aftershock sequence. The shock was felt in the Province of Santa Rosa and was particularly strong at Moyuta, where it caused great panic, but no damage.

[DES; Termer 1930]

1930 Jun 18 **Guatemala**

This further earthquake in the Province of Santa Rosa was very strong in the region between Moyuta and Cuilapa, but it caused no damage. The shock was felt in settlements along the Pacific coast and in the region of Jutiapa.

[Termer 1930]

1930 Jun 19 **Guatemala**

This was a stronger earthquake in Santa Rosa. At Tecuaco and Taxisco it damaged many houses and public buildings and caused great panic, initiating the evacuation of some settlements in the region. The earthquake was widely felt in Santa Rosa, Jutiapa and Guatemala City.

[DES, Termer 1930]

1930 Jun 24 **Guatemala**

Another very strong earthquake in Santa Rosa caused considerable damage at El Oratorio.

[Termer 1930]

1930 Jul 7 **M_S 6.2** **Guatemala**

After a short period of quiescence there was a further damaging shock in the district of Santa Rosa in Guatemala. It affected the region between Moyuta, Tecuaco, Tapalapa, Chiquimulilla Guazacapan and Ixhuatan, where almost all houses and public buildings, including many farm houses, mainly of masonry construction, were damaged beyond repair and quite a few collapsed, without causing loss of life. The shock was widely felt and caused some concern in parts of Guatemala City.

The earthquake was relatively well-recorded instrumentally and ISS reports readings from 43 stations. They adopt, however, without recalculation the epicentre of the earthquake of 30 July 1929, which is at 13.9°N, 91.2°W, about 100 km westsouthwest from our preferred macroseismic position.

[DES, Termer 1930]

1930 Jul 14 **M_S 6.9** **Figure A.38** **Guatemala**

This is the largest of the shocks that since early June had affected the Departments of Santa Rosa and Jutiapa in Guatemala.

The earthquake caused additional damage and extended the area already affected by previous shocks. The epicentral area of this shock includes Azulco, Chiquimulilla, Cuilapa, Ixpaca, Jalpatagua, Los Esclavos, La Gloria, Guazcapan, Moyuta, Oratorio, Pasaco, San Juan Tecuaco, Santa Maria Ixhuatan, Taxisco, and Tecuaburro. Within this area of about 20 km radius, almost all masonry and adobe houses, churches and public buildings in villages and farms were damaged, some beyond repair, and quite a few collapsed. Contrary to exaggerated press reports, the earthquake caused no fatalities, but made thousands of people homeless.

Landslides were numerous around the volcanoes Tecuamburro and Moyuta, where they destroyed a number of farms and killed a few people.

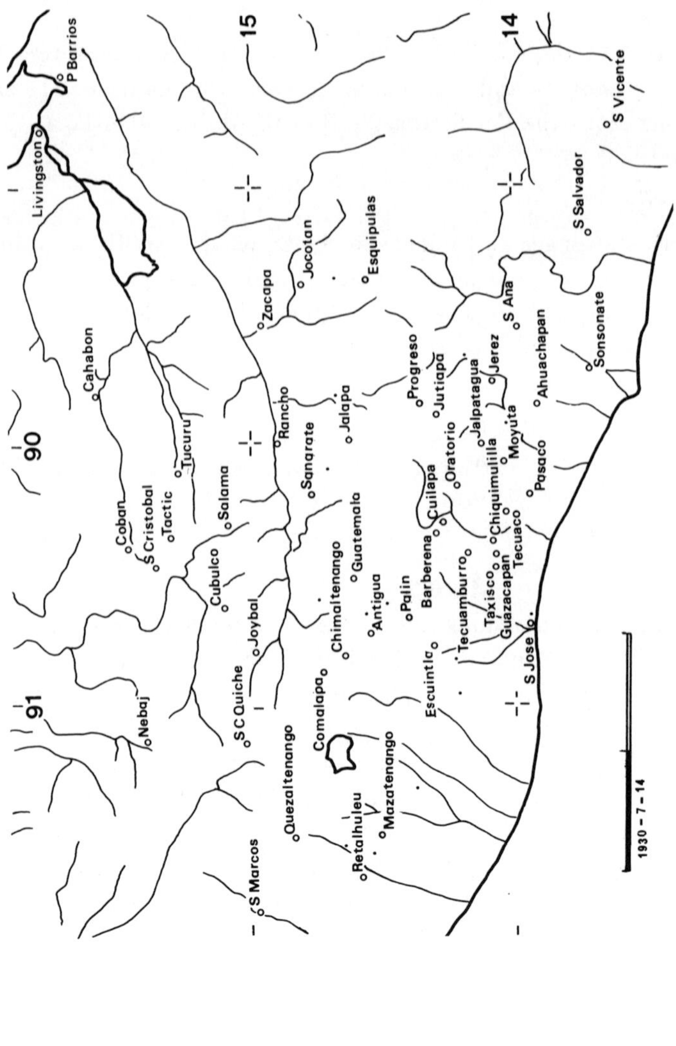

Fig. A.38 Felt effects of earthquake of 14 July 1930.

In many places the ground was cracked disrupting roads, footpaths, water mains and telegraph lines. A continuous rupture with a throw of 2 metres in places was observed extending for 12 km from Tecuaco to Ixhuatan, probably the result of slides from the steep slopes of the Volcano Quemada. More slides were triggered by heavy rains that followed the earthquake.

The shock was very strong over a relatively large area, including Rio Bravo, Comalapa, Sanarate, Jerez and Iztapa. In Guatemala City it caused some panic and water to slosh out of tanks.

The earthquake was rather strong as far as San Marcos, Cobán, Puerto Barrios, San Salvador, and it was perceptible in eastern Chiapas, Alta Verapaz and in El Salvador but not in the Department of Peten.

ISS reports readings from 101 stations, but again adopts an epicentre without recalculation. This is that of the earthquake on 24 March 1926 at 14.0°N, 89.0°W, which is about 130 km to the east of our macroseismic position.

[DES, PIMG; Termer 1930]

1930 Jul 17 M_S 5.5 Guatemala

This was the last strong aftershock of the sequence in the Department of Santa Rosa. The shock caused additional damage, mainly in the region of Moyuta, and triggered many more slides that caused some loss of life. The shock demolished a few houses at Ahuachapan in neighbouring El Salvador, and ruined a few farm houses already damaged by previous shocks. The earthquake was not felt very far.

The earthquake was poorly recorded instrumentally; ISS reports readings from 12 stations and adopts an epicentre at 15.6°N, 97.8°W, about 850 km to the west of our macroseismic position. White and Harlow (1993) report an event of magnitude 5.7 about four hours later at 22h 40m. This is likely to be a different event unless there is confusion between local time and UT.

Amplitude and period data for this event are available from short-period seismographs of the Mexican network in the distance range of 7° to 14°, but unfortunately these are unsuitable for the determination of M_S.

[DES]

1930 Jul 30 Panama

A strong earthquake occurred in central Panama.

The shock originated in the Coclé Province where it caused some minor damage, but details are lacking.

It was felt strongly along the Pacific coast, at Panama City and Balboa Heights, and it was perceptible on the Atlantic coast, at Cristobal and Colón.

[BSSA; DES]

1930 Jul 31 Honduras

A strong earthquake was felt at Tegucigalpa, causing great alarm but doing no damage.

[DES]

1930 Aug 14 **Guatemala**

Earthquake shocks of some intensity were felt at Guatemala City and caused much alarm among the inhabitants. We cannot locate them.

[DES]

1930 Aug 29 **M_S 5.7** **Panama**

A strong earthquake in Panama caused considerable panic in the District of David and at Boquete, El Volcán and David. There is no evidence that the shock, which was felt in the Osa Peninsula, caused any serious damage in the district.

It was followed by many small shocks.

[DES]

1930 Sep 9 **Panama**

An earthquake in western Panama was felt at Boquete and in the District of Chiriqui. It was preceded and followed by many small shocks.

[DES]

1931 Mar 14 **Guatemala**

A series of shocks in the Department of Amatitlan in Guatemala caused some damage at Amatitlan. The main shock caused considerable alarm in Guatemala City.

[DES]

1931 Mar 31 **M_S 6.0** **Figure A.39** **Nicaragua**

This was a locally destructive earthquake in Nicaragua.

In Managua most of the houses were destroyed by the shock, and more than 1,000 people were killed. Water mains burst, wires were down and in places near the town the railroad was destroyed. Fire from the wreckage of demolished houses burnt down 20 blocks in the centre of the city, destroying the British and American legations and causing property damage estimated at $30 million.

In Managua the earthquake caused ground ruptures. A series of ground cracks was reported in the northwest part of the city; these extended for 2 km along a narrow zone passing through the Stadium, the Barrios of El Boer, and San Sebastian, all the way to the shore of Lake Managua. Within this zone, which trends N35°E, ground cracks were discontinuous, about 5 cm wide, with the southeast side downthrown by about the same amount. This zone of deformation was not re-activated by the earthquake of December 1972.

At Masaya only a few houses suffered minor damage, and contrary to a recent report (Hansen 1975) which maintains that the shock was not felt at Granada the shock caused considerable alarm there and some minor damage.

The earthquake was generally felt as far as Leon, Rivas and San Carlos but it was not perceptible at Greytown.

Damaging aftershocks continued until late April.

[Ambraseys 1973; DES; Durham 1931; Freeman 1932; Hansen 1975; Jorgensen; Pollog 1931; Sieberg 1932; Sultan 1931; Williams 1972]

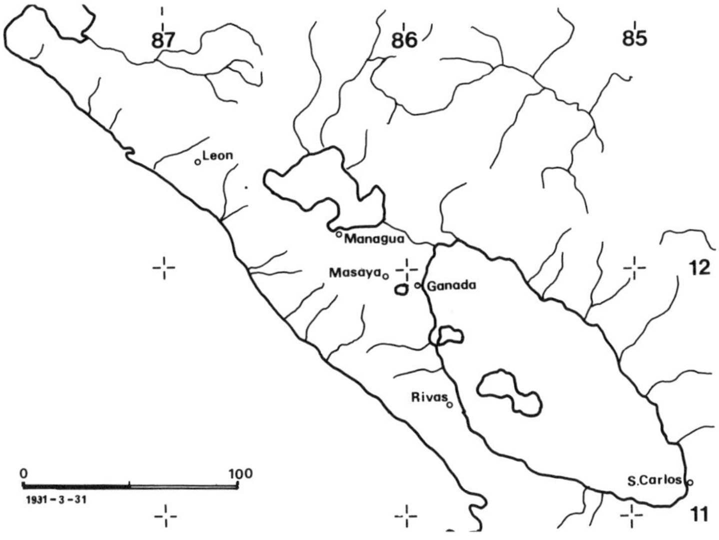

Fig. A.39 Felt effects of earthquake of 31 March 1931.

1931 Apr 12 **Panama**

A moderate shock in Panama caused minor damage at Las Palmas. The earthquake was recorded but not felt at Balboa.

[USEQ]

1931 Aug 25 **M_S 5.6** **El Salvador**

No macroseismic information is available for this earthquake.

There were few instrumental readings and our formal solution places the event about 70 km north of that of ISS, in the Pacific Ocean south of El Salvador. There is no evidence of subcrustal origin.

1931 Oct 12 **M_S 6.1** **Panama**

This earthquake was possibly felt in Panama City.

Its instrumental location, given by ISS, is in the Gulf of Chiriqui.

[USEQ]

1932 Jan 21 **M_S 4.8** **Guatemala**

This earthquake, which was preceded by a foreshock, was felt in Guatemala. It was strongly felt within a radius of about 20 km, from Chimaltenango to Escuintla and from Yepocapa to Antigua. It caused no damage.

[PRP]

1932 May 21 **M_S 7.1** **Figure A.40** **El Salvador**

This was an intermediate-depth earthquake in the Pacific with an epicentre off the coast of El Salvador near the Gulf of Fonseca. Serious damage was widespread and extended sporadically over a zone 70 km long and about 15 km wide that included Analquito, Ereguayquin, Jiquilisco, Ozantlan, Paraiso d'Osorio, San Agustin, San Juan Nonualco, San Pedro Nonualco, Santa Elena, Santa Maria Ostuma, Santiago Nonualco, Usulutan and Zacatecoluca. Most of the casualties, including a few deaths, were reported from the Districts of La Paz and Usulutan. In Zacatecoluca a few houses collapsed and others were damaged. A few casualties were also reported from the nearby town of San Juan, which was badly damaged, and from two sections of Usulutan.

Minor damage to adobe houses was reported from Chinemaca, Rosario, Olocuilta, San Antonio Masahuat, San Francisco, San Nicolas Lempa and Talpa, towns within a zone about 30 km wide and 110 km long, which extended more to the southeast, beyond Jucuaran to the coast of La Unión.

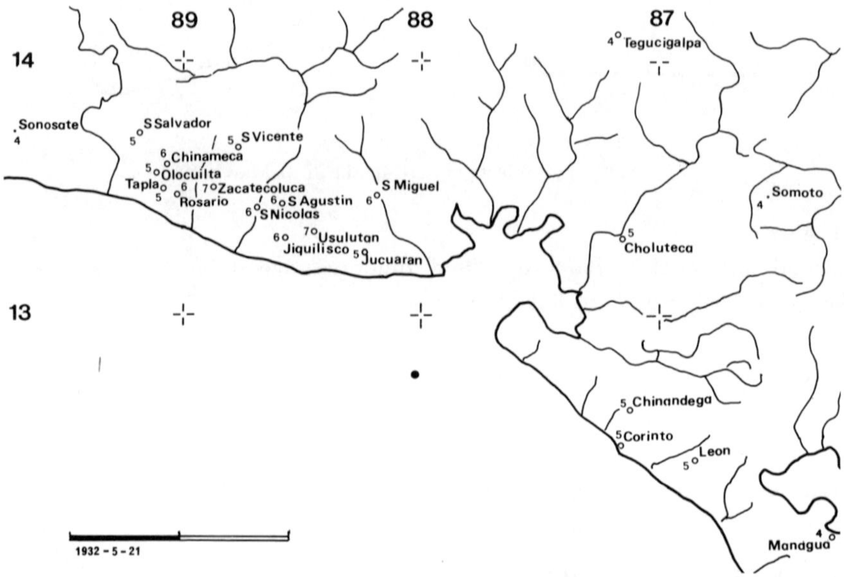

Fig. A.40 Felt effects of earthquake of 21 May 1932.

The shock was strongly felt at San Salvador, Sonosate, Choluteca and Tegucigalpa in Honduras, where in places it caused some panic, and at Chinandega, Corinto, Leon and Managua in Nicaragua.

The earthquake was felt in Guatemala City and allegedly it was perceptible in the Federal District of Mexico and in Costa Rica.

The widespread but sporadic damage over a large area did not define a clear epicentral area (Figure A.40). Perhaps the rupture propagated from the focus near Fonseca in a northwesterly direction along the sea coast of El Salvador, generating a source area that could explain the heavier damage sustained by sites in the alluvial plains along the coast and further inland. The fact that the earthquake was perceptible at large distances and was of long duration suggest an intermediate depth. This shock was extensively recorded world-wide, with many secondary phases reported. A free solution places the earthquake in the Pacific near the Gulf of Fonseca, at a depth of about 150 km. This is rather deeper than Gutenberg and Richter's value of 90 km, but is well supported by supplementary phases.

[Martinez 1978; Osegueda 1932; Schultz; TAC; US Natl.Arch.:RG 59.Box 5514(El Salvador) 20 June 1932]

1932 Jun 20 M_S 5.8 Salvador

No macroseismic information is available for this event.

This earthquake was not recorded instrumentally at many stations, and in particular there was a lack of close stations. Our calculated position off the coast of El Salvador is not in doubt, but there is no control of depth, and in particular, no instrumental reason to assume depth greater than normal.

1932 Sep 10 El Salvador

A prolonged series of shocks was felt at the port of La Unión. The people fled from their homes for several hours, but there was no damage.

We could find no evidence that the shock was recorded, even by close seismograph stations.

[DES]

1932 Dec 19 M_S 5.9 Guatemala

This earthquake was strongly felt at Socoltenango, in Chiapas, Mexico, about 250 km north of Gutenberg's instrumental epicentre, which is off the coast of western Guatemala.

[Figueroa 1973]

1933 Mar 22 Costa Rica

This locally strong earthquake in Costa Rica caused panic at Pacayas, Turialba, Santa Cruz, Cartago and Palmares, within a radius of 15 km, but it was not reported from San José.

We could find no instrumental records for this event.

1933 Apr 3 El Salvador

A strong earthquake caused some alarm in San Salvador, but it resulted in no damage. The shock was not recorded instrumentally.

[DES]

1933 May 30 M_S 5.5 Panama

This earthquake was felt widely in southern Costa Rica, in the Meseta Central, as well as along the Atlantic coast of the country, but it was barely perceptible on the Pacific coast. The shock was perceptible at Balboa in Panama and San José in Costa Rica. There are insufficient data to determine its macroseismic epicentre accurately.

[PRS; DBN; USEQ]

1933 Jul 4 **Honduras**

Press reports mention an earthquake in Honduras in the regions of Gracias, Copán and Santa Barbara, as a result of which three people were killed, but no details are given. The shock was felt at Tegucigalpa and was followed by many aftershocks.

[DES; REC: 1933]

1933 Jul 11 **Nicaragua**

An earthquake in Nicaragua caused some damage in Managua. It was strong at Tipitapa and Nagarote but hardly perceptible at Leon.

A strong aftershock on 24 August caused some additional, slight, damage in Managua.

[REC: 1933.257]

1933 Sep 30 **Honduras**

This earthquake in the region of Gracias, Santa Rosa and Copán was strong enough to cause panic, church bells to ring and to trigger a landslide at Quelacasque, which threatened to block the river.

[DES]

1933 Nov 21 **M_S 6.1** **Panama**

A strong earthquake in Panama, probably with an epicentre in the Gulf of Chiriqui, caused some damage in David. The shock was strong in the District of Chiriqui and it was felt throughout the region of El Bocas del Toro.

[USEQ]

1933 Nov 23 **M_S 6.1** **Panama**

This was the largest of the aftershocks of the earthquake of 21 November. It was strongly felt in the District of Chiriqui, but it caused no damage there.

[USEQ]

1934 Mar 7 **M_S 6.3** **El Salvador**

Macroseismic data are lacking for this earthquake.

The earthquake was well recorded in North America, but not in Europe, and its position seems well enough established in the Gulf of Fonseca. There is no reason to assume that it is other than shallow.

1934 Jul 18 **M_S 7.5** **Tsunami** **Figure A.41** **Panama**

This large magnitude earthquake with an epicentre at sea in the Gulf of Chiriqui was felt throughout Panama and Costa Rica.

In the Gulf of Chiriqui the earthquake caused a crack to form along the length of the island of Montuosa about 60 to 90 cm wide and 6 metres deep, passing near one the triangulation towers. It is not known whether this was of tectonic origin or the result of an incipient slide.

In Parida Island the shock was violent; it caused two or three landslides that obliged its few inhabitants to abandon the island.

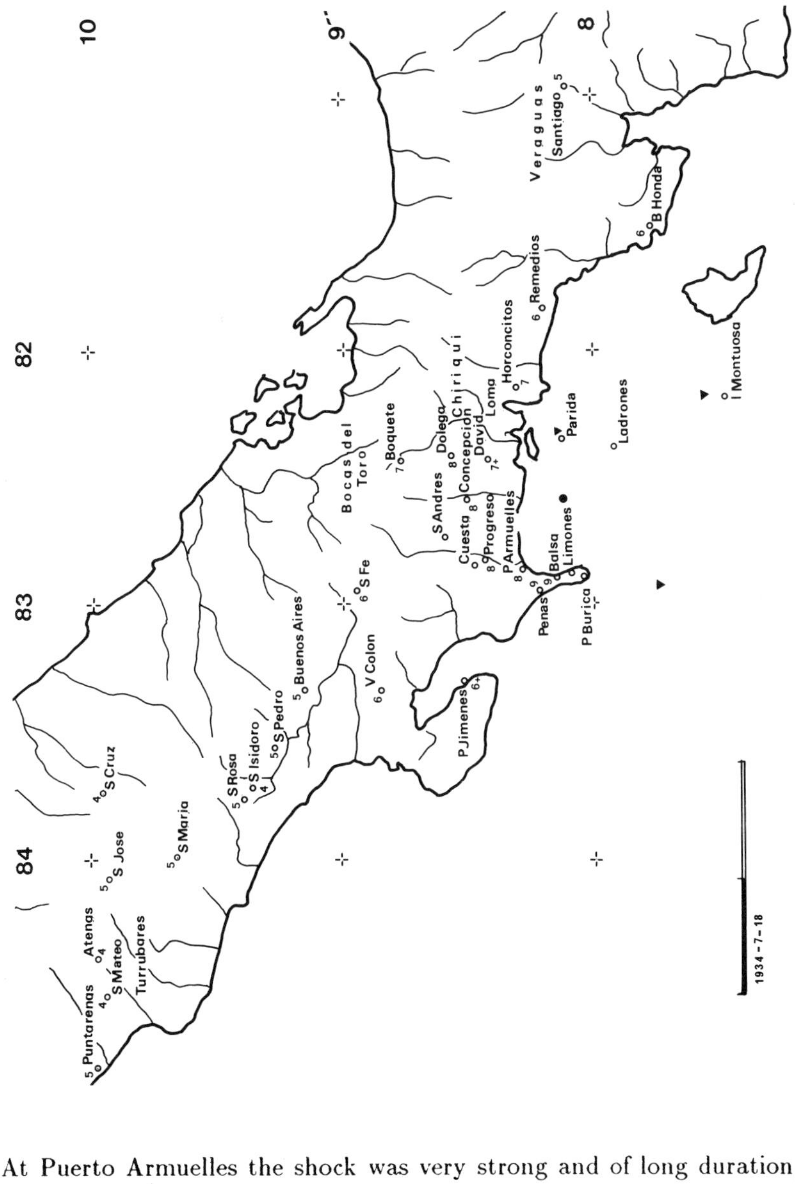

Fig. A.41 Felt effects of earthquake of 18 July 1934.

At Puerto Armuelles the shock was very strong and of long duration, causing suspended light fixtures and fans in timber houses to swing, in some cases striking the ceiling. Nevertheless only a few houses were ruined and no well-built structures in the town suffered serious damage. In contrast, the water and sewerage mains were broken, as was the pumping station and oil line in the docks.

At David the shock caused panic; walls crumbled, a few adobe houses collapsed and roofs caved in, injuring a number of people. Nearly every house in the town was damaged.

As a result of the earthquake the foundation of a triangulation tower near Balsa Point shifted about 1 cm eastward.

Damage extended to Dolega, Horconcitos, Loma and to other villages and farms in the Chiriqui district.

At Puerto Jiménez the shock was strong enough to set church bells ringing and to do considerable damage to brick houses and to the port facilities.

The shock caused panic and slight damage at Boquete, Buenos Aires, Cuesta, Remedios, and Villa Colón and it was strongly felt over a large area.

To the east the shock was felt at Panama City, 350 km from the epicentral area, and at Fort Davis. The Panama Canal was closed for a few hours for inspection. The shock was recorded on the strong-motion accelerograph of the observatory at Balboa Heights with a maximum horizontal acceleration of about 7%g.

The earthquake was felt in Costa Rica as far north as the Barra de Colorado and Los Chiles, 420 km from the epicentre.

The shock set up a seismic sea wave, which flooded the west coast of the Gulf of Chiriqui, causing minor damage. At Armuelles the seaward row of houses in the native village was washed out completely, and the main street was under water several times. At Bahia Honda the tide gauge recorded fluctuations of the sea level of 0.6 m, the disturbance lasting for over five hours. At Puerto Armuelles the tide station was destroyed and no record was obtained. The record made by the tide gauge which was fixed to the dock shows that the shock caused the structure to sink gradually into its foundation, and that the instrument ceased functioning about three minutes after the earthquake.

The earthquake was also felt on board ships in the Gulf of Chiriqui; *S.S. Nokomis* was anchored near the northeast end of Parida Island, *S.S. Tuscaloosa City* was sailing off Punta Burica and *S.S. Point Sur* was at anchor near the Islet of Montuosa. All three vessels reported very violent shaking.

The earthquake was followed by a number of relatively large magnitude aftershocks, which caused additional damage.

The macroseismic effects of this earthquake in Panama are well documented (Figure A.41). The shock is well recorded and instrumental readings give a good location in the Gulf of Chiriqui, within the epicentral area of the event.

[Chamacho 1991; Feldman 1984; Keller 1935; Kirkpatrick 1936; Miyamura 1980; PRS; Neumann 1936; USEQ]

1934 Jul 21 M_S 6.8 Panama

This was the largest aftershock of the series affecting the region between Puerto Armuelles, San Andrés and David.

At Puerto Armuelles a 130 m length of the deep-water section of the pier collapsed and sank into the sea, taking with it three banana conveyors valued at $250,000. The wharf and quay were totally destroyed and the major portion of the dock itself collapsed. The water and sewerage mains of the town were broken, as was the oil line in the dock. These had already been broken in the main shock and repaired in the intervening period. Seventeen houses and a large water tank, the main water supply of the port, that had survived the first shock, also collapsed. With the exception of timber-framed structures, almost all houses in the town were damaged beyond repair and many collapsed, injuring a number of people. Losses were estimated at $500,000.

At Armuelles waves were seen on the ground during the earthquake and there is evidence of widespread liquefaction, particularly in the native part of the town, near the waterfront, where water was ejected from cracks in the ground that run in an east-west direction. Two mooring buoys near the dock shifted about 300 m seaward. This is believed to have been due to the liquefaction and flow of sediments on the sea floor, but there was no indication of a seismic sea wave, nor was there any significant settlement of the sea bottom.

At David 90% of the houses were damaged; more than 200 were ruined and seven collapsed, but there were no casualties. The cathedral was seriously damaged, as well as many public buildings. The damage due to this earthquake at Puerto Armuelles and David alone was estimated at $750,000.

Telegraph lines between David and Puerto Aruelles were put out of commission.

Inland, cracks appeared in the mountains near Bongo, at a farm of the Chiriqui Land Co. These cracks were about 15 cm wide and quite long, extending across the border into Costa Rica for some distance.

Many structures weakened by the previous shocks collapsed entirely. At Loma and San Andrés quite a few houses were ruined and damage was serious at Las Penas, El Progreso, Cuesta, La Conceptión and Horconcentios.

There was no loss of life as practically the entire population was living outdoors.

At Balsa Point the foundation of the triangulation station mentioned in the report of the shock of 17 July was found to have shifted again about 8 cm. There were several considerable cracks in the ground, all of them about 2.5 to 5.0 cm wide, and 30 cm deep; they extended in a general SW-NE direction.

The shock was strongly felt on board *S.S. Nokomis*, which was now anchored off Balsa Point.

The earthquake was widely felt, strongly in the Canal Zone, at San José and further to the northwest.

Numerous aftershocks until the end of July added to the damage and caused additional slides, some of which again disrupted telegraph and road communications in the region.

1934 Dec 3 M_S 6.3 Figure A.42 Honduras

This damaging earthquake in Honduras was preceded by two foreshocks.

It destroyed the small towns of San Jorge, La Encarnaçion and San Fernando, and seriously damaged six other villages. Damage extended to Santa Rita, Cabanas, Esperanza, Santa Rosa and San Antonio where churches and public buildings were ruined. The earthquake caused considerable damage to the ruins at the archaeological site of Copán, to the church, which was rendered unsafe to enter and was later abandoned, and to many houses in the town.

The earthquake was severe at Metapan, Zacapa, Sinuapa, Gracias and Naranjito, and at Comayagua the shock set church bells ringing. The shock was strong at Ilama, and at Teguchigalpa it caused panic and some minor damage. The shock was felt at Yoro and Puerto Truxillo.

In the regions of Copán and Cerro Azul the earthquake, which occurred during a period of heavy rains, triggered landslides that continued to occur during the aftershock period.

In the Districts of Ceiba and Truxillo streams were dammed by landslides, which were later overtopped and breached, causing floods which damaged bridges and the railway line. It is not clear whether these slides were triggered by the earthquake or whether they were the result of heavy rains.

In all about 50 people were killed and hundreds were injured.

The earthquake was followed by many aftershocks, which were widely felt, in places causing additional damage.

[Bois 1934; DES, REC 1934; Estrada and Flores 1934; Von Hagen 1945.174]

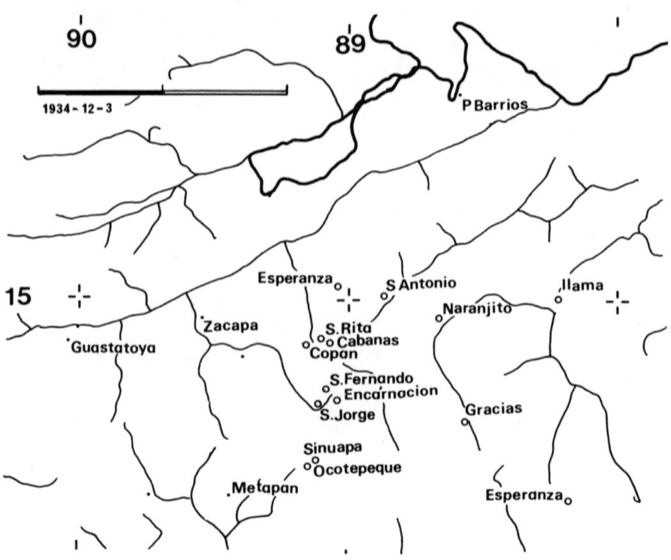

Fig. A.42 Felt effects of earthquake of 3 December 1934.

1935 Apr 24 M_S 5.8 Mexico

This earthquake on the coast of Chiapas, in Mexico, was strongly felt in the region of Mapastepec, but it caused no damage there.

[Figueroa 1973]

1935 Jul 26 **M_S 5.7** **Panama**

An earthquake in western Panama caused some damage in the northern part of the District of Chiriqui, near the Lago di Chiriqui, but details are lacking.

[USEQ]

1935 Aug 1 **M_S 6.1** **Figure A.43** **Costa Rica**

This earthquake in the Guanacaste region in Costa Rica caused extensive damage, mainly to adobe houses, at Bagaces, Salitral and Mogote, but caused no loss of life. Sporadic but minor damage extended to Salto, Liberia and Las Canas.

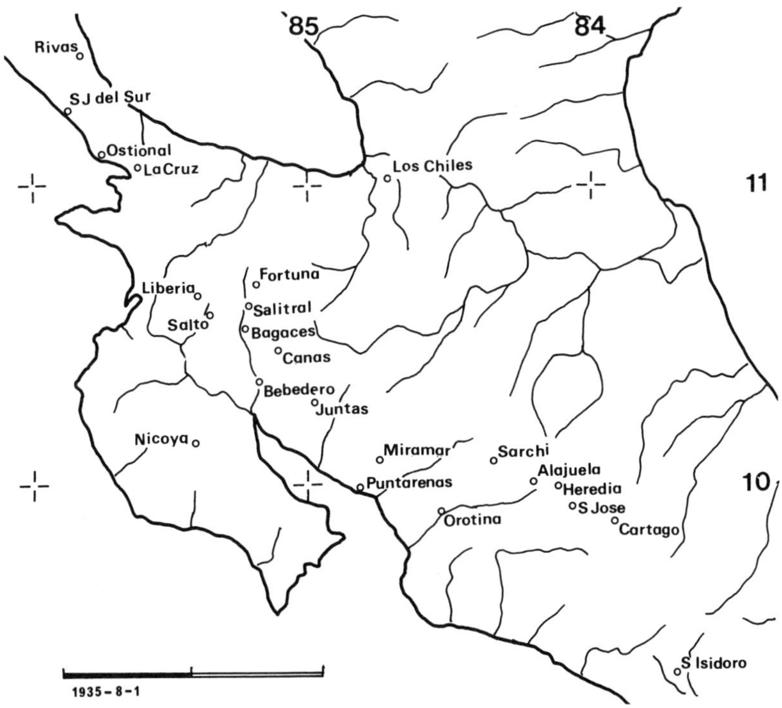

Fig. A.43 Felt effects of earthquake of 1 August 1935.

The shock was strongly felt at Los Chiles, La Cruz and Puntarenas and it was perceptible as far as Managua in Nicaragua and La Cuesta on the border between Costa Rica and Panama.

[Miyamura 1980; Montero and Alvaro 1988; PRS]

1935 Oct 31 **Honduras**

This was a strong shock in Honduras.

At Morazon the church was split through the centre by the earthquake and at El Negrito houses were damaged.

The shock was very strongly felt at Moraita and San Pedro Sula. We could find no instrumental data for this event.

[DES; Von Hagen 1945.91]

1935 Nov 30 **M_S 6.2** **Panama**

This earthquake was felt throughout Panama.

It was particularly strong in the Canal Zone. At Balboa it caused panic and the damage of a few buildings. No serious damage was done to the Panama Canal.

[USEQ]

1935 Dec 14 **M_S 7.3** **Mexico**

This major earthquake with an epicentre off the coast of the Guatemala-Mexico border was felt over a large area.

Damage was reported from as far as Tuxtla Gutierrez, Ciapa de Corzo and San Cristobal Las Casas, but details are lacking. The shock was less strong in Chicoasen.

Felt reports confirm the general location of this event, which is determined from instrumental readings as being off the coast of Mexico.

[Figueroa 1973]

1935 Dec 21 05h **M_S 5.6** **Mexico**

This small aftershock was strong at Acapetahua, Escuintla and Motozintla, in Chiapas, where it caused considerable concern, but little or no damage. This location is about 100 km to the north of the instrumental position given by ISS.

[Figueroa 1973]

1935 Dec 21 11h **M_S 5.8** **Mexico**

Another strong aftershock was felt in the same area, again about 150 km to the north of the instrumental location.

[Figueroa 1973]

1936 Mar 7 **M_S 5.3** **Costa Rica**

A relatively small earthquake was felt on board a ship sailing 70 km off Point Guiones, Costa Rica.

[BSSA]

1936 Apr 2 **Nicaragua**

A strong shock was felt in Managua, Nicaragua; there was no damage. We cannot locate it.

[DES]

1936 May 6 **Panama**

This earthquake occurred in the district of Colón in Panama.

At Santa Rita houses were damaged and at Anton the walls of a few dwellings were cracked. The shock was strong at San Carlos, Chame and at Balboa, where it caused a few cracks and some loosened tiles. The earthquake caused general alarm throughout the Isthmus and was perceptible on the Atlantic coast of Colón.

Balboa Heights was the only seismograph station to record the shock, and at Miraflores a strong-motion instrument registered a peak acceleration of 6%g.

[USEQ]

1936 Sep 3 **M_S 5.6** **Mexico**

This relatively small magnitude earthquake in Chiapas caused some minor damage at Chiapilla and Riviera de Cangui. The macroseismic position is 320 km to the north of the instrumental location given by ISS.

[Figueroa 1973]

1936 Oct 29 **M_S 6.1** **Panama**

An earthquake with an epicentre in the Gulf of Chiriqui was felt at David and Santiago. The shock was perceptible at Balboa and Panama City.

[USEQ]

1936 Nov 1 **El Salvador**

A strong shock was felt at San Salvador.

[STR]

1936 Nov 10 **El Salvador**

This was a further shock at San Salvador.

[STR]

1936 Nov 15 **El Salvador**

This was a strong earthquake at San Salvador.

[STR]

1936 Nov 19 **M_S 6.9** **Guatemala**

This relatively large, subcrustal, earthquake in Guatemala was strongly felt, particularly in the Department of Escuintla.

The shock, which was of long duration, was felt over a large area, from La Grandeza and Ciudad Cuauhtemoc in Chiapas in Mexico to Guatemala City, and to central and western El Salvador. There is no evidence that it caused any damage.

The earthquake was recorded at many stations, regional, in North and South America, and in Europe. The most likely instrumental location is in south Guatemala, in the Department of Escuintla, but there is an uncertainty of depth arising from uneven station distribution and particularly the lack of close stations. We have chosen a subcrustal focus that is consistent with the macroseismic observations and supported by observed depth phases.

[DES; Figueroa 1973; STR]

1936 Nov 22 **M_S 6.4** **Guatemala**

This was a strong aftershock of the earthquake of 19 November in Guatemala. It caused some damage at Escuinlta, but no details are available. It was felt in central and western El Salvador.

[STR]

1936 Nov 23 **Honduras**

This was a strong earthquake in Honduras in the region of La Esperanza. It is not known whether it caused any damage. The shock was perceptible in San Salvador. We could find no instrumental data for this event.

[STR]

1936 Nov 26 **M_S 6.0** **Costa Rica**

This was a damaging shock in northeast Costa Rica, presumably in the region of San Carlos. The shock was strong at San José, and it was followed by many aftershocks.

[DES; STR]

1936 Nov 27 **El Salvador**

A strong earthquake in El Salvador was reported from Chicotepec and San Vicente; it was not felt very far. It was followed by many aftershocks until 19 December.

[STR]

1936 Dec 20 **M_S 6.1** **Figure A.44** **El Salvador**

This damaging earthquake in El Salvador was preceded by numerous foreshocks that began a month earlier.

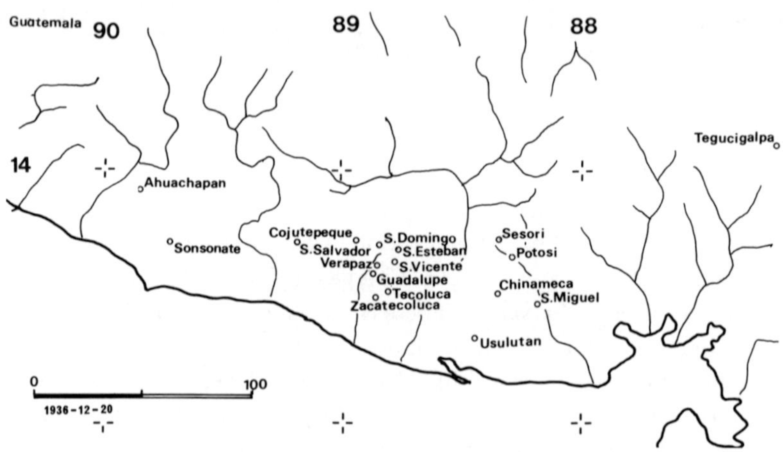

Fig. A.44 Felt effects of earthquake of 20 December 1936.

The shock almost totally destroyed San Esteban Catarina, San Vicente, Iztepeque and Tecoluca.

At San Esteban 219 houses were destroyed, 10 people were killed and 98 injured; in places the ground opened up and slides added to the damage, particularly at the foothills of Santa Rita. Springs near the base of the Volcano of San Vicente became sulphurous, making large supplies of drinking water unfit for use.

Damage extended to Mercedes, Verapaz, Tepetitan, Guadalupe and Apastepeque, and communications were temporarily disrupted.

The shock was felt throughout El Salvador.

An aftershock on 21 December at 01h 35m added to the damage at San Vicente.

This earthquake was recorded reasonably well at seismograph stations in Central and North America, but there were few readings from Europe. With the closest station more than 8° away there is little direct control of depth, but all solutions give subcrustal depths (between 55 and 75 km), with confirmation from depth phases. The macroseismic position is well established, and should be preferred, for the seismological evidence is not well enough constrained to override it.

There is some uncertainty about the depth, with the felt effects strongly suggesting that the event was shallow, despite the instrumental evidence for a focus rather deeper than normal.

[BCIS 1936: 132-3; DES; Jordan and Martinez 1980; Levin 1940; Martinez 1976; PRS; Schultz 1963]

1936 Dec 29 **M_S 4.9** **El Salvador**

Poor instrumental evidence suggests the occurrence of a small event in the Pacific Ocean off the coast of Guatemala, close to the position assumed by ISS for an event on 14 November 1936. The only felt reports at this time are for a shock allegedly felt at Sihuatepeque in El Salvador, 450 km to the east, which may be associated with the aftershock sequence occurring in that area following the event of 20 December.

[STR]

1937 Jan 11 **M_S 5.6** **Mexico**

This earthquake in Mexico, in the District of Chiapas, was very strongly felt at Chicosen, Tuxtla Gutierrez, Chiapa de Corzo and San Cristobal las Casas. It caused no damage and was not felt very far.

[Figueroa 1973]

1937 Jan 12 **Nicaragua**

A series of local shocks in Nicaragua caused the collapse of five buildings at Chinandega and Corinto.

[REC]

1937 Mar 9 **M_S 6.4** **Costa Rica**

This earthquake is relatively well located on the southwest coast of Costa Rica. The shock was strongly felt in the region of Ciudad Cortes and to a lesser extent at David. Allegedly the same event was also perceptible at Balboa Heights and throughout the Pacific section of the Panama Canal, 430 km away.

It was followed by a number of strong aftershocks.

[Jorgensen 1966; USEQ]

1937 Mar 29 **M_S 5.6** **Costa Rica**

A strong aftershock was reported from David. It was perceptible in Costa Rica and in the Canal Zone.

[STR; USEQ]

1937 Apr 3 **M_S 5.5** **Guatemala**

This rather small earthquake was very strongly felt in the western part of the District of San Marcos in Guatemala, and at Mazapa de Madero and Motozintla in Chiapas, where it caused some minor damage, of which there are no details. The instrumental position of this event given by ISS is 410 km to the southsoutheast of its macroseismic location, which we prefer.

[Figueroa 1973]

1937 May 28 **M_S 5.6** **Mexico**

An intermediate-depth earthquake in Chiapas caused some small damage at Chiapa de Corzo, 20 de Noviembre and at Tuxtla Gutierrez. It was less strong at Chioasen. We adopt Gutenberg's position.

[Figueroa 1973]

1937 Jun 8 **M_S 5.8** **Mexico**

This was an intermediate-depth earthquake with an epicentre in Chiapas, Mexico. The shock was particularly strong at Pichucalco, Ocotepec and Tapalapa, but it caused no damage there. It was perceptible throughout the State of Chiapas, in Oaxaca and Vera Cruz.

[Figueroa 1973; TAC]

1937 Jun 24 **M_S 6.0** **Costa Rica**

An earthquake with an epicentre in the Pacific, off the Osa Peninsula, was felt on board a vessel sailing at 8.22°N, 84.13°W. It was preceded by a strong foreshock two minutes earlier.

[DBN]

1937 Jun 30 **El Salvador**

A strong earthquake was felt at San Salvador.

[STR]

1937 Sep 15 **M_S 6.5** **Guatemala**

This earthquake was relatively well-located macroseismically in western Guatemala, 220 km from the instrumental location given by ISS. It was felt in Chiapas, particularly at Socoltenango, but there are no macroseismic details from there.

[Figueroa 1973]

1937 Sep 28 **M_S 6.0** **Guatemala**

An earthquake in western Guatemala caused damage in the District of San Marcos and across the border in the District of Chiapas in Mexico, but details are lacking. The earthquake, which was followed by strong aftershocks, was felt in Guatemala City.

[STR]

1937 Sep 29 **El Salvador**

A strong shock was felt at San Salvador at the same time that the Volcano Santa Ana became active.

[DES]

1937 Dec 18 **Ms 5.4** **Panama**

This earthquake was felt by many people at Balboa Heights in Panama. Its instrumental epicentre is 210 km away in the Gulf of Montijo in the Azuero Peninsula, but we have no macroseismic information from there.

[USEQ]

1937 Dec 26 **Ms 5.8** **Figure A.45** **El Salvador**

A relatively small magnitude earthquake in El Salvador caused considerable damage within a small area of about 9 km radius. At Ahuachapan and Atiquizaya, not counting adobe houses, four buildings were totally destroyed and four others were substantially damaged, killing one person and injuring 11 others.

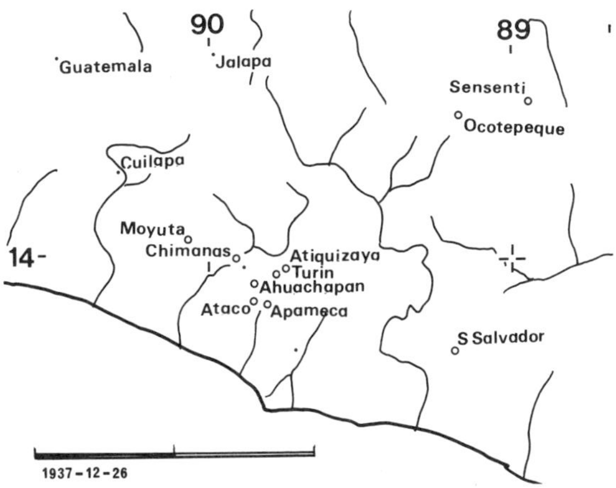

Fig. A.45 Felt effects of earthquake of 26 December 1937.

In the surrounding villages of Ataco, Apaneca, Chimanas, Junquillo, Palo Pique, Las Puertas and Turin many adobe houses were damaged beyond repair.

Sporadic damage extended into Guatemala in the regions of Chimanas and Moyuta, but details of damage are lacking.

The shock disrupted telegraph and power services and triggered landslides.

The earthquake was felt at Jalapa, Cuilapa, San Salvador, and was perceptible throughout the country as well as in southwestern Honduras, at Sensenti and in the District of Ocotepeque.

[BCIS: 1937.115; DES; Jordan and Martinez 1980; Martinez 1976; PRS; Schultz 1963; US Natl.Arch.:Rec.G.84/C8.8 vol.7(S.Salvador)]

1938 Mar 5 **El Salvador**

An earthquake in El Salvador damaged a few houses and a hospital at San Salvador.

We could find no instrumental information about this earthquake.

[REC: 1939]

1938 Apr 25 **M_S 6.0** **Nicaragua**

This was a damaging shock in Nicaragua.

At Leon and Chinandega, and within a radius of about 15 km from them, many houses were damaged, but there was no loss of life. The shock was felt along the entire Pacific coast of Nicaragua.

[REC: 1939.238]

1938 May 6 **M_S 6.0** **Nicaragua**

Another damaging shock in Nicaragua added to the damage already caused by the earthquake of 25 April. At Telica many people were injured and the village was almost totally ruined. Considerable non-structural damage was also done at Leon, but details are lacking.

[REC: 1939.238)

1938 Sep 1 **M_S 6.1** **El Salvador**

An earthquake with an epicentre off the coast of El Salvador was felt in the capital.

[REC]

1939 Jan 20 **M_S 6.2** **Guatemala**

This subcrustal event occurred in the Pacific Ocean. It was widely felt in the central and western parts of El Salvador, but caused no damage.

[REC]

1939 Jan 27 **M_S 5.7** **El Salvador**

This was a strong foreshock of the earthquake of 29 January. Its epicentre was in the Pacific, off the coast of El Salvador.

It was felt strongly in the western part of El Salvador and caused some panic in San Salvador, but no damage.

The instrumental relocation of this event suggests a focal depth of about 120 km.

[BSSA: 1919.416]

1939 Jan 29 **M_S 5.7** **El Salvador**

This earthquake was felt in El Salvador; very strongly in the western part, strongly in the central part, and lightly in the eastern part. It caused some small damage and a few injuries, but details are lacking. The shock was strong at San Salvador.

There is seismological evidence that this was an intermediate depth earthquake, very similar to its foreshock.

[DES; REC 1929.416]

1939 Jun 18 **M_S 5.8** **Costa Rica**

An earthquake in central Costa Rica was particularly strong in the Province of Heredía, where it did some damage, about which we have no details. It was less strong at San José.

[Boschini and Montero 1993; DES; Miyamura 1980; REC]

1939 Aug 16 **M_S 6.0** **El Salvador**

This earthquake had an epicentre in the Pacific off the coast of El Salvador. It was very strong in the west and central parts of the country, but caused no damage there.

[REC]

1939 Sep 28 **m_B 6.2** **Guatemala**

This was an intermediate-depth earthquake in Chiapas in Mexico. It was widely felt, and was particularly strong at Coatepeque and San Marcos in Guatemala, and at Ciudad Hidalgo, Puerto Madero and Tapachula in Chiapas.

[Figueroa 1973]

1939 Oct 20 **M_S 5.6** **Panama**

The epicentre of this event is relatively well located instrumentally in the Gulf of Chiriqui in western Panama, but we have no macroseismic data from there. The shock was allegedly widely felt at Balboa Heights, 360 km away.

1939 Nov 28 **M_S 5.2** **Panama**

This intermediate-depth earthquake in the Gulf of Panama was generally felt throughout the Isthmus of Panama.

[USEQ]

1939 Dec 5 **M_S 6.9** **Guatemala**

A relatively large magnitude earthquake with an epicentre off the coast of Chiapas was very strongly felt along the Pacific coast of southwestern Chiapas and Guatemala. It caused panic and minor damage at Acapetahua, Las Casas, Escuintla, Pueblo Nuevo Retalhuleu, and Tapachula.

[Figueroa 1973; STR]

1939 Dec 12 **m_B 5.5** **Mexico**

This was an intermediate-depth earthquake.

We have no reports from Guatemala, but in Chiapas the shock was reported as strong at Angostura, Chiapas de Corzo and San Cristobal Las Casas.

[Figueroa 1973]

1939 Dec 13 **El Salvador**

A series of shocks in El Salvador culminated in a violent shock at San Vicente, which allegedly caused some damage.

[DES; REC]

1939 Dec 21 — Ms 6.9 — Figure A.46 — Costa Rica

This was the first of two strong earthquakes, eight hours apart, that shook much of Costa Rica, causing some damage.

The shock, at 20h 54m, caused panic and some damage at Bagaces, Chomes, Grecia, Santa Cruz, San Mateo, San Pablo and set church bells ringing at Liberia. At San José a few old houses suffered minor damage, and the cross on the dome of the cathedral fell off.

The earthquake was widely felt, allegedly as far as Panama, and it was followed by an equally large shock (M_S 6.9) the following day at 04h 43m.

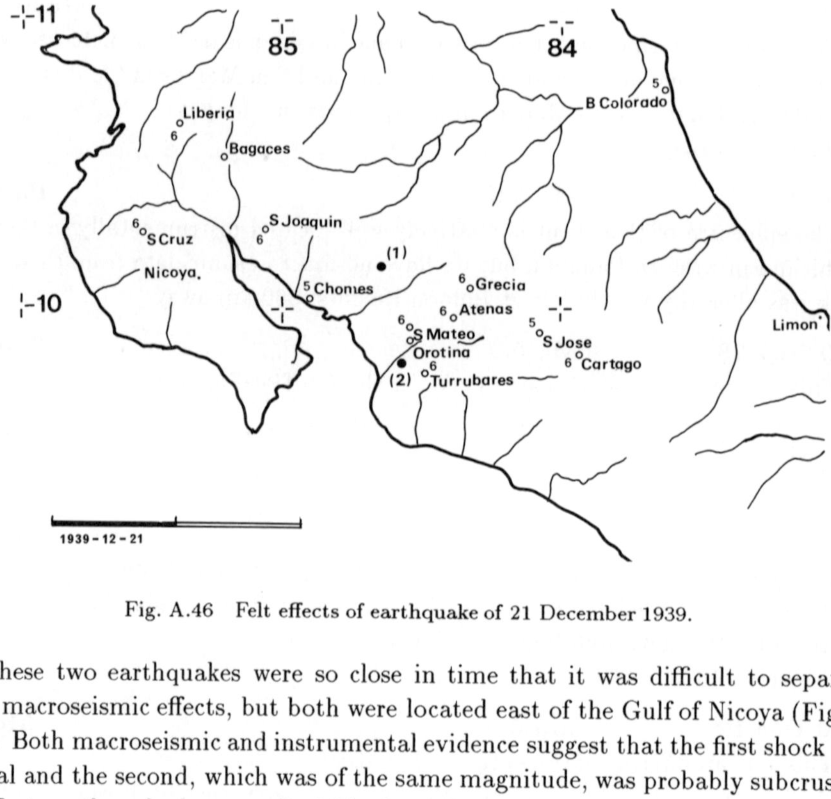

Fig. A.46 Felt effects of earthquake of 21 December 1939.

These two earthquakes were so close in time that it was difficult to separate their macroseismic effects, but both were located east of the Gulf of Nicoya (Figure A.46. Both macroseismic and instrumental evidence suggest that the first shock was crustal and the second, which was of the same magnitude, was probably subcrustal. The first earthquake happened within five minutes of a magnitude 8.0 earthquake in the Celebes, making it difficult to separate surface-wave amplitudes of the two events.

[DES; Feldman 1984; Miyamura 1980; REC]

1939 Dec 22 — Ms 6.9 — Costa Rica

Eight hours later a second earthquake followed, affecting the same region and causing some additional small damage. Its effects are difficult to separate from those of the first shock.

1939 Dec 26 M_S 5.7 El Salvador

A subcrustal earthquake shook the Pacific coasts of Nicaragua and El Salvador. It was strong at Corinto, Leon, Chinandega and La Unión, but caused no damage. The shock was felt in Managua and San Salvador.

The instrumental evidence suggests a depth of about 60 km.

[DES, REC]

1940 Feb 6 El Salvador

An earthquake shook San Salvador, but caused no damage. It was preceded by shocks which were felt in northern El Salvador. Instrumental information about these shocks is lacking.

[DES; REC]

1940 Mar 28 M_S 5.4 El Salvador

A strong earthquake was felt in a number of unspecified towns along the Pacific coast of El Salvador. It was preceded and followed by many shocks.

The main shock caused some alarm at San Salvador, but caused no damage.

[DES; REC]

1940 Apr 4 El Salvador

An earthquake in El Salvador caused panic at San Vicente.

[REC]

1940 Apr 7 El Salvador

A strong earthquake in El Salvador caused some minor damage in the region of the delta of Rio Lempa, at San Augustin and San Marcos. The shock was felt at San Salvador. Instrumental data for this earthquake are lacking.

[REC]

1940 May 17 M_S 5.8 Panama

This earthquake in the Gulf of Montijo in western Panama is relatively well located instrumentally, but there is no local macroseismic information.

The shock was perceptible at Balboa Heights, 270 km from the instrumental position.

[USEQ]

1940 Jul 13 M_S 6.2 Panama

We could find no macroseismic information for this event, whose instrumental position is in the Gulf of Chiriqui.

It is said that this shock was perceptible at Balboa Heights, 350 km away.

[USEQ]

1940 Oct 27 M_S 6.8 Costa Rica

In spite of its relatively large magnitude, this earthquake in Costa Rica caused disproportionately small damage in the region of Turrubares. The shock was felt strongly at Esparta, Puntarenas, Orortina, Grecia and Alajuela.

[Miyamura 1980]

1941 Mar 10 03h **Panama**

This small earthquake was felt by many at Balboa Heights, Panama.

[USEQ]

1941 Mar 10 04h **M_S 5.2** **Panama**

This earthquake in the Peninsula of Azuero in Panama was preceded and followed by many small shocks,.

The shock was particularly strong at Flores and Pedasi, but caused no damage there. It was perceptible in Central Panama, and as far as Balboa Heights and the Canal Zone.

[LPZ; USEQ]

1941 Mar 23 **M_S 5.6** **Honduras**

A strong earthquake was felt on Swan Island, which is about 200 km north of the coast of Honduras. On the island buildings swayed visibly and heavy objects were displaced.

A sailing ship about a mile off the west end of the island was shaken violently.

Aftershocks continued to be felt intermittently until November.

[USEQ]

1941 Apr 6 **M_S 5.2** **El Salvador**

This was a damaging earthquake in El Salvador. Ten villages and farms on the coastal region of Rio Lempa suffered considerable damage, but details are lacking. The shock was strong at San Salvador.

[REC]

1941 Dec 5 **M_S 7.6** **Tsunami** **Figure A.47** **Costa Rica**

A major earthquake in the Golfo del Dulce was felt throughout Costa Rica and in much of Panama.

The earthquake caused severe damage in the Osa Peninsula. Most affected were the alluvial plains along the Golfo del Dulce and the Bahia de Coronado, where soil liquefaction was the cause of severe damage.

At Golfito the port facilities, the water supply, the electricity system and many buildings of all types of construction were destroyed, and four people were killed. The railway line was also destroyed along a length of about one kilometre due to excessive ground deformations, most probably caused by lateral spreading of the ground. Landslides near the port and slumping of the ground near the coast added to the damage.

Puerto Jiménez was also ruined and its landing pier and many of its timber houses were destroyed.

Palmar Sur, Piedras Blancas, Puerto Cortés, La Angelita and Pozos Sur were almost totally destroyed, and even timber houses collapsed or became uninhabitable, mainly due to the liquefaction and excessive deformation of the ground.

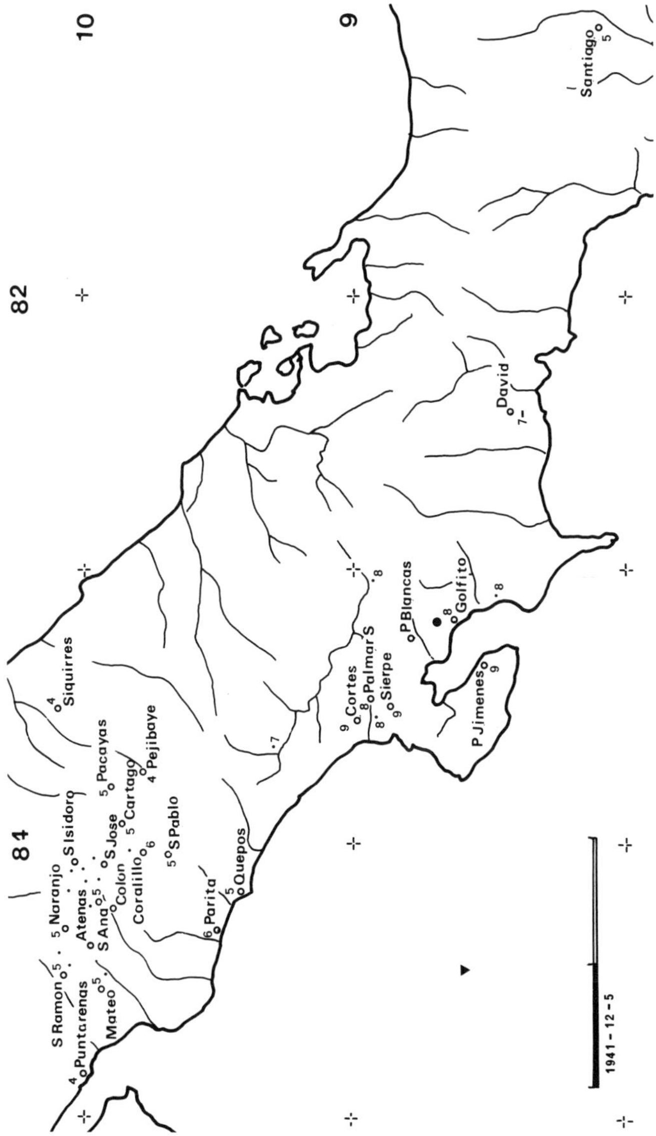

Fig. A.47 Felt effects of earthquake of 5 December 1941.

Ground movements were very strong; at La Angelita animals staggered and at Puerto Cortés trees were uprooted. The landing pier of Puerto Cortés was put out of commission and in the town the church was damaged beyond repair, the school collapsed and many houses were destroyed. Two people were killed and many were injured.

Segments of the railroad track between Golfito and Puerto Cortés, totalling a few kilometres, slumped due to excessive settlements of the ground and they were destroyed.

At Santiago, Panama, a number of houses, the municipal building and a school were toppled and at David a few houses were destroyed.

The shock caused great panic and minor damage over a large area that extended to Quepos, Paritta, San Pablo and to the densely populated Central Plateau.

The earthquake was strongly felt as far as Bagaces, Guanacaste, Puntareas and San Isidoro in the northwest and in the Peninsula de Azuero at Pedasi and Panama City in the east. At Balboa Heights the seismograph pendulum clock stopped.

The shock was experienced on board a vessel steaming near 8.58°N, 84.45°W, 160 km from the epicentre.

The earthquake was recorded on a tide gauge at Puntarenas in Costa Rica. The disturbance began at 21h 45m, about one hour after the main shock, and continued for about six hours with an average range of 23 cm.

The earthquake was followed by many strong aftershocks that caused additional damage and triggered new slides in the region between Puerto Cortes and Golfito.

Both macroseismic and instrumental data confirm a location for this earthquake east of the Golfo del Dulce in Costa Rica (Figure A.47). Much of the macroseismic information comes from far-field sites in the Central Plateau, with relatively little from the near-field to define clearly its epicentral area and its effects in Panama.

[Boschini 1989; Feldman 1984; Miyamure 1980; USEQ]

1941 Dec 6 01h Ms 5.9 Costa Rica

This earthquake occurred in Costa Rica, in the region of Guanacaste.

Maximum damage was observed at Bagaces, a site already affected by the earthquake of 1935, where almost all the houses, most of them old, and public buildings were damaged beyond repair. No one was killed or injured. At Salitar and Montano damage was less serious.

The shock was felt at Liberia and Nicoya, and was perceptible in San José.

[Feldman 1984; Miyamura 1980; Montero and Alvaro 1988; PRS]

1941 Dec 6 21h Ms 7.0 Tsunami Figure A.48 Costa Rica

This was the largest aftershock of the earthquake of 5 December, with an epicentre off the Osa Peninsula.

The shock added to the damage in the region of the Golfo del Dulce and elsewhere, but its effects are difficult to separate from those of the main shock.

It was felt throughout most of Costa Rica and in western Panama.

A long series of aftershocks followed, some of them causing minor damage.

The earthquake was recorded on a tide gauge at Puntarenas, Costa Rica. The disturbance began at 22h 20m and continued for about eight hours with an average range of 8 cm.

[USEQ]

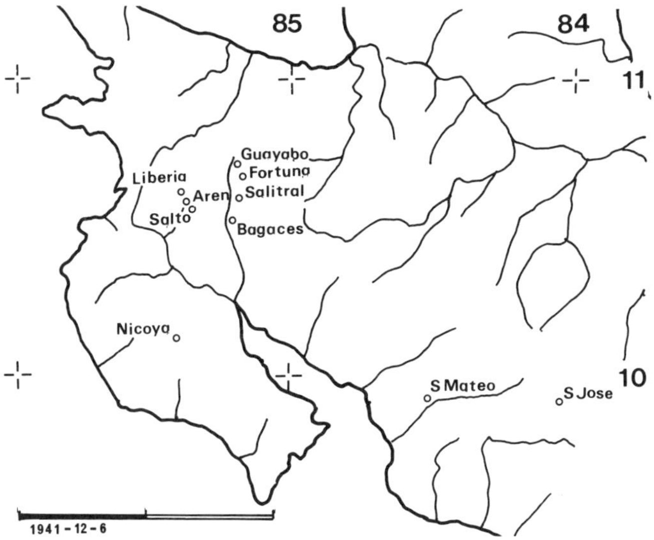

Fig. A.48 Felt effects of earthquake of 6 December 1941.

1942 Aug 6 M_S 7.9 Figure A.49 Guatemala

This major lower crustal earthquake in western Guatemala was the largest in our area for the period of our study, being rivalled only by the Costa Rican earthquake of 5 October 1950.

The shock, which was of long duration, caused widespread damage and some destruction along the west-central highlands in Guatemala, where 38 people were killed. Material losses amounted to over one million dollars. While the damage was heavy and severely affected most of Guatemala, save the eastern seaboard, the catastrophe may not fairly be characterised as one of first magnitude. In general, the destruction was confined to old adobe buildings and modern houses seemed to have fared well. The combination of the earthquake and the heavy seasonal rains added to the destruction and caused landslides which at numerous places destroyed roads, the Inter-American Highway and telegraph lines.

In Tecpan more than 60% of the houses were demolished and in Totonicapan and Chimaltenango 50% were destroyed. In San Marcos the losses amounted to 45% of the building stock and in Amatitalan 33%. About half of Patzicia and one-third of Escuintla were totally ruined. Acatenango, Antigua, Pochuta, Yepocapa and Zaragoza were shattered, and access roads were blocked by landslides. The entire water supply of Antigua was destroyed. Damage was serious at Mazatenango, where the ice plant and cemetery were totally destroyed, as well as at Quetzaltenango and La Esperanza, where all public buildings and most of the houses were seriously damaged. Many settlements, especially in the western Pacific coast region, suffered some damage with 10% to 30% losses in places.

Sporadic damage was reported from as far as Huehuetenango, El Progresso, Rabinal, Salama and Santa Rosa. In Guatemala City damage was widespread but not very serious.

The earthquake alarmed the population throughout Guatemala. It was strongly felt as far as Puerto Barrios, Alta Varapaz, southern Peten, El Salvador and in the State of Chiapas, as well as in Vera Cruz and Tabasco in Mexico.

Because of the Second World War press reports about this earthquake are few and limited in detail.

Instrumental relocation of the earthquake confirms the general position in central Guatemala indicated by the macroseismic information, and a focus in the lower crust.

[TAC; USEQ]

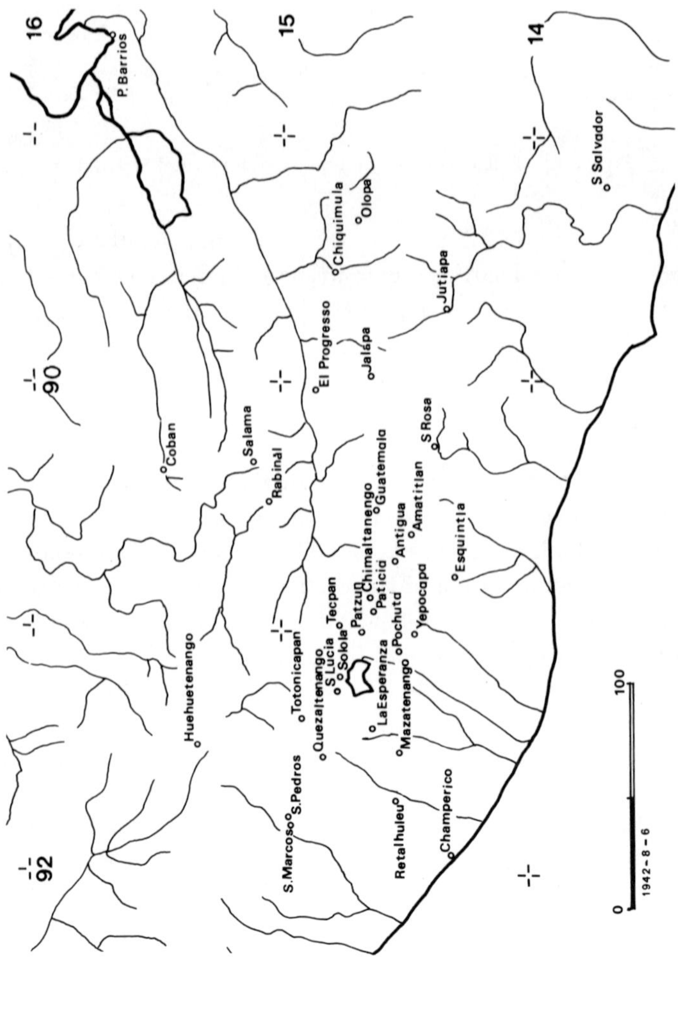

Fig. A.49 Felt effects of earthquake of 6 August 1942.

1942 Nov 11 M_S 5.7 Costa Rica

This shock was strongly felt in central Costa Rica, where it did no damage, but details are lacking.

[BSSA]

1942 Dec 22 M_S 5.4 Panama

An earthquake in the Gulf of Panama was felt by a few people in places on the Pacific coast.

It was followed by aftershocks, which were more widely felt at Balboa and Cristobal.

[USEQ]

1943 Apr 19 M_S 5.6 Honduras

This earthquake in the Caribbean Sea was felt in Swan Island, 260 km west of its instrumental position.

[USEQ]

1943 May 2 M_S 7.0 Panama

This earthquake was located instrumentally south of the Peninsula of Azuero, Panama.

The shock was strongly felt in the District of Los Santos, where it caused some minor damage, and in the Canal Zone, where many people were alarmed.

At Balboa five out seven pendulum clocks in the canal administration building were stopped.

The shock was felt throughout almost all Panama and in western Colombia.

The small amount of macroseismic information available for this earthquake confirms the general location determined by ISS.

[USEQ]

1943 Jun 15 M_S 6.5 Guatemala

This earthquake was located off the southwest coast of Mexico. It was widely felt in Chiapas, but caused no damage there.

It was followed by numerous aftershocks.

1943 Sep 23 M_S 6.1 Guatemala

This earthquake was located instrumentally in western Guatemala and assigned an intermediate depth.

The shock caused damage at Grandeza in Chiapas and to other nearby settlements and villages across the Mexican border, but details of damage in Guatemala are lacking.

The shock was very strong at Amatenango, Mazapa and Motozintla and was felt in the State of Chiapas.

[Figueroa 1973]

1944 May 27 **El Salvador**

A strong earthquake was felt at San Salvador. No instrumental data are available for this shock.

[BSSA]

1944 Jun 28 **M_S 7.2** **Guatemala**

This was a major lower crustal earthquake with a relocated position just off the coast of Guatemala.

We could find no macroseismic information.

1944 Aug 24 **M_S 5.9** **Mexico**

This strong earthquake in Chiapas, Mexico, was very strong at Arriaga and Tonala.

[Figueroa 1973]

1945 Jan 12 **M_S 6.0** **Mexico**

This earthquake in Chiapas, Mexico, has been assigned a focus of intermediate depth.

The shock was very strong at Arriaga, Tonala, Villa Corzo and Villa Flores, and strong at Tuxtla Gutierrez.

[Figueroa 1973; TAC]

1945 Jun 3 **M_S 6.3** **Panama**

The earthquake was felt strongly in the Province of Chiriqui, Panama, where it caused minor damage to buildings, particularly at David and Puerto Armuelles.

The shock was felt as far as the Canal Zone, where some people were awakened.

[USEQ]

1945 Jul 5 **M_S 4.9** **Panama**

A strong earthquake in central Panama caused some damage in the Department of Veraguas, where roofs were damaged, and a few houses in Santa Fé collapsed, without causing any casualties.

The shock was generally felt at Balboa Heights, and was perceptible in the Canal Zone, 180 km away.

[BPH; USEQ]

1945 Aug 10 **M_S 6.0** **Figure A.50** **Guatemala**

This was a damaging shock in northeast Guatemala.

At Quirigua considerable damage was done to houses, to the hospital and church, and a few people were injured.

In places, which are not named, the ground was fissured and elsewhere new hot springs appeared. It is said that the earthquake also cracked the Mayan stelae near Los Amatos in Quirigua; it was also strongly felt in the Districts of Chiquimula, Zacapa and parts of Izabal.

The earthquake was followed by a strong aftershock, about three hours later,

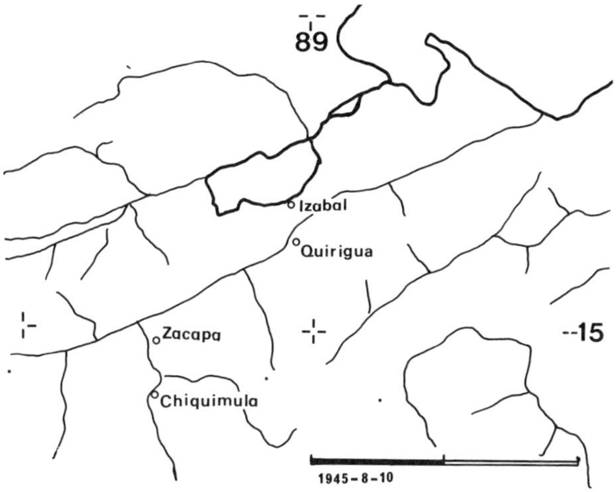

Fig. A.50 Felt effects of earthquake of 10 August 1945.

which at Quiringa destroyed the hospital and a number of buildings already damaged by the main shock.

[REC: 1949.25; White 1980]

1945 Oct 27 **M_S 6.0** **Guatemala**

This earthquake in western Guatemala has been assigned an intermediate depth. It was felt in Chiapas, and caused some damage in the region of El Soconusco and Tapachula. Details of the effects of this event in Guatemala are lacking.

[Figueroa 1973]

1946 Jan 5 **M_S 5.7** **Guatemala**

This was an intermediate-depth earthquake with an instrumental epicentre located in western Guatemala. It was felt very strongly in Chiapas, at La Concordia, Socoltenango and Trinitaria, about 200 km southeast from its instrumental position.

[Figueroa 1973]

1946 Feb 22 **m_B 6.0** **Mexico**

An intermediate-depth earthquake with an instrumental epicentre in Mexico was strongly felt in Chiapas at Cilla Corzo and Villa Flores.

[Figueroa 1973]

1947 Jan 25 03h **M_S 6.1** **Figure A.51** **Honduras**

An earthquake in the Gulf of Fonseca was strongly felt in eastern El Salvador, Honduras and Nicaragua.

In the town of La Unión a number of adobe houses, the hospital and the bell towers of the church collapsed, killing one person. Damage to the more substantial buildings in the port was negligible. Damage extended to El Carmen, San Alejo and to farms in the region of Choluteca in Honduras, where many houses were badly damaged and a few collapsed; precise details are lacking.

Causeways and waterways in Honduras, within a radius of about 80 km around the Gulf of Fonseca, suffered differing degrees of damage.

Telegraph and telephone comunications between El Salvador and Nicaragua were interrupted.

The shock was very strong at Chinandega, Somotilo and in some parts of the District of San Miguel in El Salvador, but it did no damage there.

The earthquake was also felt at Chichigalpa, Corinto, Esteli, Leon, Ocotal, Pueblo Nuevo, Somoto and Villanueva, as well as in the eastern provinces of El Salvador. The shock was perceptible in Managua but not in Granada.

[REC; Mayer-Abich 1956.92;]

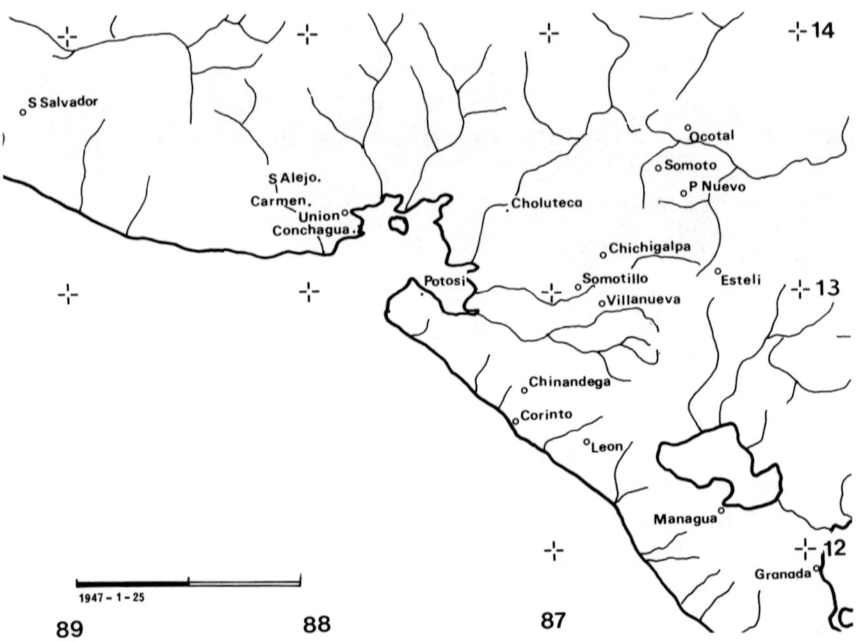

Fig. A.51 Felt effects of earthquake of 25 January 1947.

1947 Jan 25 04h — Honduras

A second shock about one hour later was felt strongly in the region of Choluteca, but caused no damage. Allegedly this shock was associated with the emission of vapour from the Volcano of Conchagua.

The effects of this earthquake are often wrongly attributed to the shock of 26 January.

1947 Jan 26 M_S 6.7 Nicaragua

The macroseismic data indicate an intermediate-depth earthquake in Nicaragua not far from the region affected by the shallow shock of 25 January.

The earthquake was strong at Managua, Leon and Chinandega, where it caused no damage, and it was felt in Granada and in the region of La Unión, where ground motions were slow and lasted for a long time. Damage in this area attributed to this event was in fact caused by the shallow earthquake of 25 January.

This was a well recorded event which gave an opportunity to test the determination of depth in this region. Over 100 stations gave useful readings, and although none was closer than 7°, there was a good distribution of stations out to distances of *PKP* phases. Stations reported from Central, North and South America and from Europe, with more distant readings from India and New Zealand. There was also a profusion of later phases. Many of these fitted as depth phases *pP* and *sP*, which give a depth of about 160 km. There is no doubt that the earthquake is of intermediate depth.

[Jordan and Fernandez 1980; Fernandez; REC; Schultz 1963; US Natl. Arch.:RG 59 B.5169(El Salvador)]

1947 May 26 Guatemala

This small shock in Chiapas was strongly felt at Angel Albino Corzo and Jaltenango.

[Figueroa 1973]

1947 Jun 5 M_S 5.4 Guatemala

The instrumental location given by ISS for this relatively small event is in southeast Guatemala.

The shock was perceptible in Guatemala City, where the maximum acceleration recorded by a strong-motion instrument was about 0.5%g.

[USEQ]

1947 Jul 15 Nicaragua

A strong earthquake was felt at Leon, Nicaragua; we could find no instrumental data.

[REC]

1947 Dec 23 Guatemala

A small shock in Chiapas was felt at Motozintla.

[Figueroa 1973]

1947 Dec 30 M_S 5.8 Costa Rica

This earthquake was felt in the central part of Costa Rica, but details are lacking. At San José a strong-motion instrument recorded an acceleration of 1.2%g.

[USEQ]

1948 Jul 16 **M_S 6.2** **Guatemala**

Two intermediate-depth earthquakes, seven minutes apart, shook Guatemala. They were felt at Guatemala City and San Salvador, and were followed by many aftershocks.

The aftershocks of 17, 18 and 19 July were recorded by a strong-motion instrument at Guatemala City and showed maximum ground accelerations of a few gals.

[BSSA; USEQ]

1948 Jul 28 **M_S 5.5** **Panama**

Two shocks, 44 minutes apart, caused some small damage on Coiba Islandindex Coiba, in the Gulf of Chitriqui, Panama.

The second and larger shock was felt at Coco Solo and Balboa.

[BSSA; USEQ]

1948 Aug 13 **Mexico**

A local shock in Chiapas was strongly felt at Cintalapa de Figueroa and Jiquipiles. It was perceptible at Tuxtla Gutierrez.

[Figueroa 1973]

1948 Sep 16 **Costa Rica**

The location of this event is uncertain. Its rough instrumental position is off the southwest coast of Costa Rica.

The shock was recorded at San José by a strong-motion instrument, which showed a peak acceleration of 0.3%g. It is not known whether the shock was felt.

[USEQ]

1948 Nov 19 **M_S 6.3** **Costa Rica**

This was an intermediate-depth earthquake in Costa Rica.

Shaking was strong at Cartago, Santa Cruz, Tuirrialba, in other places in the Central Valley, and further south at Puerto Cortés and Puerto Jiménez, with no damage resulting. Slight damage was reported from San José, where an accelerograph recorded a peak ground acceleration of 3%g, and from a few other places. There was no indication of a specific epicentral area.

The shock was felt throughout Costa Rica, but there is no information from Nicaragua and Panama.

[Boschini 1989; Boschini and Montero 1993; USEQ]

1949 Jun 9 **Guatemala**

This small earthquake in Chiapas was probably a subcrustal event. It was felt strongly at Las Margaritas.

[Figueroa 1973]

1949 Aug 18 **M_S 6.1** **Panama**

This earthquake in western Panama was very strong in the Department of Chiriqui, where it caused some small damage at David and considerable concern at

Golfito and Puerto Armuelles.

The earthquake was felt by a few Canal Zone residents at an epicentral distance of 390 km.

[BSSA; REC; USEQ]

1949 Dec 22 M_S 6.2 Mexico

This was a subcrustal earthquake in Chiapas, which was strongly felt over a large area including Chiapa de Corzo, Rivera de Cangui, San Cristobal las Casas, Tuxtla Gutierrez, Villa Corzo, and Villa Flores.

The shock was felt throughout the Department of Chiapas.

It was recorded by a strong-motion instrument at Guatemala City 310 km away, with a peak ground acceleration of 0.2%g.

[Figueroa 1973; USEQ]

1950 Mar 24 Panama

This strong earthquake in Coclé, Panama, was felt at Santiago and in the Canal Zone.

[STR]

1950 Oct 5 M_S 7.9 Tsunami Figure A.52 Costa Rica

This major lower crustal earthquake in Costa Rica caused disproportionately little damage for its size.

At Nicoya many houses and the cathedral church were seriously damaged. Equally serious damage was reported from Filadelfia, Santa Cruz, Paraiso and from settlements on either side of the Nicoya Peninsula.

At Puntarenas, the port facilities and many houses were damaged on account of their poor foundations, and a few were destroyed, but there were no casualties. In places the ground slumped, causing damage to the port, to the railway lines and water mains.

The railroad to the east of Puntarenas, between kilometres 45 and 48, and at kilometre 59, was damaged and in places the tracks were displaced and blocked by slides.

In Aluejela, San Ramón, San Isidoro, Santa Barbara and Tres Rios and in the villages to the south and east, the shock caused great panic and damage to many houses and to a few public buildings and churches, mainly old constructions, but there were no casualties.

At San José the shock caused widespread damage and the injury of a few people. A strong-motion instrument in the city recorded an acceleration of just under 6%g.

The shock was very strong at Cartago, Orotina, Oreamuno, Liberia and Quepos and was felt at Rivas, Virgen, Guapiles and Cortes.

There was a series of strong aftershocks.

The earthquake was associated with a small seismic sea-wave which was reported from the Pacific coast of the Nicoya Peninsula with periods of about 45 minutes. The tide station at Puntarenas was also shaken, and soon after recorded a seiche or possible seismic sea-wave. Small oscillations that may have resulted from this earthquake were also recorded on tide gauges at San Juan del Sur, Nicaragua, and at La Unión, El Salvador. We could find no evidence to substantiate the statement by Cruz and Wyss (1983) that this earthquake set up a seiche at La Libertad, El Salvador, reaching an amplitude of 8.9 metres.

It is not known how far the earthquake was felt; reports from neighbouring Panama and Nicaragua are lacking.

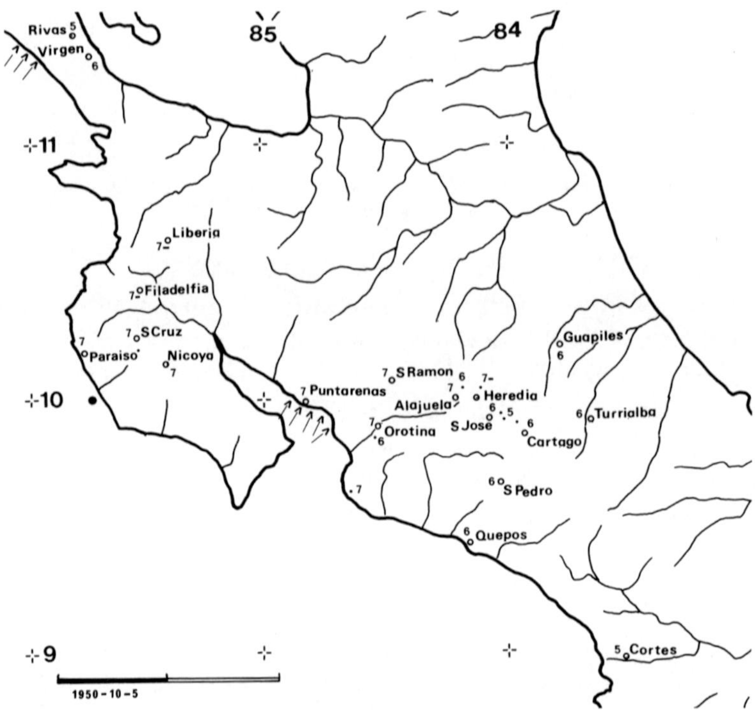

Fig. A.52 Felt effects of earthquake of 5 October 1950. Arrows offshore show areas where tsunami effects were reported.

This was the second largest earthquake in our area during the period of our study, being only slightly smaller than the Guatemalan earthquake of 6 August 1942. The overall macroseismic observations are good, but lacking in detail for the epicentral area (Figure A.52). They support a location of the epicentral area partly on the west coast of the Nicoya Peninsula and partly at sea, and a focus in the lower crust. Various instrumental relocations of this event give epicentres further to the east and a focus in the crust. The seismic sea-waves reported from two areas of the coast, 200 km apart, are likely to be due to the local slumping of submarine slopes.

[Barquero and Rojas 1994b; Cruz and Wyss 1983; Feldman 1984; Güendel 1986; Miyamura 1980; REC; USEQ]

1950 Oct 23 M_S 7.3 Tsunami Guatemala

This was a major earthquake in southwestern Guatemala.

At San Marcos not a single house was left undamaged; some were destroyed and several people were injured.

The shock was strongly felt in central Guatemala, and in Chiapas at Puerto Madera, Tapachula and in the Department of Soconusco.

In Guatemala City people were awakened. The earthquake was recorded with a peak acceleration of 4%g by a strong-motion instrument.

The earthquake was followed by many aftershocks.

The tide gauge at San José, Guatemala, recorded a wave of about 30 cm amplitude, but at La Unión, El Salvador, it was barely noticeable on the tide gauge. The wave was not recorded by other tide gauges in Mexico and Central America.

In spite of the large size of this earthquake the available macroseismic data are sufficient only to confirm the general area of the instrumental relocation in southwestern Guatemala.

[Figueroa 1973; REC; TAC; USEQ]

1950 Oct 24 M_S 5.8 Guatemala

This strong aftershock was felt at Puerto Madero, and to a lesser degree at Tapachula, and in western Guatemala.

It was followed by equally strong shocks in the following two days.

[Figueroa 1973]

1950 Oct 28 M_S 5.4 Guatemala

This aftershock, probably in the lower crust, was reported as strong from Mazapa, Motozintla and other places in Chiapas.

[Figueroa 1973]

1950 Nov 1 Guatemala

Strongly felt in Chiapas, at Purto Madero and Ciudad Hidalgo, this aftershock was less strong at Tapachula.

[Figueroa 1973]

1950 Nov 5 M_S 5.8 Guatemala

More strong shocks were felt at Purto Madero, Ciudad Hidalgo and Tapachula. The shock was equally strong across the border in Guatemala, but details are lacking.

[Figueroa 1973]

1951 Jan 5 M_S 5.9 Panama

This was a strong foreshock of the earthquake in central Panama that followed a day later; its focus was probably deeper than normal.

At Penonome and Santa Fé the shock caused some alarm, and it was strongly felt in the Azuero Peninsula. It was perceptible throughout the Canal Zone.

[USEQ]

1951 Jan 6 M_S 6.6 Panama

This earthquake in central Panama caused great panic in the Districts of Colcé, Veraguas and to a lesser extent in Los Santos, particularly at Calbre and Penonome, where people ran into the streets.

There is no evidence that the shock caused any damage.

In the Canal Zone the shock was very strong. Most residents were awakened, and in a few places it produced plaster cracks and caused old cracks in walls to re-open.

The peak ground acceleration recorded at Balboa Heights reached 1%g.

[USEQ]

1951 Jan 29 Guatemala

This shock was reported as strong from Motozintla, Chiapas.

[Figueroa 1973]

1951 Feb 13 Guatemala

This shock was very strongly felt at Motozintla and Mazapa de Madero, Chiapas, and was also reported from Pueblo Nuevo and from a number of places in Guatemala.

[Figueroa 1973]

1951 May 6 23h 03m M_S 5.9 Figure A.53 El Salvador

Two successive earthquakes five minutes apart, followed by a third shock twenty-one hours later, caused localised destruction and loss of life in eastern El Salvador.

The first shock was widely felt but caused little damage.

At Jucuapa (pop. 6700) and Chinameca (pop. 7000) the earthquake consisted of a violent vertical shock. It was strong enough to throw people to the ground and to damage a few houses of the "bahareque" type of construction. No adobe houses collapsed and there was no loss of life.

At the coffee factory of San José the shock was very strong. It caused no damage to the buildings of the plant, but its free-standing adobe fence wall was thrown down.

At San Buenaventura (pop. 1540) only the adobe church was ruined.

At Nueva Guadalupe (pop. 1720) the shock was strong and distinctly vertical in motion, but it caused no serious damage.

At Santiago de Maria (pop. 6880) the shock was strongly felt.

At Berlin (pop. 5120) the shock was distinctly felt.

At Nueva Guadelupe (pop. 1720) the motion was vertical.

At Mercedes Umana (pop. 4430) and Lolotique (pop. 2760) it was felt.

At Triunfo (pop. 1240) the shock was strong, and caused trees to sway.

The shock was felt at San Salvador and probably at San Miguel.

ISS used readings from 84 stations to place the epicentre of the first shock at 13.0°N, 87.8°W and the focus at a depth of 100 km. Other instrumental positions determined by Strasbourg and USCGS give similar locations and depths between 100 and 150 km. Our re-determination of these events confirms intermediate depths from both free solutions and recordings of depth phases. The instrumental locations are about 16 km southwest of the macroseismic epicentre.

[Meyer-Abich 1952; PHRG; PIMG; PPRN; REC: 1954.196-7; Schultz 1963; Wallstrm 1953]

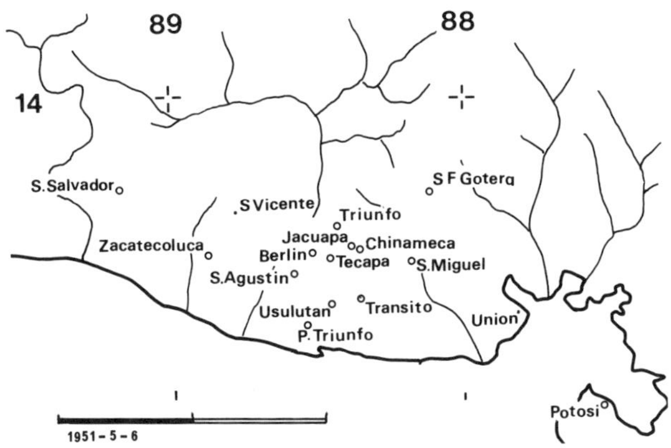

Fig. A.53 Felt effects of earthquake of 6 May 1951.

1951 May 6 23h 08m M_S 6.0 Figure A.53 El Salvador

This second shock occurred about five minutes later.

At Jucuapa the earthquake consisted of a powerful horizontal jolt that completely destroyed the town, throwing houses to the east and killing about 350 people. The only structure left standing, badly damaged, was an unfinished market-building. The shock triggered slides of alluvial banks of dry water courses, the collapse of the abutments of a small concrete bridge at the entrance to the town and of a number of stone masonry retaining walls.

At Chinameca the earthquake killed about 50 people and destroyed or damaged beyond repair 80% of the houses; most of them fell to the east. The rest of the houses were badly broken. The steeple of the church fell, together with the walls of the choir, and its concrete cupola was left hanging in the air. Tombstones in the nearby cemetry were overturned.

About 800 m to the south of Chinameca the ground was badly cracked. Individual ground ruptures were found running in a NW-SE direction but the exact location of these cracks is not given.

On the Plan de las Mesas, an extensive basaltic lava flow situated about one kilometre to the northeast of Chinameca, the shock caused some concern but damage was negligible. Dilapidated old adobe farmhouses on the Mesa were left standing and the school building suffered only plaster cracks and fallen roof tiles.

The coffee factory of San José was almost totally destroyed and most of its buildings collapsed to the east. Here, heavy machinery was displaced on its foundations, concrete tanks were fractured and light roofs were thrown off their supporting walls.

The buildings of the plantation of Los Naranjos were also ruined and its stables made of timber collapsed.

Of the 300 adobe houses of San Buenaventura 270 collapsed, together with three churches, killing three people.

Of the 320, mainly adobe, houses in Nueva Guadalupe, 50 collapsed to the east, 20 became uninhabitable, and those remaining were ruined.

At Santiago de Maria there was widespread but minor damage consisting of cracks in plaster and ruptures of the water supply.

At Berlin adobe houses suffered plaster cracks.

At Mercedes Umana the shock caused panic. Adobe houses were badly cracked and tiles fell off roofs. The shock made the majority of houses uninhabitable.

At Lolotique the church and about 20% of the houses, mostly adobe, were either destroyed or damaged beyond repair. Only 10% of the houses survived intact and the rest were damaged. A 60 m deep water well was blocked by the collapse of its wall.

At Triumfo adobe houses sustained some damage, tiles fell off roofs and a pendulum clock was stopped.

The shock was felt in San Salvador, San Francisco Gotera at Potosi and probably at La Unión.

The second shock, which was reported in ISS by 66 stations, occurred about four and a half minutes later. Its depth is also established as intermediate at about 70 km. Because of the closeness in the times of the two shocks, it is difficult to separate the phases, and most of the amplitude readings available are for the first shock.

1951 May 7 M_S 5.5 El Salvador

This strong aftershock of the earthquakes of May 6 caused additional damage outside the area of maximum effects of the main shocks.

At Santiago de Maria the shock was violent, particularly in the east part of the town, where the foundation materials are not strong. In all, about 40% of the houses were destroyed, 40% ruined; the rest were damaged and the electricity supply was cut. A reinforced concrete structure was damaged.

At Berlin the shock was very strong. It set church bells ringing and water sloshed from reservoirs. The covered market of the town was ruined and 10% of the houses were destroyed and about 40% damaged, particularly those of adobe construction.

Starting from the farm of Llanes, about 1.6 km northwest of Berlin, a 10 m wide zone of ground cracks was found running for about 2 km, and striking, regardless of topography, between N305°E and N330°E. Individual cracks were about 20 cm wide with no evidence of vertical displacement.

At Mercedes Umana new cracks appeared in walls but no house collapsed.

At Jucuapa ground movements were slow and of long duration; they were not as strong as those of the second main shock, but strong enough to overturn free-standing objects and ruined walls.

In the region Chinameca and San José the shock was strongly felt, without causing additional damage.

At Nueva Guadalupe the shock was weak and of a much longer duration than the previous shocks, causing the fall of a number of ruins.

At Lolotique a few houses ruined by the previous shocks collapsed.

At Triumfo the shock was felt, but caused no damage.

The shock was strong at San Salvador, Puerto El Triunfo, and San Miguel.

ISS reports readings from 57 stations, and adopts the same epicentre as for the main shock and an intermediate depth.

[Jordan and Martinez 1980; Meyer-Abich 1952; PIMG; PPRG; PPRN; REC; Stoiber and Carr 1973; USNA; Wallstrm 1953]

1951 May 11 **El Salvador**

A belated aftershock was felt at Jucuapa and surroundings, in San Miguel and other parts of the central volcanic zone. It was particularly strong at San Salvador.

[Mayer-Abich 1952]

1951 Jun 18 **M_S 5.3** **Costa Rica**

The shock was perceptible in San José, Costa Rica, where it was recorded by a strong-motion instrument with an acceleration of 1%g.

[USEQ]

1951 Jun 25 **El Salvador**

A strong earthquake in El Salvdor was felt in the region of San Marcos Lempa and Zacatecoluca. The shock was felt also at Alegria, Berlin, Jucuapa, San Vicente, and Santiago de Maria, within a radius of about 40 km.

It was followed by a few aftershocks on 26 and 27 June, which were reported from San Marcos.

The shock was recorded at San Salvador but we could find no instrumental data from other stations.

[Mayer-Abich 1952]

1951 Aug 2 **M_S 5.6** **Nicaragua**

This was the first of two shocks four hours apart, which affected the region around the Volcano of Cosigüina in Nicaragua. It was preceded by a foreshock on 1 August.

The shock did considerable damage and allegedly there was some loss of life in this sparsely populated part of the south coast of the Gulf of Fonseca, but details are lacking.

The shock was perceptible in Jucuapa in El Salvador.

White and Harlow list an earthquake at 20h 30m to which they assign a magnitude 5.8 on the basis of readings from De Bilt. They add that this earthquake caused 1,000 deaths on the Nicaraguan side of the Gulf of Fonseca. We could find no evidence that De Bilt reported a surface-wave amplitude for this event.

[Mayer-Abich 1952; White and Harlow 1993; REC]

1951 Aug 3 **M_S 5.9** **Nicaragua**

This was the second and larger earthquake of the pair and added to the damage.

As a result of this shock the flanks of the cone of the Volcano of Cosigüina were in places split open, releasing large volumes of water and debris, which virtually destroyed the nearby port of Potosi.

This shock was also felt at Jutiapa in El Salvador.

[Mayer-Abich 1952; REC]

1951 Aug 6 **M_S 5.1** **Nicaragua**

The facts about the effects of this earthquake, a belated aftershock of the series that started in the beginning of the month in Nicaragua, are not clear.

Press reports mention that in this earthquake Jinotega, a town 160 km east of Cosigüina, was damaged. It is probable that it referred to Chinandenga, a town in the vicinity of the Gulf of Fonseca.

[REC]

1951 Aug 22 **M_S 5.4** **Figure A.54** **Costa Rica**

This was a small, damaging earthquake in Costa Rica.

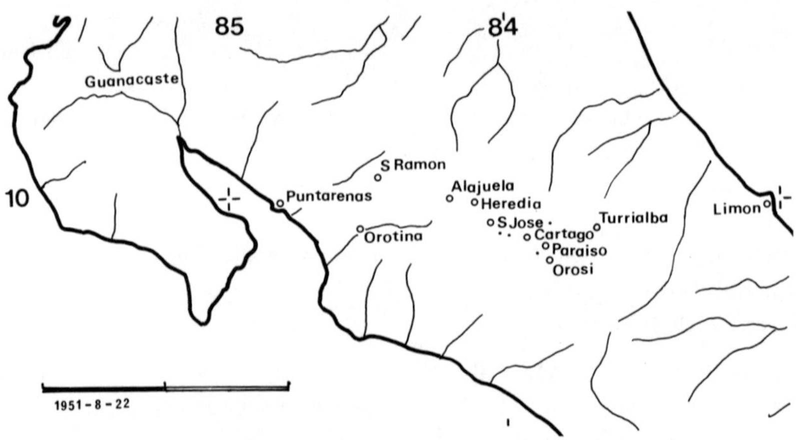

Fig. A.54 Felt effects of earthquake of 22 August 1951.

At Orosi a few houses were totally destroyed and the rest were damaged; there was no loss of life. The church and its bell tower were seriously damaged.

At Paraiso the shock caused the collapse of about 30 houses and damage to twice as many; it injured five people and made 200 homeless. The church, school and a bridge suffered some damage, and the water mains were ruptured.

Minor damage extended to Cachi.

At Cartago about 10 houses were damaged beyond repair and other buildings suffered some minor damage.

The shock was strongly felt at Oreamuno, Taras and also at San José, but it was not recorded by the strong-motion accelerograph in operation in that city. It was felt at Heredía, Alajuela, San Ramón and Orotina, but not in Guanacaste. It was also felt at Turrialba and Limón.

The shock was followed by two strong aftershocks.

[Boschini 1989; Miyamura 1980]

1951 Oct 25 M_S 5.3 Mexico

A strong earthquake in Chiapas, Mexico, was felt strongly at San Cristobal las Casas, Teopisca, Venustiano Carranza, and Chiapa de Corzo, causing some damage at Chiapilla.

The earthquake was followed by aftershocks, the strongest of which occurred on 11 November.

[Figueroa 1973]

1952 Jan 15? Guatemala

A long series of earthquakes occurred in eastern Guatemala.

The seismic activity began at the end of December 1951 in the region of Ixhuatan and intensified. During mid-January 1952 a series of moderately strong earthquakes occurred; the village of Ixhuatan was the most damaged. Cracks in the ground were observed, which may have been caused by slumping.

The activity ended at the beginning of February.

[White and Sanchez 1980]

1952 Feb 4 Mexico

This earthquake in Chiapas, Mexico, was very strong at Acapetahua, Escuintla and Pueblo Nuevo. It was less intense at Huixtla.

[Figueroa 1973]

1952 Feb 26 M_S 6.4 Nicaragua

This subcrustal earthquake occurred in the Pacific off the coast of Nicaragua.

It was strong at Managua, but caused no damage there.

[BSSA]

1952 May 10 M_S 5.1 Mexico

A rather strong earthquake in Chiapas was reported from Chiapilla, San Cristobal las Casas and Teopisca.

[Figueroa 1973]

1952 May 13 M_S 6.4 Tsunami Costa Rica

This lower crustal event in northwestern Costa Rica was strongly felt around the Gulf of Nicoya, at Espatza, Palmares, San Ramón and Santa Cruz.

The shock was felt at San José, where it lasted almost one minute.

McNally and Minster refer to a seismic sea-wave associated with this event, recorded at Puntarenas; we cannot authenticate its occurrence.

[Cruz and Wyss 1983; McNally and Minster 1981; PNCN]

1952 May 16 **M_S 5.6** **Guatemala**

This earthquake in western Guatemala was preceded by an equally strong shock. It was possibly of intermediate depth. It was reported in Chiapas from Mazatan, Puerto Madero, Suchiate and Tapachula. Details of its effects in Guatemala are lacking.

[Figueroa 1973]

1952 Jul 9 **M_S 5.8** **Panama**

This earthquake occurred off the south coast of Panama.

It was felt strongly by m/s *Agamemnon* at 7.33°N, 82.33°W, by a few people at Balboa and in the Canal Zone, and it was followed by an almost equally strong aftershock about two hours later.

[USEQ]

1952 Jul 15 **M_S 5.6** **Mexico**

An earthquake in Chiapas was felt rather strongly at Ciudad Hidalgo, Puerto Madero, Suchiate and Tapachula.

[Figueroa 1973]

1952 Aug 20 **m_B 6.0** **Mexico**

An intermediate-depth earthquake in Chiapas was very strongly felt at Las Margaritas and Trinitaria. It was less strong at Comitan.

[Figueroa 1973]

1952 Sep 9 **M_S 6.8** **Figure A.55** **Costa Rica**

This large earthquake with an epicentre in the Pacific off the southwest coast of Costa Rica was felt throughout the country.

It was very strong at Parrita and Quepos, where the shock lasted 35 seconds. It caused considerable damage to houses and schools at San Marcos and San Lorenzo. At Puertos Cortés and Jiménez, the damage was small, mainly because houses were of timber construction. Here the shaking lasted more than 30 seconds, causing structures to sway. At San Isidoro del General the earthquake was very strong, causing some panic. There is some evidence that the shock was also strong at Buenos Aires, but details are lacking.

Further inland damage was minor, varying from place to place. At Turrucares the shock was strong but did no damage. At Atenas and San Mateo, shaking was of long duration. In Rio Secundo the shock was strong enough to damage the walls of the school, and at Heredía to cause panic. In San José the earthquake lasted almost a minute and caused some alarm, in a few houses displacing furniture; a broken water main was the principal damage. At Alajuelita the church of Nuestra Señora de Esquipules, the school and a few adobe houses suffered some damage. At Tres Rios the earthquake was violent and did some minor damage. At Cartago it overturned objects in houses and caused some damage to merchandise. There was no damage at El Tejar, Paraiso or Cot. At Tierra Blanca only a few adobe houses were cracked. At El Yas, Pacayas and Turrialba, where shaking lasted for over 30 seconds, there was panic but no damage. In Puriscal and San Nicolas de Tacas the shock caused some alarm.

In the south the earthquake was equally strong at Cuesta and David in Panama, and it was felt at Aruelles.

At San Ramón, Grecia, San Pedro de Poas and Sabanilla de Alajuela shaking was not as strong as in the Central Plateau.

The earthquake was felt at Upala but not at Liberia nor La Cruz, but it was perceptible in the southern part of Guanacaste and as far as Rivas in Nicaragua. It was rather strong at the Bara del Colorado and at the Boca del Rio San Carlos; at Siquirres and Limón the shock was of long duration, but it did no damage.

The shock was not reported felt from Panama City, Managua, Tegucigalpa or San Salvador.

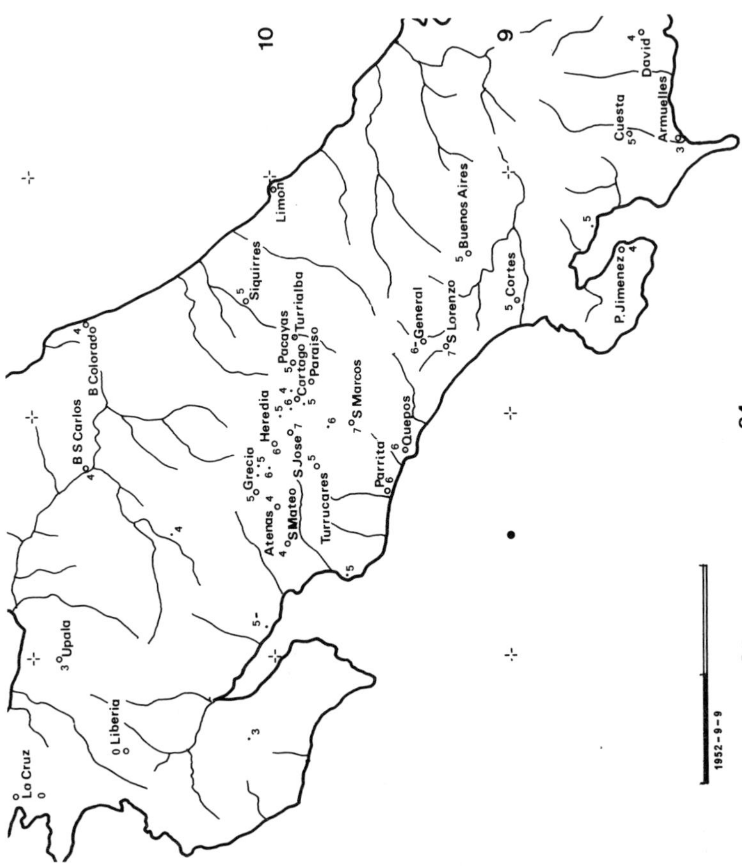

Fig. A.55 Felt effects of earthquake of 9 September 1952.

This earthquake is relatively well documented macroseismically with an epicentre off the Pacific coast of Costa Rica (Figure A.55). We have adopted the computer relocation given by Güendel (1986).

[Barquero and Rojas 1994b; BSSA; Feldman 1984; Güendel 1986; Montero and Climente 1990]

1952 Oct 21 M_S 5.3 Costa Rica

This aftershock of the earthquake of 9 September was felt at San José, but was not reported felt from Quepos.

[BSSA]

1952 Dec 1 Mexico

An earthquake in Chiapas was strongly felt at Acapetahua, Escuintla and Pueblo Nuevo.

[Figueroa 1973]

1952 Dec 30 M_S 5.7 Figure A.56 Costa Rica

This was a damaging shock in Costa Rica.

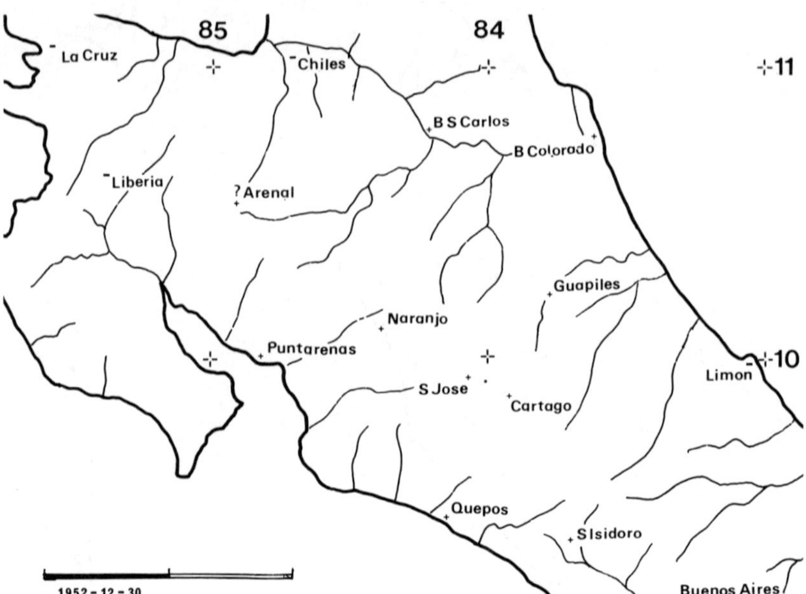

Fig. A.56 Felt effects of earthquake of 30 December 1952.

Damage was concentrated within the Canton of Coronado, where livestock and about 30 people were killed in villages and farms on the slopes of the Irazu Volcano. The villages of Patillos, Cascajal and Tabacales suffered most. The shock triggered landslides and rockfalls, which blocked streams and were responsible for much of the destruction caused at Guayabillos and Piedra de Tres Filos. Less damage was done in some parts of the Canton of Cartago.

The shock was strong at nearby San José, Tres Rios and Cartago, but it caused no damage. At San José the peak acceleration recorded was 9.8%g.

The shock was felt at Arenal, Boca San Carlos and Colorado, San Isidoro, Quepos and Puntarenas, but not at Liberia, Los Chiles, Limón or Buenos Aires.

[BSSA; Feldman 1984; REC: 1954.197]

1953 Jan **Nicaragua**

Press reports mention an earthquake in Nicaragua, which was strongly felt in the region of Somoto.

[REC]

1953 Jan 6 **M_S 5.3** **Costa Rica**

This was a medium-magnitude earthquake in southern Costa Rica. It caused no damage and was felt within an area of 90 km radius encompassing Batan, Cartago, San José, Quepos, Puerto Jiménez and Almirante.

1953 Jan 7 **M_S 5.6** **Figure A.57** **Costa Rica**

This earthquake caused widespread but minor damage over a relatively large area in eastern Costa Rica.

In the port area of Limón (pop. 11,000) the shock ruined 50 houses and shops, damaged a few brick houses and injured two people. Timber-frame structures suffered little or no damage, and the shock had no effect on the port installations. The electricity and power supplies were disrupted temporarily, and two fires started by the earthquake in the commercial part of the town added to the damage, which was estimated at one million colones.

Minor damage extended into the alluvial plain around Puerto Limón, at Bomba, Ramal, Moin, Sandoval and Liverpool. Here the shock damaged two small bridges near Moin and Bufalo, and cracked a number of causeways.

The shock was strongly felt further away from Puerto Limón. It caused some damage at Copalci and triggered landslides which destroyed a section of the road between Santa Maria and Corralillo. Unconfirmed reports add that damage, probably due to slides, extended southwest of Limón, and affected a number of settlements in the Matama Range of Talamanka, and in the region of El Tablazo.

The earthquake caused some panic at Batan, San Joaquin, in the region of Turrialba, at Santa Maria, in the district of Cartago, in Jerico and at Desamparados.

At San José the shock caused some concern; the peak acceleration recorded was about 2%g.

The shock was perceptible as far as Upala, Boca San Carlos, Barra del Colorado, Almirante, and across the border in Panama.

There is no evidence that this earthquake caused an uplift of the Caribbean coast as suggested by some authors.

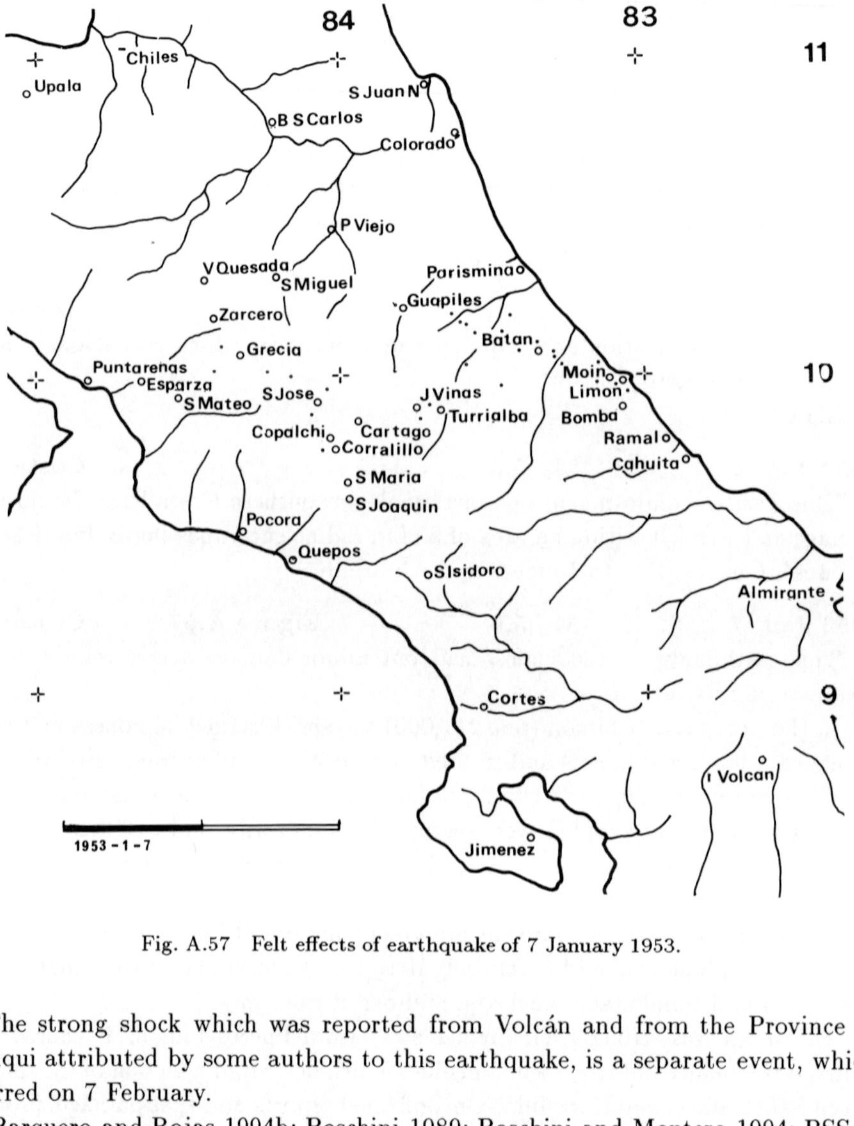

Fig. A.57 Felt effects of earthquake of 7 January 1953.

The strong shock which was reported from Volcán and from the Province of Chiriqui attributed by some authors to this earthquake, is a separate event, which occurred on 7 February.

[Barquero and Rojas 1994b; Boschini 1989; Boschini and Montero 1994; BSSA: 1953.182; Feldman 1984; Güendel 1986; Miyamura 1975; USEQ]

1953 Feb 7 **Panama**

An earthquake was felt at El Volcán and in the Province of Chiriqui in Panama. [USEQ]

1953 Feb 25 **El Salvador**

This was a sharp shock in San Salvador. There were no reports of damage, and we could find no instrumental data for this earthquake.

[BSSA]

1953 May 21 **m_B 5.6** **Guatemala**

This intermediate-depth earthquake had an epicentre off the coast of Guatemala.

It was felt strongly in Chiapas, at Ciudad Hidalgo, Puerto Madero, Suchiate and Tapachula. We have no information from Guatemala.

[Figueroa 1973]

1953 Aug 24 **M_S 5.6** **Guatemala**

This intermediate-depth earthquake was felt in southwestern Guatemala and Mexico.

It was strong in Soconusco, where it caused some small damage at Ciudad Hidalgo and Puerto Madero, and considerable concern at Suchiate and Tapachula. The effects of this shock in Guatemala are not known except that in Guatemala City the recorded peak ground acceleration was 1%g.

[Figueroa 1973; USEQ]

1953 Sep 5 **Guatemala**

The shock was strongly felt at Ciudad Hidalgo, Puerto Madero and Suchiate in Chiapas.

[Figueroa 1973]

1953 Oct 22 **Mexico**

This earthquake was strong at San Cristobal las Casas and at Teopisca.

[Figueroa 1973]

1953 Nov 17 **M_S 6.9** **Guatemala**

We could find no macroseismic data for this earthquake, which was located off the coast of Guatemala.

We have adopted the location calculated by Kelleher *et al.* (1973).

[Kelleher, Sykes and Oliver 1973]

1954 Jan 13 **El Salvador**

This shock was very strongly felt in southeastern El Salvador, causing some panic at Usulutan. It was widely felt at San Salvador.

[Martinez 1978]

1954 Feb 19 **M_S 6.8** **Nicaragua**

An intermediate-depth earthquake with an epicentre off the west coast of Nicaragua was strongly felt along the entire Pacific area of Nicaragua, El Salvador and northern Costa Rica.

The only damage reported was cracked walls in Managua and in other towns along the coast of Nicaragua, but further details are lacking.

The shock was reported as strong from San Salvador in the northwest, and from Guanacaste and Naranjo in the southeast. It was perceptible as far as San Joaquin in Costa Rica.

The earthquake was followed, 21 hours later, by an intermediate-depth event of almost equal magnitude, and by many smaller shocks.

[BSSA; Hansen 1975; Martinez 1978; Miyamura 1980; REC]

1954 Feb 20 **M_S 5.8** **Nicaragua**

This was an aftershock of the earthquake of 19 February, which was strong at Managua. It was also perceptible at San Salvador.

1954 Mar 8 **M_S 4.9** **El Salvador**

A subcrustal earthquake with an epicentre off the coast of El Salvadorindex caused considerable alarm in San Salvador and in other parts of the country. There are no reports of damage or casualties.

[BSSA; Martinez 1978]

1954 Mar 11 **M_S 5.3** **Guatemala**

Another subcrustal event in Guatemala was very strong in the region of Armenia, and caused some concern at San Salvador.

[Martinez 1978]

1954 Mar 27 **M_S 4.7** **Costa Rica**

This subcrustal earthquake was felt in southwestern Costa Rica, at San Joaquin de Flores, and was reported from Portero Grande and San Vito.

[Miyamura 1980]

1954 Apr 3 **Mexico**

This small earthquake was very strong at Las Margaritas, Chiapas.

[Figueroa 1973]

1954 Apr 23 **Costa Rica**

An earthquake with an epicentre off the Pacific coast of Costa Rica was felt on board *S.S. Horace Irving* sailing at 8.21°N, 83.48°W; violent tremors shook the ship for about five seconds.

[USEQ]

1954 Jul 6 **Guatemala**

This intermediate-depth earthquake was rather strong at Ciudad Hidalgo and Puerto Madero, Chiapas.

[Figueroa 1973]

1954 Aug 23 **Panama**

A local shock in Panama was felt strongly at Puerto Armuelles. It was preceded and followed by a few shocks.

[USEQ]

1954 Oct 21 **M_S 6.0** **Guatemala**

A subcrustal earthquake in eastern Guatemala was strongly felt in the region of Santa Rosa and throughout western El Salvador. At San Salvador electric services were disrupted, but no other damage was reported.

[BSSA; Martinez 1978]

1954 Dec 21 **Guatemala**

This was a widely felt shock, probably subcrustal, that was reported felt from as far as San Salvador.

[Martinez 1978]

1955 Mar 5 **M_S 5.1** **El Salvador**

A relatively small intermediate-depth earthquake off the coast of Guatemala was perceptible at San Salvador.

[Martinez 1978]

1955 Mar 20 **Guatemala**

A subcrustal earthquake with an epicentre in western Guatemala was felt over a large area. It was very strong at Mazapa and Motozintla de Madero in Chiapas, and in the District of San Marcos in Guatemala, but caused no damage.

The shock was felt in western El Salvador as far as San Salvador.

[Figueroa 1973; Martinez 1978]

1955 Mar 28 **El Salvador**

A strong earthquake was felt throughout El Salvador. Although it was rather strong at San Salvador there are no reports of damage.

[BSSA; Martinez 1978]

1955 Apr 4 **M_S 6.0** **Nicaragua**

An earthquake in northwestern Nicaragua caused considerable damage in the region of the Estero Real, northwest of the Volcano Viejo, where a number of houses were ruined at Puerto Morazan, San Cristobal and in other settlements up to the Honduras border. No one was killed, but details regarding damage are lacking.

The shock did some minor damage at Leon and Managua, and was felt at San Salvador.

[Johansson 1988; Martinez 1978]

1955 Apr 14 **Guatemala**

This intermediate-depth earthquake had an instrumental epicentre in northwestern Guatemala. The shock was reported as very strong at Concordia in Chiapas.

[Figueroa 1973]

1955 Apr 23 **El Salvador**

An earthquake in El Salvador was felt at San Salvador, but there are no reports of damage or casualties. We could find no instrumental data for this event.

[BSSA; Martinez 1978]

1955 Apr 26 **M_S 5.8** **El Salvador**

Very strong shaking was felt at San Salvador, originating from an earthquake in the Pacific off the coast.

Shocks continued to be felt intermittently for ten hours. No damage was reported.

[BSSA; Martinez 1978]

1955 Apr 28 **El Salvador**

A sharp shock was felt at San Salvador. No injuries or damage were reported and no instrumental data are available.

[BSSA]

1955 Apr 30 **M_S 5.9** **Nicaragua**

Two consecutive shocks of almost the same magnitude, eleven minutes apart, occurred in the northwestern part of Lake Managua.

They caused some damage in the Momotombo area, and were felt at San Salvador.

[Martinez 1978]

1955 Aug 12 **Mexico**

This intermediate-depth earthquake was felt strongly in Chiapas at Las Margaritas and Trinitaria.

[Figueroa 1973]

1955 Aug 28 **M_S 6.6** **El Salvador**

A subcrustal earthquake was strongly felt at San Salvador and at San José in Guatemala, where it caused some concern.

It was preceded by a strong foreshock.

[REC; Martinez 1978]

1955 Sep 1 **M_S 5.5** **Figure A.58** **Costa Rica**

This locally damaging earthquake in Costa Rica was preceded by small foreshocks.

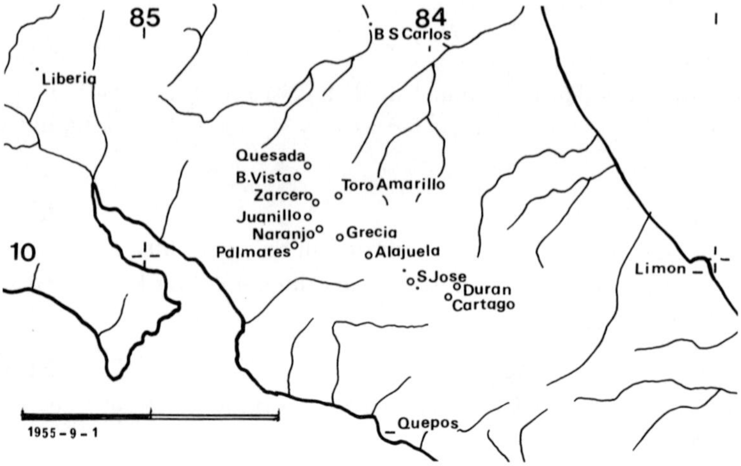

Fig. A.58 Felt effects of earthquake of 1 September 1955.

Toro Amarillo, on the west flank of the Volcano Poas, was almost totally destroyed; one person was killed and many were injured by collapsed houses and landslides. Hundreds of people were made homeless, and the village was evacuated.

A number of houses were also damaged in Zarcero, San Juanillo, Naranjo, and Crecia, and a few people were injured.

The shock was strong at Quesada and Alajuela and to a lesser extent at Cartago.

The earthquake caused cracks along the Rio Segundo near Toro Amarillo, probably due to slumping of the ground, in places 1.5 m wide, running for almost 500 m.

Near Bajos del Toro landslides temporarily blocked small streams and caused flooding when their dammed sections were overtopped.

The earthquake was probably felt at Liberia, and it was perceptible at Boca San Carlos, but not in Limón or Quepos.

[Alvarado *et al.* 1988; Miyamura 1980; REC]

1955 Sep 3 M_S 6.5 El Salvador

This was an intermediate-depth earthquake with an epicentre in the Pacific.

It was strong at San Salvador. No casualties or damage were reported.

[BSSA; Martinez 1978]

1955 Sep 26 M_S 6.4 Mexico

An intermediate-depth earthquake in Chiapas caused some damage at Angel Albino Corzo and at Jaltenango; it was strong at Chicomuselo, Mapastepeque, Pijijiapan and Socoltenango.

It was felt throughout southeast Chiapas and in the isthmus of Tehuantepec.

[Figueroa 1973]

1955 Dec 28 El Salvador

A strong earthquake was felt at San Salvador; no instrumental data are available.

[BSSA]

1956 Jan 9 13h Mexico

This earthquake was strong at Ciudad Hidalgo and Puerto Madero, in Chiapas.

[Figueroa 1973]

1956 Jan 9 17h m_b 5.6 Mexico

This intermediate depth-earthquake in Chiapas was very strong at Altamiran and Ocosingo.

[Figueroa 1973]

1956 Mar 13 M_S 6.0 Panama

This earthquake had an epicentre in the Gulf of Chiriqui, off the south coast of Panama. It was widely felt and was strong at Balboa Heights and Panama City.

The shock reported felt at San Salvador about the same time was a different local event which occurred at 09h 02m UT.

[Martinez 1978; USG]

1956 Mar 24 **El Salvador**

A strong shock was felt in the region of Puerto San José; it was also felt at San Salvador, but caused no damage there.

[BSSA; Martinez 1978]

1956 May 7 **El Salvador**

An intermediate-depth earthquake was very strongly felt in western El Salvador and eastern Guatemala.

[Martinez 1978; REC]

1956 Jun 16 **Mexico**

This intermediate-depth earthquake in Chiapas was very strongly felt at Angel Albino Corso.

[Figueroa 1973]

1956 Jul 5 **El Salvador**

A prolonged and rather strong earthquake was felt at San Salvador and reported from the coastal area. There were no reports of casualties or damage.

[BSSA; Martinez 1978]

1956 Jul 19 **M_S 5.7** **Costa Rica**

Two intermediate-depth eathquakes in Costa Rica were felt eleven minutes apart. They were particularly strong at San Ramón, Palmares, Orotina, and San José, causing some panic in places, but no damage.

The shock was reported from as far as Portero Grande, San Vito and Golfito.

[Barquero and Rojas 1994b; Miyamura 1980]

1956 Aug 10 **Costa Rica**

This earthquake in Costa Rica was probably subcrustal. It was felt over a relatively large area of the Central Valley, and was particularly strong at San Juan Vinas, Santa Ana and San José; it was felt from Palmares to Portero Grande and San Vito.

[Miyamura 1980]

1956 Aug 20 **M_S 5.5** **Panama**

An earthquake with an epicentre off the east coast of the Peninsula of Azuero in Panama was very strong in Los Santos, and was also felt at Balboa Heights.

It was followed by two aftershocks which were perceptible in the Canal Zone.

[Miyamura 1980; USG]

1956 Sep 10 **El Salvador**

An earthquake was felt along the coast of El Salvador. It was very strong at San Salvador.

[BSSA; Martinez 1978]

1956 Sep 25 **M_S 4.9** **Panama**

This earthquake in the District of David in Panama was accompanied by foreshocks and aftershocks. It was felt at Puerto Armuelles and David, and was perceptible at Coto in Costa Rica. The shock was not felt at Balboa Heights.

[Miyamura 1980; USG]

1956 Oct 23 **El Salvador**

This earthquake was felt strongly on the coast of El Salvador and at San Salvador.

[Martinez 1978]

1956 Oct 24 **M_S 7.2** **Nicaragua**

This earthquake was very strongly felt along the Pacific coast of Nicaragua. It caused some minor damage at coastal towns as well as further inland at Managua. The shock was felt at San Salvador, in southeastern El Salvador and in northwestern Costa Rica.

It was followed by two relatively large subcrustal aftershocks, which were perceptible at San Salvador.

We adopt the location in the Pacific recomputed by Dewey and Algermissen (1974).

[BSSA; Hansen 1975; Martinez 1978; REC]

1956 Oct 25 **M_S 6.2** **Nicaragua**

An aftershock was felt in Nicaragua and El Salvador.

[REC]

1956 Nov 1 **Panama**

This earthquake was instrumentally located near the north coast of Panama. It was felt rather strongly at Balboa Heights and in the Canal Zone.

[USG]

1956 Dec 4 **M_S 5.8** **Guatemala**

This subcrustal earthquake with an epicentre in western Guatemala was widely felt. It was very strong at Mazapa de Madero, Motozintla, San Cristobal las Casas and in other places in southeastern Chiapas. It was also felt throughout southern Guatemala, but details are lacking. It was strong in western El Salvador and at San Salvador.

[Figueroa 1973; Martinez 1978; USG]

1957 Feb 1 **El Salvador**

This earthquake in El Salvador was very strong at Acajutla, and was felt at San Salvador.

[Martinez 1978]

1957 Feb 4 **M_S 4.7** **Figure A.59** **Costa Rica**

A small subcrustal earthquake was perceptible over much of central and southern Costa Rica.

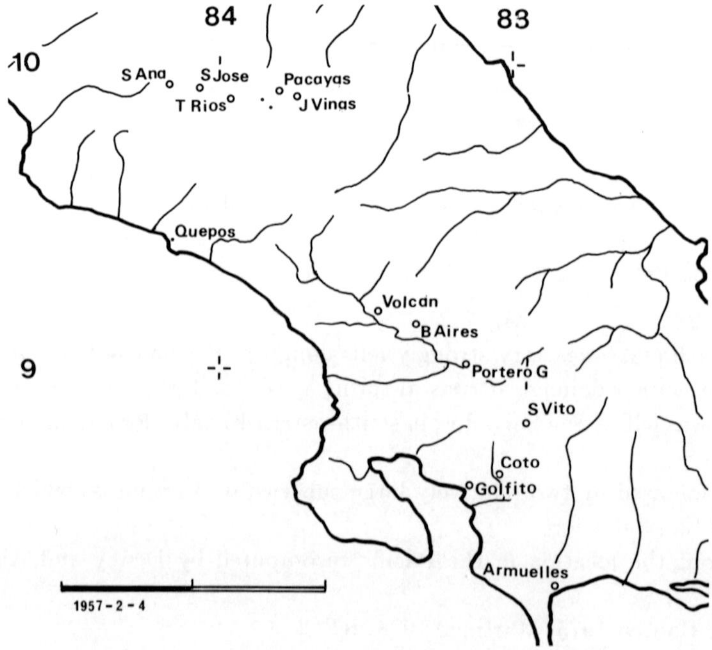

Fig. A.59 Felt effects of earthquake of 4 February 1957.

The shock was felt with low intensities, which varied erratically from place to place. It was strong at Quepos, Tres Rios, Pacayas, Juan Vina, and was reported felt at Santa Ana, San José, Volcán, Buenos Aires and Portero Grande. The earthquake was perceptible as far southeast as Golfito and Puerto Armuelles.

Contrary to later reports we could find no evidence that the shock caused damage or undue concern anywhere in the country.

[Miyamura 1980, USG]

1957 Apr 5 **Nicaragua**

A strong earthquake off the Pacific coast of Nicaragua was felt in the eastern part of El Salvador and in San Salvador; there were no reports of damage.

[Martinez 1978; REC]

1957 Apr 8 **M_S 5.8** **Panama**

This earthquake was widely felt in southeastern Costa Rica and Panama. It was rather strong at Coto, Golfito and Puerto Armuelles, and was felt at San José, in the Canal Zone and at Balboa Heights.

[REC; USG]

1957 Apr 17 **M_S 5.6** **Mexico**

This subcrustal earthquake near the Mexico-Guatemala border was reported as very strong from Ciudad Hidalgo and Tapachula in Chiapas.

[Figueroa 1973]

1957 Jul 8 **M_S 5.4** **Guatemala**

This was an intermediate-depth earthquake in Guatemala.

The most severe shaking was reported from the regions of San Marcos and Totonicapan, where there was moderate damage. The shock was strong at Mazapa, Motozintla and Talisman in Chiapas, and was felt throughout almost all southern Guatemala and western El Salvador.

[Figueroa 1973; Martinez 1978; REC]

1957 Jul 10 **M_S 6.2** **Panama**

This earthquake in the Gulf of Chiriqui was felt on board a ship sailing near 7.37°N, 82.80°W, and was perceptible at Balboa Heights.

[REC; USG]

1957 Jul 21 **M_S 5.6** **Guatemala**

This earthquake in western Guatemala was very widely felt.

At Ciudad Hidalgo and Puerto Madero it was very strong and allegedly caused some damage in the region of Retalhuleu, but details are lacking.

The shock was felt all along the Guatemala coast and as far as San Salvador.

[Figueroa 1973; Martinez 1978]

1957 Aug 15 **Panama**

An earthquake near the north coast of Panama was felt in the Canal Zone, Panama City and Balboa Heights.

[USG]

1957 Sep 12 **M_S 5.8** **Honduras**

This earthquake has been located near the Islas de la Bahia, north of Honduras, but we have no information from there.

The shock is said to have been felt in Havana, Cuba, but not in Jamaica.

[USG]

1957 Sep 17 **El Salvador**

An earthquake was felt in the region of Sonsonate, El Salvador. It was perceptible at San Salvador.

[BSSA; Martinez 1978]

1957 Sep 23 **El Salvador**

This earthquake was strongly felt on the El Salvador side of the Gulf of Fonseca and in the eastern part of the country.

The shock was perceptible at San Salvador.

[BSSA; Martinez 1978]

1957 Sep 26 **Mexico**

An intermediate-depth earthquake in Chiapas was felt rather strongly at Ciudad Hidalgo and Puerto Madero. We could find no evidence that the shock was felt in Guatemala.

[Figueroa 1973]

1957 Oct 27 **M_S 4.0** **Honduras**

This earthquake has been relocated close to the north coast of Honduras, but we could find no felt reports from there.

The earthquake was allegedly felt at San Salvador.

[Martinez 1978]

1957 Dec 2 **El Salvador**

A subcrustal earthquake in the Gulf of Fonseca was very strongly felt in south-eastern El Salvador. It was strong at San Salvador, but it caused no damage.

[Martinez 1978]

1957 Dec 23 **Nicaragua**

An intermediate-depth earthquake was felt at Managua.

[BSSA]

1958 Jan 8 **El Salvador**

A strong earthquake shook the south coast of El Salvador, causing some alarm. At San Salvador cracks appeared in many old buildings and the city's water supply was interrupted for several hours.

[BSSA; Martinez 1978]

1958 Jan 15 **El Salvador**

An earthquake was strongly felt in the western part of El Salvador and at San Salvador, but caused no damage.

[BSSA; Martinez 1978]

1958 Feb 1 **Nicaragua**

A strong earthquake shook the Nicaragua-Costa Rica border. Residents of Rivas fled from their homes, but no damage or casualties were reported.

[BSSA]

1958 Feb 3 **El Salvador**

A local earthquake in El Salvador shook San Miguel Tepezontes and the region southeast of Lake Ilopango. The shock was not felt at San Salvador, and was not recorded by the world netowrk.

[Martinez 1978]

1958 Feb 9 **M_S 5.1** **Panama**

This earthquake had an epicentre in the Gulf of Panama.

The shock was strongly felt along the southwest coast, at Rio Hato, Santa Clara and in the Canal Zone.

[USEQ]

1958 Mar 1 **El Salvador**

This earthquake was felt in southwest El Salvador and at San Salvador, where it was rather strong.

[Martinez 1978]

1958 Mar 3 **Guatemala**

An intermediate-depth earthquake in western Guatemala was reported as strong from Soconusco in Chiapas, Ciudad Hidalgo and Puerto Madero in Mexico. We have no information from Guatemala.

[Figueroa 1973]

1958 Mar 11 **Guatemala**

This intermediate-depth earthquake located in central Guatemala was perceptible at San Salvador. Felt information from Guatemala is lacking.

[Martinez 1978]

1958 Mar 21 **Mexico**

An intermediate-depth earthquake in Chiapas was felt in Soconusco, and was strong at Mazapa de Madero and Motozintla.

[Figueroa 1973]

1958 Apr 8 **El Salvador**

A lower crustal earthquake off the coast of El Salvaror was felt strongly at San Salvador; there are no reports of damage.

[BSSA]

1958 Apr 11 **El Salvador**

An intermediate-depth earthquake off the southeast coast of El Salvador was felt rather strongly at Santiago de Maria, Verapaz and San Salvador.

[Martinez 1978]

1958 Apr 15 **M_S 6.3** **Costa Rica**

This earthquake was relocated off the Pacific coast of Costa Rica. It was rather strong in the Osa Peninsula at Golfito and Coto, and was felt as far north as the Central Valley at San José, Santa Ana, and San Juan Vinas.

[Miyamura 1980]

1958 May 17 **El Salvador**

An earthquake was felt throughout the southeastern part of El Salvador and was perceptible at San Salvador. It was followed by a few aftershocks.

[Martinez 1978]

1958 Jun 6 **M_S 6.5** **Costa Rica**

Relocated in the Pacific off the coast of Costa Rica, this earthquake was reported with varying degrees of intensity from a number of places in southwestern Costa Rica. It was rather strong at San Vito and Buena Vista and was perceptible at Santa Ana, San José and Juan Vinas.

[Miyamura 1980]

1958 Jun 16 **El Salvador**

The earthquake was rather strong in the southern part of El Salvador.

[Martinez 1978]

1958 Jun 19 **Mexico**

An intermediate-depth earthquake in Chiapas was felt strongly at Mazapa de Madero and Motozintla.

[Figueroa 1973]

1958 Jun 27 **M_S 5.6** **El Salvador**

This lower-crustal earthquake was located off the southeastern coast of El Salvador.

It caused slight damage in the district of Usulutan and was very strong at Santiago de Maria, as well as in the region of Chinandega in Nicaragua.

[BSSA, Hansen 1975; Martinez 1978]

1958 Jul 3 **El Salvador**

An earthquake was felt in western El Salvador in the region of Sonsonate. It caused no damage.

[Martinez 1978]

1958 Jul 15 **Panama**

An earthquake was felt at Boquete and El Volcán in Panama.

[USEQ]

1958 Jul 19 **El Salvador**

An earthquake off the east coast of El Salvador was felt in the Department of La Unión, at El Carmen.

[Martinez 1978]

1958 Sep 2 **M_S 5.6** **Mexico**

This earthquake off the coast of Chiapas was very strong at Puerto Madero, Mazatan and Ciudad Hidalgo. It was also strong at Tapachula.

It was followed by a number of aftershocks that were felt more widely.

[Figueroa 1973]

1958 Nov 3 **Nicaragua**

An earthquake originating in the Pacific off the coast of Nicaragua was felt in Chinandega and Corinto in Nicaragua, where it caused alarm. It was also felt at the Gulf of Fonseca, El Salvador.

[Hansen 1975; Martinez 1978]

1958 Nov 14 05h **Guatemala**

An earthquake was felt in southeast Guatemala as far as San Salvador. In Guatemala City it caused some damage to houses and injured a number of people.

[Martinez 1978; REC]

1958 Nov 14 15h **M_S 5.1** **Nicaragua**

This earthquake in Nicaragua caused some alarm in the region of San Francisco del Carnicero. It was strong at Managua and to the north of Lake Managua at Casas Viejas, Dario, El Jicaral, Las Maderas and Sebaco. It was widely felt, as far as Waspan; it caused no casualties or damage.

[BSSA; Hansen and Chiavez 1973; Hansen 1975]

1958 Nov 22 **El Salvador**

An intermediate-depth earthquake in El Salvador was strongly felt in the regions of Nahuizalco and Juayua. It caused no damage.

[Martinez 1978]

1958 Dec 4 12h **Guatemala**

An intermediate-depth earthquake in southern Guatemala was reported from southwestern El Salvador, where it was rather strong, as well as from Ciudad Hidalgo and Puerto Madero, Chiapas. We have no felt reports from Guatemala.

[Figueroa 1973; Martinez 1978]

1958 Dec 4 19h **M_S 5.3** **Nicaragua**

A subcrustal earthquake off the Pacific coast of Nicaragua was widely felt along the south coast of Nicaragua, where it caused some alarm.

[Hansen 1975]

1959 Jan 9 **Guatemala**

This subcrustal earthquake was located in southeastern Guatemala. It was reported as strong from southwestern El Salvador.

[BSSA; Martinez 1978]

1959 Jan 13 **M_S 5.5** **Costa Rica**

This subcrustal earthquake in central Costa Rica was strong at San Juan Vinas, Pacayas, San José and Santa Ana, and was widely felt in Esparza.

It was felt at Golfito, Coto and San Vito, and reported from as far as El Volcán in Panama.

[Barquero and Rojas 1994b; Miyamura 1980; USEQ]

1959 Jan 15 **Costa Rica**

This earthquake located in Costa Rica was reported felt in northwestern Panama.

[USEQ]

1959 Feb 20 **M_S 5.2** **Guatemala**

This lower-crustal earthquake occurred in Alta Verapaz in northern Guatemala, from where there is evidence that the shock caused some damage, without loss of life, to a number of farms situated in northern Quiche Department; details about the sites affected are lacking.

The shock was rather strong in Chiapas at La Independecia, and was felt in Guatemala City and at Tenosique in Mexico.

[BSSA; Figueroa 1973; White and Harlow 1993]

1959 Apr 11 **Panama**

A subcrustal earthquake on the Caribbean coast of Panama was perceptible at Volcán in Panama and at San Vito, Coto, Golfito and Juan Vinas in Costa Rica.

[Miyamura 1980; USEQ]

1959 Apr 18 **El Salvador**

This earthquake was widely felt in southern El Salvador, with varying degrees of intensity. It was strong in San Salvador and in the southeast of the country.

[BSSA; Martinez 1978]

1959 Apr 22 **M_S 5.4** **Nicaragua**

This earthquake was felt along the coast of Nicaragua and at Managua, and caused some panic, but no damage.

[BSSA; Hansen 1975]

1959 Apr 28 **M_S 6.5** **Mexico**

The recalculated location of this subcrustal event is off the coast of Chiapas. It was very strong at Angel Albino Corzo, La Concordia and Jaltenango in Chiapas, but information about its effects in Guatemala is lacking.

The shock was generally felt at San Salvador.

[Figueroa 1973; Martinez 1978; REC]

1959 Apr 29 **Mexico**

This subcrustal earthquake in Chiapas caused some minor damage at Ciudad Hidalgo and Puerto Madero, and was strong at Mazatan and Tapachula.

[Figueroa 1973]

1959 May 3 **M_S 5.4** **Nicaragua**

This earthquake off the coast of northwestern Nicaragua was strong at San Miguel and felt to a lesser degree at San Vicente.

The shock was felt at San Salvador and in northern parts of Nicaragua.

[Martinez 1978; REC]

1959 May 12 **Guatemala**

This subcrustal earthquake was relocated in southern Guatemala, from where we have no felt reports. The shock was perceptible in San Salvador.

[Martinez 1978]

1959 Sep 17 **m_b 5.7** **El Salvador**

A subcrustal earthquake in El Salvador caused slight damage throughout the southern part of El Salvador. At San Salvador a few buildings suffered slight damage, and one person was killed by a falling wall.

[BSSA, Martinez 1978; REC]

1959 Nov 14 **El Salvador**

A subcrustal earthquake near the Guatemala-El Salvador border caused slight damage in the Department of Ahuachapan.

[Martinez 1978; USEQ]

1960 Jan 26 **Nicaragua**

This shock was felt at San Salvador.

[BSSA]

1960 Feb 21 **El Salvador**

A small earthquake in El Salvador caused some slight damage at Tacuba. The shock was not felt in San Salvador.

[REC]

1960 Feb 29 **Panama**

An earthquake in the Gulf of Panama was perceptible at Balboa Heights.

[USEQ]

1960 Mar 28 **M_S 5.9** **Panama**

This earthquake was relocated in the Gulf of Chiriqui.

It was felt on Coiba Islandindex Coiba, and in the Districts of Veraguas and Chiriqui, but details are lacking.

The shock was perceptible at Santa Ana in Costa Rica, and at Balboa Heights.

[Miyamura 1980; USEQ]

1960 Apr 13 **M_S 5.7** **Guatemala**

This was a damaging earthquake in northwestern Guatemala.

In the District of Huehuetenango damage was widespread, but there are no details. In the District of San Marcos at least one person was killed and 14 injured, and the town of San Marcos suffered some damage.

Minor damage extended to Chiapas at Motozintla and Mazapa de Madero.

[Figueroa 1973; USEQ]

1960 Apr 25 **El Salvador**

This shock was felt in southwest El Salvador.

[USEQ]

1960 May 12 **M_S 5.9** **Panama**

This earthquake in the Peninsula of Azuero, Panama, caused some small damage at Ocu and Penonome. It was strong at La Mesa, Tonosi, Las Tablas and Chitré, and was felt at David and Balboa Heights.

The earthquake was followed by strong aftershocks.

[USEQ]

1960 May 23 **Tsunami**

The seismic sea-wave associated with the large earthquake off the coast of Chile was recorded at La Unión in the Gulf of Fonseca, at an epicentral distance of 5,900 km, with a maximum amplitude of about one metre.

[Gonzales 1960]

1960 Jun 5 **Guatemala**

This intermediate-depth earthquake was located in central Guatemala. No felt reports are available from there.

The shock was very violent at Ciudad Hidalgo and Puerto Madero in Chiapas in Mexico, and was strong at Tapachula.

[Figueroa 1973]

1960 Jul 10 **Nicaragua**

An intermediate-depth earthquake in Nicaragua was widely felt.

[REC]

1960 Jul 19 **Mexico**

This intermediate-depth earthquake was relocated 450 km northeast of the position calculated by ISS.

The shock was strong near the southwest coast of Chiapas at Ariaga and Tonala.

[Figueroa 1973]

1960 Sep 21 **El Salvador**

This lower crustal earthquake was relocated 370 km north of the position calculated by USCGS.

The shock caused some slight damage at San Salvador.

[USEQ]

1960 Sep 29 **m_b 4.8** **Guatemala**

This small lower crustal earthquake in northeast Guatemala was perceptible in San Salvador.

[USEQ]

1960 Oct 17 **M_S 5.5** **Guatemala**

An earthquake with an epicentre off the coast of Chiapas was strongly felt in the District of Soconusco. At Ciudad Hidalgo, Puerto Madero and Tapachula the shock caused considerable alarm.

It was preceded by a foreshock.

[Figueroa 1973]

1960 Oct 23 **El Salvador**

This was a small earthquake which was felt in the southeast of El Salvador.

It caused some damage in the District of Usulutan, particularly at Jiquillisco, where two people were injured.

There is no evidence that the shock was widely felt at San Salvador.

The instrumental location of this event determined by USCGS is 170 km to the south of the macroseismic epicentre found by us.

[REC]

1960 Dec 4 **El Salvador**

This earthquake was felt in El Salvador.

[USEQ]

1960 Dec 6 **M_S 6.0** **Panama**

An earthquake in the border region between Panama and Costa Rica was felt in the District of Chiriqui in Panama, and at Coto, Golfito and San Vito in Costa Rica.

It was perceptible in the Central Plateau at Santa Ana and the San José area.

[Miyamura 1980; REC]

1960 Dec 14 **El Salvador**

A local earthquake in El Salvador was very strong around the Volcano Tacapa and at Alegria, Berlin and Santiago di Maria.

There is no evidence that the shock was felt very far.

[Martinez 1978]

1961 Jan 24 **M_S 4.7** **Panama**

A strong earthquake in the District of David, Panama, caused no damage, and there is no evidence that it was felt to any great distance.

It was preceded and followed by a number of shocks.

[USEQ]

1961 Feb 5 **M_S 5.5** **Panama**

An earthquake with an epicentre in the Gulf of Chiriqui, Panama, was strongly felt in the Region of David and was perceptible at Balboa Heights, and in Costa Rica at Helechales, Golfito and Coto.

[Miyamura 1980; USEQ]

1961 Feb 7 **El Salvador**

An intermediate-depth earthquake in western El Salvador was felt in Guatemala and San Salvador.

[Martinez 19858; USEQ]

1961 Feb 16 **Nicaragua**

This shock was felt at Chinandega, Nicaragua.

[USEQ]

1961 Feb 23 **Mexico**

This lower crustal earthquake in Chiapas was strong at Ciudad Hidalgo and Puerto Madero.

[Figueroa 1973]

1961 Feb 27 **m_b 5.8** **Costa Rica**

This was a subcrustal earthquake in Costa Rica.

It was reported from the central part of the country, where it caused some alarm. At Pacayas, San José and Duran the shock caused some minor damage.

[Miyamura 1900; USEQ]

1961 Mar 2 **Mexico**

This intermediate-depth earthquake was felt strongly at Motozintla in Chiapas.

[Figueroa 1973]

1961 Mar 4 **Nicaragua**

This earthquake was felt strongly in the southeastern part of El Salvador.

[Martinez 1978]

1961 Mar 20 **M_S 5.8** **Nicaragua**

This earthquake was located off the southwest coast of Nicaragua.

It was felt at Managua and in towns along the coast, but it caused no damage.

It was followed by a strong aftershock.

[Hansen 1975; USEQ]

1961 Apr 12 **M_S 5.8** **El Salvador**

This subcrustal earthquake was felt strongly in the southern part of El Salvador and at San Salvador, where hundreds of houses were cracked.

The shock was felt throughout the country, and allegedly it was perceptible as far as Hueheutenango in Guatemala.

[Martinez 1978; REC]

1961 May 13 **El Salvador**

An earthquake located off the coast of El Salvador caused widespread but minor damage in the southern part of the country. At San Salvador it cracked walls in modern reinforced concrete office buildings, and caused the collapse of some walls in old houses.

[REC]

1961 May 23 03h **M_S 5.3** **Costa Rica**

An intermediate-depth earthquake was widely felt in Costa Rica.

It was rather strong in the Central Vally at Cartago, Duran, San Juan Vinas, San José and particularly at Santa Ana and San Marcos. It was perceptible at Golfito.

It was followed by a few aftershocks.

[Miyamura 1990; REC]

1961 May 23 16h **M_S 5.4** **Nicaragua**

This subcrustal earthquake was located off the northwest coast of Nicaragua.

The shock was very strong at Puerto Morazan, Chinandega and Corinto, where it caused alarm and some minor damage.

In Managua plaster fell off some houses, and at Leon and Masaya a number of old houses were damaged.

The shock was felt in southeastern El Salvador, at San Miguel and in the Gulf of Fonseca, but details are lacking.

[Hansen 1975; REC]

1961 Jun 17 **M_S 5.9** **Guatemala**

The earthquake was felt in southwestern Guatemala, and in Chiapas at Ciudad Hidalgo and Puerto Madero, where it was strong.

It was followed by a strong aftershock.

[Figueroa 1973]

1961 Jun 21 **El Salvador**

This intermediate-depth earthquake was felt at San Salvador.

[USEQ]

1961 Jun 28 **El Salvador**

This small earthquake was located 190 km south of San Salvador, where it is said that the shock was felt.

[USEQ]

1961 Sep 1 **Honduras**

This earthquake was felt in northern Honduras; no details are available.

[USEQ]

1961 Sep 3 **Nicaragua**

An intermediate-depth earthquake in Nicaragua was widely felt at Managua and in the western coast region, where it caused some slight damage.

[Hansen 1975; USEQ]

1961 Sep 19 **M_S 5.6** **Panama**

This earthquake was located off the coast of western Panama.

It was perceptible in the central provinces as far as Balboa Heights, and it was rather strong at Aguadulce and Chitré.

[USEQ]

1961 Nov 11 **Guatemala**

This intermediate-depth earthquake on the eastern coast of Guatemala was felt strongly at San Salvador.

[STR]

1961 Nov 22 **M_S 5.1** **Guatemala**

An intermediate-depth earthquake was reported from western Guatemala and southern Chiapas. At Motozintla and neighbouring sites the shock caused some minor damage.

[Figueroa 1973]

1961 Dec 20 **Guatemala**

This earthquake was located instrumentally in central Guatemala.

The shock caused some minor damage at Ciudad Hidalgo and Puerto Madero in Chiapas, and was very strong in Soconusco, at Tapachula and elsewhere, at places

160 km from the instrumental epicentre. The true epicentre may be closer to the felt area.

[Figueroa 1973]

1962 Jan 30 **M_S 5.0** **Nicaragua**

This shock was felt in Managua; there are no reports of damage.

[BSSA]

1962 Feb 19 **El Salvador**

This event, originating from an epicentre off the coast of El Salvador, was felt at San Salvador.

[USEQ]

1962 Feb 24 **El Salvador**

This earthquake was located off the coast of El Salvador and felt in the southern part of the country.

[USEQ]

1962 Mar 12 09h **M_S 5.4** **Panama**

A foreshock of an earthquake in western Panama caused some small damage at Puerto Armuelles, Conceptión and David.

It was strong in the District of Chiriqui and in the Osa Peninsula. The shock was felt as far as Balboa Heights, where it awakened many people.

[USEQ]

1962 Mar 12 11h **M_S 6.5** **Tsunami** **Panama**

The main shock of this sequence occurred in the Golfo del Dulce near the south coast of Panama and Costa Rica.

It caused some slight damage to David, Conceptión and Puerto Armuelles, and was strongly felt at Puerto Jiménez, Golfito and Coto.

The shock was felt in the Vale del General at San Vito and Helechales and was reported from San José, Santa Ana and other places in the Central Plateau. The earthquake was rather strong at Balboa Heights.

The shock was felt on board a vessel at 7.73°N, 82.95°W, sailing 33 km off Punta Burica, and was associated with a seismic sea-wave of about 0.3 m recorded at Puerto Armuelles and San Cristobal, Galapagos Islands.

The earthquake was followed by a number of aftershocks.

[Miyamura 1980; USGS]

1962 Jun 4 **m_b 4.4** **Panama**

This earthquake was located in the Peninsula de Azuero, from where there is no detailed macroseismic information except that the shock was strongly felt.

Allegedly the shock was very strong at Balboa Heights and Panama City.

[USEQ]

1962 Jul 24 **M_S 5.4** **Mexico**

An intermediate-depth earthquake in Chiapas was strongly felt at Chichares.

[Figueroa 1973]

1962 Jul 26 **M_S 7.3** **Panama**

We could find little macroseismic information for this major earthquake with an epicentre in the Gulf of Chiriqui.

The shock was widely felt in western Panama and in places such as Golfito, Volcán, Veraguas and Aguadulce it was strong, causing some panic.

The earthquake was perceptible as far as Coco airport and San José to the northwest and in Panama City in the east. At Balboa Heights the shock awakened many people.

The main shock was followed by a long sequence of aftershocks.

We adopt the position in the Gulf of Chiriqui recomputed by Molnar and Sykes (1969), which is consistent with the felt area on land.

[Miyamura 1980; USEQ]

1962 Aug 29 **m_b 4.4** **Panama**

This shock near the south coast of Panama was perceptible at Balboa Heights.

[USEQ]

1962 Sep 4 **m_b 5.4** **Guatemala**

We have no macroseismic data from Guatemala for this intermediate-depth earthquake located near the Mexico-Guatemala border. The shock was strong in Chiapas in Mexico at Angel Albino Corzo and Jaltenango.

[Figueroa 1973]

1962 Sep 14 **Costa Rica**

This earthquake was felt strongly in the region of La Fortuna de San Carlos in Costa Rica.

We can find no instrumental readings for this event.

[Miyamura 1980]

1962 Sep 18 **M_S 6.7** **Panama**

This was a relatively large magnitude earthquake with an epicentre 80 km south of the Gulf of Chiriqui, Panama.

The shock caused some slight damage over a large area in the Province of Chiriqui, at Puerto Armuelles, David, Horncocitos, Guabala and probably in the Island of Coibaindex Coiba, but details are lacking.

The shock was felt in southeastern Costa Rica and in Panama as far as Balboa Heights.

Three ships in the vicinity of the epicentre, at 7.33°N, 82.33°W, about 90 km southeast of Punta Burica, reported violent shaking.

The earthquake was followed by a long series of aftershocks.

[USEQ]

1962 Oct 25 **M_S 5.4** **Panama**

A strong aftershock of the event of 18 September, with an epicentre in the Gulf of Chiriqui, was felt at Volcán, Panama and Coto, Costa Rica.

[Miyamura 1980; USEQ]

1962 Oct 27 **M_S 4.7** **Guatemala**

This subcrustal earthquake was located in southern Guatemala, from where we have no macroseismic information. It was reported felt from western El Salvador.

[USEQ]

1962 Oct 30 **M_S 5.3** **Nicaragua**

A subcrustal earthquake was located south of the Gulf of Fonseca. It was reported felt in San Salvador.

[USEQ]

1963 Feb 3 **M_S 3.8** **Guatemala**

A subcrustal earthquake in western Guatemala was reported as strong from Chicharas, Chiapas.

[Figueroa 1973]

1963 Feb 24 **M_S 5.2** **Guatemala**

An intermediate-depth earthquake located in southwestern Guatemala was reported as strong from Mazapa de Madero and Motozintla in Chiapas, and from the western part of El Salvador.

[Figueroa 1973; Martinez 1978]

1963 Mar 10 **M_S 3.5** **El Salvador**

A strong earthquake in western El Salvador caused some small damage at Ahuachapan and Atiquizaya, but it was not felt beyond Jalpatagua, Santa Ana and Sonsonate.

It was preceded by foreshocks, and followed by a very long sequence of aftershocks.

[STR]

1963 May 9 **M_S 5.2** **Nicaragua**

This was a locally damaging earthquake in Nicaragua, followed by a strong aftershock two minutes later.

The shock caused some damage at Leon and Telica, but we could find no evidence that it was felt beyond Managua.

The earthquake was accompanied by a reactivation of the volcano Cerro Negro.

ISS give an intermediate depth, but we adopt the shallow depth given by Dewey.

[STR]

1963 Jun 25 **M_S 4.2** **Panama**

A small earthquake in western Panama was felt at Boquete.

[USG]

1963 Jul 9 **M_S 5.7** **Costa Rica**

An earthquake in the Golfo del Dulce, Costa Rica was felt in the Osa Peninsula, at Golfito and Coto, as well as at El Volcán, Bambito and other unspecified places in the Chiriqui Province of Panama.

[BPH; Miyamura 1980]

1964 Jul 6 **Costa Rica**

A small earthquake in Costa Rica was strongly felt at Juan Vinas and Pacayas, and was perceptible at Duran.

[Miyamura 1980]

1964 Jul 30 **M_S 5.9** **Nicaragua**

An earthquake off the Pacific coast of southern Nicaragua was felt rather strongly along the coast and at Peñas Blancas.

[Miyamura 1980]

1964 Aug 10 **Costa Rica**

A small local earthquake in Costa Rica was very strong at Pacayas.

It was felt at Duran and was perceptible at Turrialba.

[Miyamura 1980]

1965 Jan 23 **Costa Rica**

An earthquake in southeastern Costa Rica was perceptible at Coto and Helechales.

[Miyamura 1980]

1965 Jan 30 **m_b 4.2** **Costa Rica**

A small earthquake in Costa Rica was felt at Cartago and was perceptible at Juan Vinas, San Isidoro Coranado and Tres Rios.

[Miyamura 1980]

1965 May 3 **M_S 5.9** **Figure A.60** **El Salvador**

This was a locally destructive earthquake in El Salvador preceded by foreshocks and followed by a long series of aftershocks.

Earthquakes began on 2 February from a focus close to San Salvador. The number of shocks increased rapidly from eight on 2 February to 210 on 3 February, 323 on 4 February, 538 on 5 February and 562 on 6 February. After that, seismic activity decreased, to be suddenly resumed shortly before the destructive shock of 3 May.

Much of the destruction and most victims were in the northeast outskirts of San Salvador, and in the villages of Ilopango, Soyapango, Villa Delgado, Cuscatancingo, Mejicanos, and Ayutuxtepeque, within an area of 7 km radius, where intensities ranged between *VI+* and *VIII+* (MM). The worst hit area was between Ilopango airport and the western shore of Lake Ilopango, an area covered by deposits of pumice (tierra blanca). Much of the destruction was caused to houses of local type, but some reinforced concrete buildings, in particular the airport, the prison and the

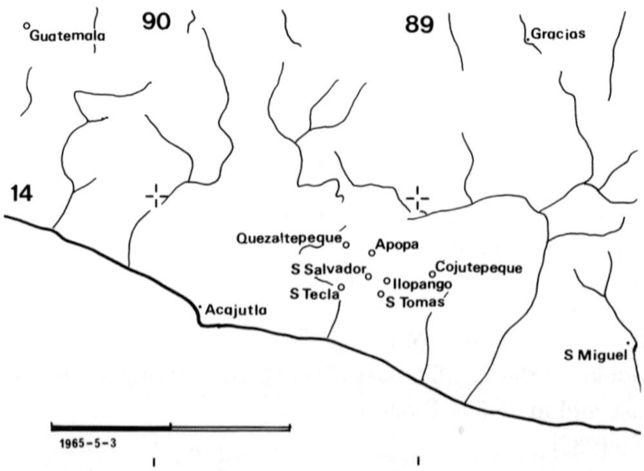

Fig. A.60 Felt effects of earthquake of 3 May 1965.

university were damaged. There were more than 110 killed and 400 injured; 30,000 people were made homeless.

There was some minor damage at Santa Tecla, Santo Tomas, Cojutepeque, Apopa and Quezaltepeque.

The shock was felt throughout El Salvador, including Acajutla and San Miguel, and as far as Gracias in Honduras and Guatemala City, 170 km away.

[Dengo *et al.* 1965; Jordan and Martinez 1980; Lomnitz and Schultz 1966; Martinez 1977, 1978]

1965 Jul 19 m_b 4.7 Costa Rica

An intermediate-depth earthquake in Guanacaste, Costa Rica, was perceptible at Canas and Tilaran.

[Miyamura 1980]

1965 Oct 4 m_b 4.9 Costa Rica

A small earthquake in southeast Costa Rica was perceptible at Rincon de Osa, Pacayas and Rancho Redondo.

[Miyamura 1980]

1965 Nov 12 m_b 4.6 Costa Rica

An earthquake in northern Costa Rica was felt strongly at Quebrada Azul, and was perceptible at Puerto Viejo Sarapiqui, Santa Clara, San Carlos and Zarcero.

[Miyamura 1980]

1966 Mar 13 m_b 4.2 Honduras

A small earthquake in Honduras was felt rather strongly in El Salvador, and recorded at San Salvador with a peak acceleration of 0.5%g.

1966 Mar 27 M_S 5.6 Costa Rica

This earthquake in southeastern Costa Rica was strongly felt at Rincon de Osa, and was perceptible in the Central Valley.

[Miyamura 1980]

1966 Apr 9 02h 34m m_b 5.3 Costa Rica

This was a large foreshock of the earthquake in Costa Rica that followed seven minutes later.

It caused no damage but was widely felt, from Santa Ana to Rincon, and from the Pacific coast to Heleches.

[Miyamura 1980]

1966 Apr 9 02h 42m M_S 5.6 Figure A.61 Costa Rica

An earthquake in Costa Rica was very strongly felt in the region between Jaco and San Isidoro de El General, where it caused minor damage.

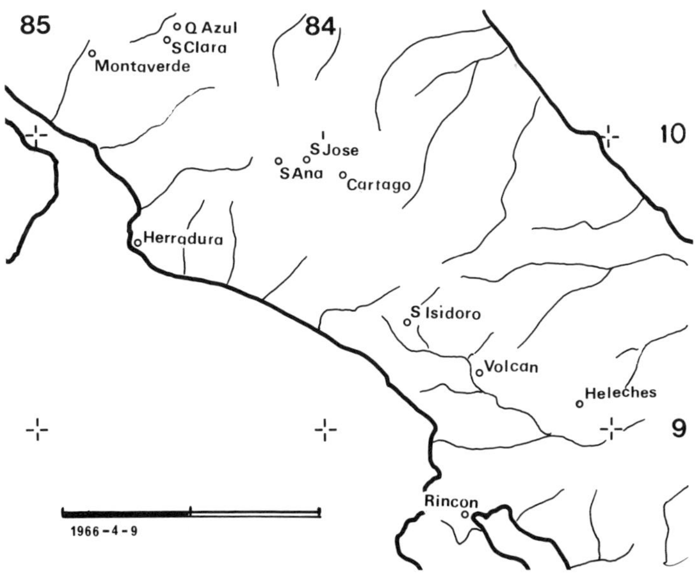

Fig. A.61 Felt effects of earthquake of 9 April 1966.

The shock was perceptible as far as Quebrada Azul in the northwest and to Rincon de Osa in the southeast. It was not reported from Limón.

[Miyamura 1980]

1966 May 11 m_b 4.2 Costa Rica

A locally damaging shock in Costa Rica destroyed four houses at Puerto Gonzalez Viquez.

[Miyamura 1980]

1966 Jul 10 **m_b 4.1** **Costa Rica**

This small earthquake in Costa Rica was rather strong at Pacayas, about 60 km from its instrumental location, which is likely to be in error.

[Miyamura 1980]

1966 Aug 9 **M_S 5.0** **Costa Rica**

A small earthquake located in the valley of El General was reported felt at Rincon de Osa.

[Miyamura 1980]

1966 Dec 11 **m_b 4.4** **Costa Rica**

A small intermediate-depth earthquake in central Costa Rica was perceptible within a radius of 40 km, at Santa Ana, Quebrada Azul, Buena Vista and Monteverde.

[Miyamura 1980]

1966 Dec 27 **m_b 5.2** **El Salvador**

An intermediate-depth earthquake off the coast of El Salvador was strongly felt in the southern central part of the country and at San Salvador.

1967 Apr 22 **M_S 5.0** **Panama**

A small earthquake in the Chiriqui District was reported felt at Rincon de Osa.

[Miyamura 1980]

1967 Aug 1 **Costa Rica**

A small event in Costa Rica was very strong at San José; it was perceptible only as far as Santa Ana and San Rafael Ojo Agua, within a radius of 13 km.

[Miyamura 1980]

1967 Oct 3 **M_S 6.1** **Costa Rica**

An earthquake in the Gulf of Papagayo in Costa Rica was strongly felt at Puerto Humo, and was perceptible at Canas, Monteverde and Quebrada Azul. Information from the coastal area of Costa Rica and Nicaragua is lacking.

[Miyamura 1980]

1967 Oct 15 **M_S 6.5** **Nicaragua**

A large intermediate-depth earthquake in Nicaragua caused panic in Managua, and was felt in Costa Rica at Canas, Tilaran, Monteverde, Quebrada Azul and Santa Ana.

[Miyamura 1980]

1967 Nov 18 **m_b 5.0** **El Salvador**

A subcrustal event off the coast of El Salvador was recorded at San Salvador with a peak ground acceleration of 2%g.

It was followed by many aftershocks that registered larger accelerations at San Salvador.

1968 Jan 4 **m_b 4.6** **Figure A.62** **Nicaragua**

A small magnitude earthquake in Nicaragua caused some damage in the southern outskirts of Managua.

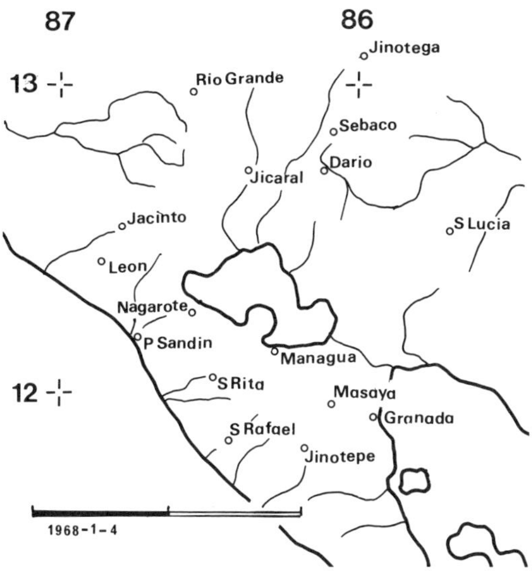

Fig. A.62 Felt effects of earthquake of 4 January 1968.

Much of the damage was concentrated within 2.5 km from the city in the Districts of the Colonia Centroamericana, Universidad Centroamericana, Colonia Nicaro and Colonia 14 Septiembre, where about 300 houses were rendered uninhabitable, and the remaining were damaged to some degree. The shock caused some injuries and moderate property damage in the southern part of Managua.

The earthquake was strongly felt at Masaya and Nagarote, and was felt as far as Jinotega, Leon, to the west of Boaco, and to the southeast of Granada over an area of 70 km radius.

We adopt the position of Engdahl *et al.* (1998), which is about 25 km from the area of highest intensity, at a depth of 77 km.

[Algermissen *et al.* 1974; Brown 1968; BSSA; Hansen 1975; Karim *et al.* 1968; Williams 1972]

1968 Mar 1 **m_b 4.6** **Nicaragua**

An intermediate-depth earthquake in Nicaragua was perceptible at Monteverde in Costa Rica.

[Miyamura 1980]

1968 Mar 3 **m_b 4.2** **Costa Rica**

A small earthquake in Costa Rica was perceptible at Limón.

[Miyamura 1980]

1968 Mar 28 M_S 5.5 Mexico

An intermediate-depth earthquake near the Mexico-Guatemala border was strong in Chiapas at Chicharras and Motozintla, and in the San Marcos area in Guatemala. It caused no damage there.

It was perceptible at Tehuantepec.

[Figueroa 1973]

1968 May 6 m_b 5.0 Guatemala

An intermediate-depth earthquake was felt in much of Guatemala, and was perceptible at San Salvador.

1968 Jun 15 M_S 6.1 Panama

A relatively large earthquake off the Pacific coast of Panama was perceptible at Golfito.

[Miyamura 1980]

1968 Jul 2 m_b 4.2 Costa Rica

A small earthquake in Costa Rica was felt at Golfito.

[Miyamura 1980]

1968 Jul 4 m_b 4.2 Costa Rica

An intermediate-depth earthquake in Costa Rica was perceptible at San Rafael Ojo Agua.

[Miyamura 1980]

1968 Jul 17 M_S 5.1 Costa Rica

This earthquake was located near the Caribbean coast of Costa Rica.

It was very strong at Tortuguero and Barra del Colorado, but details are lacking.

The shock was perceptible further inland at Puerto Viejo, Vara Blanca, San Rafael, San Isidoro and Limón.

[Miyamura 1980]

1968 Sep 25 M_S 6.3 Tsunami Mexico

This was an intermediate-depth earthquake in the Mexico-Guatemala border region.

Much of the damage was done along the coast from Ciudad Hidalgo to Mapastepec, where 48 people were killed, 507 were injured and 333 houses, mostly new constructions, were destroyed; colonial buildings suffered relatively little. Worst damage was at Acapetahua, Ejido Cuatimoc, Iturbide, Laguna del Carmen, Comatitlan, Huehuetan, Mapastepec, Mazatan, Tuzantan and Ulapa where material losses exceeded $2.6 million.

Power failures were reported from many places in Chiapas.

The Inter-American Highway and other roads were blocked in several places by landslides, rail tracks were bent, embankments slumped and a train was derailed.

A seismic sea-wave 1.5 m high flooded the coast.

There is no information about the effects of this earthquake in neighbouring Guatemala, except that damage extended to the region of Suchiate, and that the shock was felt in Guatemala City.

[Figueroa 1973]

1968 Dec 27 **m_b 4.8** **Panama**

A small earthquake in Chiriqui, Panama, was perceptible at Coto.

[Miyamura 1980]

1969 Feb 25 **M_S 5.1** **Honduras**

An earthquake in northern Honduras was felt at Tegucigalpa.

1969 Mar 10 **m_b 5.2** **Nicaragua**

An earthquake off the coast of Nicaragua was felt along the central part of the coast of Nicaragua, and was perceptible at San Salvador.

1969 Mar 14 **M_S 5.6** **Nicaragua**

An intermediate-depth earthquake in Nicaragua was perceptible in El Salvador and Nicaragua.

1969 Apr 1 **m_b 5.0** **El Salvador**

A subcrustal earthquake off the coast of El Salvador was perceptible at San Salvador.

1969 Apr 21 **M_S 5.9** **Guatemala**

An intermediate-depth earthquake in Guatemala was felt at San Salvador.

1969 May 13 **M_S 6.5** **Nicaragua**

A large earthquake off the Pacific coast of Nicaragua caused considerable concern in Managua, Masaya and Rivas, and was very strong at Peñas Blancas.

The shock was strong in Guanacaste and at Liberia, and was felt at Puntarenas, San Ramón and other places in northwestern Costa Rica, but not in San José.

1969 Jun 28 **M_S 5.1** **El Salvador**

A subcrustal earthquake off the coast of El Salvador was felt at San Salvador.

1969 Jul 11 **M_S 5.2** **Costa Rica**

A small, intermediate-depth earthquake in Costa Rica was perceptible in Nicoya.

1969 Aug 15 **m_b 4.4** **Costa Rica**

A small earthquake in Costa Rica was perceptible at Santa Ana, Turrucares, Volcán and San Vito.

[Miyamura 1980]

1970 Jan 1 **M_S 5.5** **Costa Rica**

This relatively small earthquake was located in the Osa Peninsula, Costa Rica and was reported felt from San Vito.

[Miyamura 1980]

1970 Apr 29 11h Ms 6.5 Mexico

This was a large foreshock of the event at 14h 01m off the coast of southern Chiapas, Mexico.

The earthquake was widely felt in Chiapas, Guatemala and as far as El Salvador, but its reported effects are difficult to separate from those of the main shock, which occurred two and a half hours later.

It caused some minor damage at Mazapa de Madero, and was very strong at Comitan, Motozintla. Tapachula and Mazapa. Information about the effects of this event in Guatemala is lacking.

The shock was perceptible throughout Chiapas, and as far as Mexico City and San Salvador.

1970 Apr 29 14h Ms 7.3 Mexico

This was a major earthquake located instrumentally off the Pacific coast of Mexico and Guatemala.

In Mexico maximum damage consisting of the ruin of old houses was done at Amatenango de Frontiera, Mazapa de Madero and Motozintla, on the Mexico-Guatemala border. Damage extended into the San Marcos area, but there are no details of the effects of this shock in Guatemala.

In the rest of the District of Chiapas in Mexico the earthquake caused widespread but minor damage. At Cacahoatan, Ciudad Hidalgo and Tapachula, towns damaged by earlier shocks, a few new buildings were badly cracked. There were no reported victims.

The shock was widely felt. It was strong at San Salvador and perceptible in the lower parts of Mexico City, but its epicentral area was diffuse with disproportionately small damage for the magnitude.

The earthquake was followed by many aftershocks.

[Figueroa 1973; P]

1971 Oct 12 Ms 5.4 Guatemala

This earthquake was in northwestern Guatemala.

At Barillas two people were killed and a few hundred were made homeless.

The shock was felt as far as Huixtla in Mexico.

1972 Jan 5 Ms 3.9 Nicaragua

This small earthquake was part of a series in Nicaragua. It was strongly felt in Managua, where a few houses in the western part of the city were slightly damaged. Its epicentral area is believed to have been located near Hacienda Santa Justina.

It was preceded by foreshocks and followed by many aftershocks, which continued for a few days.

[Williams 1972]

1972 Dec 23 Ms 6.3 Figure A.63 Nicaragua

This was a very shallow, locally destructive earthquake in Nicaragua.

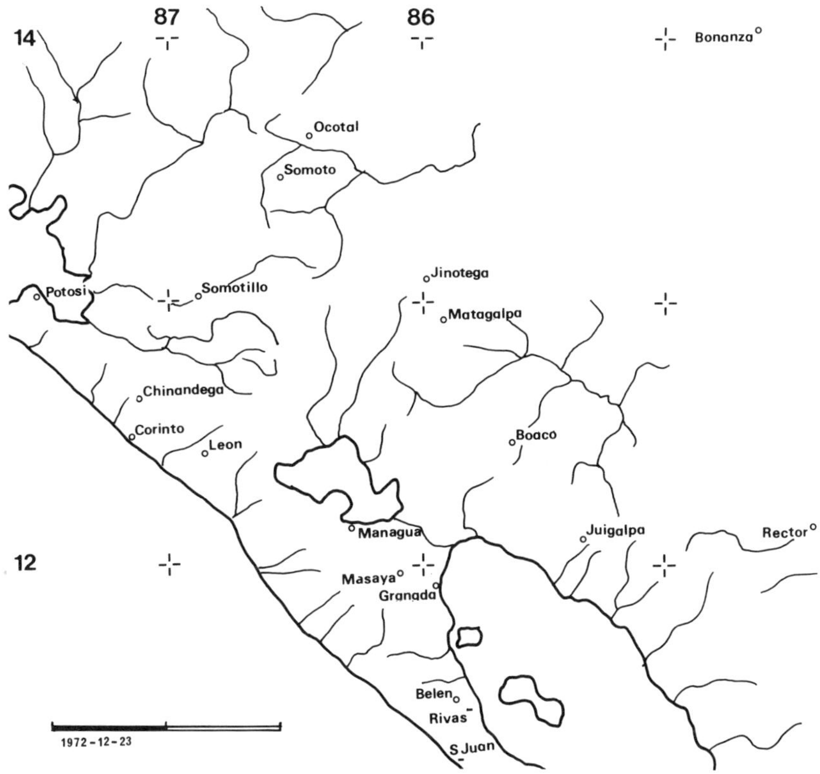

Fig. A.63 Felt effects of earthquake of 23 December 1972.

Heavy damage was confined within a radius of only 5 km around the centre of Managua. The city, of about 400,000 inhabitants, was almost totally destroyed by the shock and resultant fires, with great loss of life. About 10,000 people were killed and as many injured. There was damage to virtually every building in Managua and its outskirts and to all houses both of reinforced concrete and local construction. About three quarters of the city's housing units were destroyed or rendered uninhabitable, leaving 200,000 people homeless.

An accelerograph record obtained in Managua shows that maximum ground acceleration was 39%g, while seismoscope records suggest that accelerations exceeding 50%g may have occurred within the zone of surface faulting.

The earthquake was associated with a zone of ground ruptures passing through Managua and showing 5 to 30 cm left-lateral displacements. These faults were about 6 km long, extending for at least another 6 km northeast of the city south of Lake Managua. Damage was negligibly small at epicentral distances beyond 12 km.

The shock was felt over much of Nicaragua, as far as Rivas, El Recreo, Bonanza, Somoto, Somotillo and Potosi. It was not reported from Honduras or Costa Rica.

[Algermissen *et al.* 1974; Ambraseys 1973; Brown *et al.* 1973; Hansen and Chavez 1973; Dewey and Algermissen 1974; Knudson *et al.* 1974; Langer *et al.* 1974; Matthiesen 1975; Proceedings Managua 1973; Sozen 1975; Uribe 1973; Ward *et al.* 1974; Wyllie *et al.* 1974]

1973 Apr 14 M_S 6.5 Costa Rica

An earthquake in northwestern Costa Rica caused significant damage in the Region of Arenal, at Rio Chiquito, Tronadora, Arenal and Tilaran, rural settlements confined within a radius of about 9 km.

Twenty-three people were killed at Rio Chiquito, where a landslide buried a farmhouse. Ten deaths were also caused by landslides south of Silencio. No surface faulting was observed.

The relatively small structural damage sustained in the epicentral area is attributable to the predominance of timber frame houses and lightweight roofs.

The electric and water supply systems were put out of order by numerous landslides.

The shock was strongly felt as far as Puerto Viejo, Cartago, San José, Santa Maria and Herradura, but it is not known how far it was felt.

The earthquake was followed by many aftershocks.

[Barquero and Rojas 1994b; BSSA; Plafker 1973]

1973 Jun 7 M_S 6.4 Guatemala

An intermediate-depth earthquake on the coast of Guatemala was felt in southeastern Mexico, Guatemala and El Salvador.

1974 Mar 6 M_S 6.0 Nicaragua

This intermediate-depth earthquake in Nicaragua caused some slight damage at Managua, and was felt in much of Central America.

1974 Apr 10 M_S 5.1 Guatemala

An intermediate-depth earthquake in Guatemala was felt in Chiapas, Guatemala and much of El Salvador.

1974 Dec 6 M_S 5.5 Figure A.64 Panama

This was an earthquake in the Gulf of Chiriqui, Panama.

At Puerto Armuelles the shock caused some minor damage, water sloshed from containers and suspended objects swung.

The shock was strong at Esperanza, Progresso, Coto and David, where it caused considerable concern.

The earthquake was perceptible over much of Costa Rica and western Panama.

[Miyamura 1980]

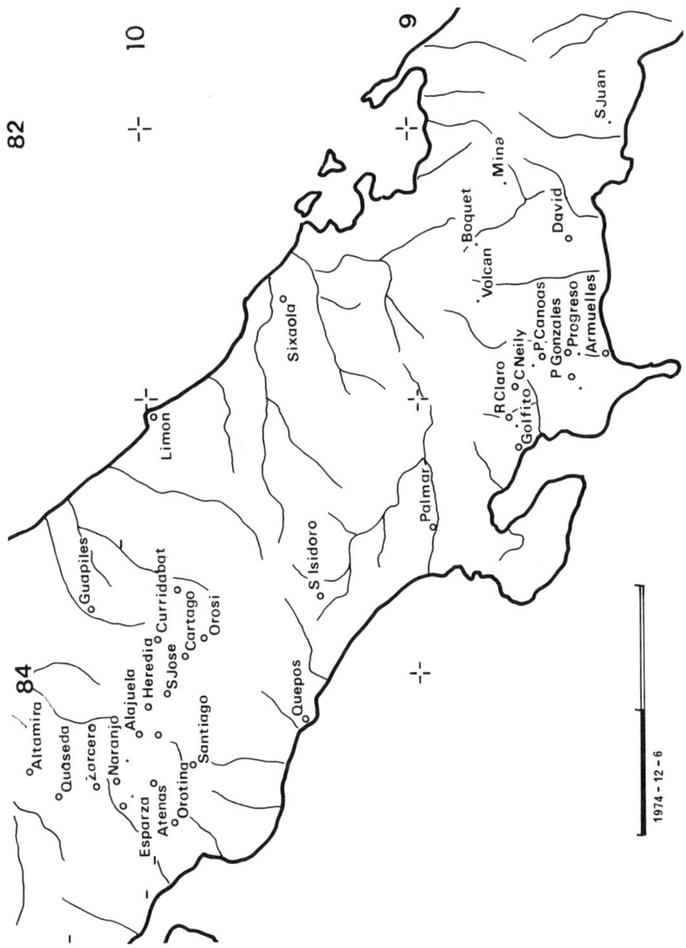

Fig. A.64 Felt effects of earthquake of 6 December 1974.

1975 Feb 3 **M_S 5.2** **Guatemala**

An intermediate-depth earthquake in the Guatemala-Mexico border area was felt in the region of Comitan.

1975 Jul 17 **M_S 3.9** **Figure A.65** **El Salvador**

A small earthquake in El Salvador caused some minor damage in the District of San Vicente. It was preceded by a foreshock. Guadelupe, Mercedes, La Ceiba, Paraiso de Osorio, and Santa Maria Ostuma, villages confined within a radius of about 4 km, suffered some damage but no casualties.

The shock was strongly felt at Cojutepeque and was perceptible at San Salvador and possibly at Usulutan.

The earthquake was followed by many small aftershocks.

[Jordan and Martinez 1980]

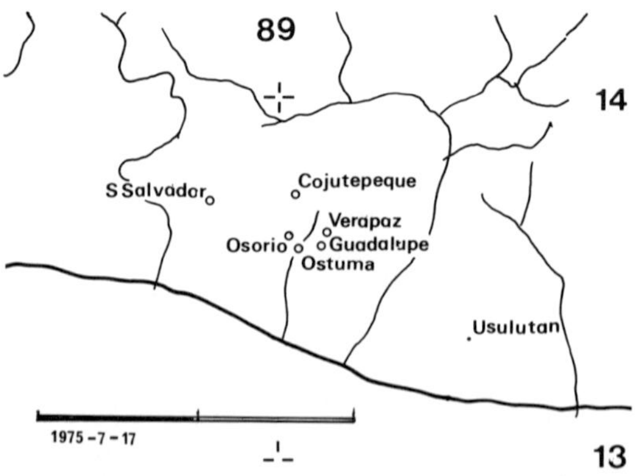

Fig. A.65 Felt effects of earthquake of 17 July 1975.

1976 Feb 4 M_S 7.5 Tsunami Figure A.66 Guatemala

This was a major earthquake in Guatemala.

The earthquake was associated with a left-lateral surface break of the Motagua fault, 230 km long, extending from near Quebradas in the lower Motagua Valley on the east to about 10 km east of Patzaj on the west. Maximum horizontal slippage was 325 cm with an average value of about 200 cm. Slippage also occurred on a number of secondary normal faults in the vicinity of Guatemala City, implying substantial intraplate deformation adjacent to the principal boundary in central and western Guatemala.

The distribution of damage within the epicentral area was erratic. Communities and small towns that suffered total damage covered an area of 1,700 km^2. The towns and villages which were almost totally destroyed were: Cabañas, Comalapa, Gualán, El Jicaro, Joybal, Morazan, Parramos, Patzicia, El Progreso, Rabinal, San Andreas Itzapa, San José Poaquie, Progreso Cabecera, San Martin Jilotepeque, San Pedro and San Juan Sacatepeque, San Pedro Ayampuc, Sumpanjo, Tecpan, Yepocapa, and Zaragoza. At some villages adobe houses near the fault break were essentially undamaged, while at greater distances from the fault in the highlands there was widespread collapse of adobe buildings. Guatemala City suffered severe damage in the poorer districts and along the deep canyons on the west side of the city. Modern earthquake-resistant structures in Guatemala City and elsewhere were damaged, including several hospitals. Several modern buildings collapsed with loss of life.

Two seismoscope records from the University station show a maximum relative displacement of 5.3 cm.

The bridge of Agua Caliente collapsed and that of Benque Viejo was almost totally destroyed.

From a total population of 3.2 million, there were 22,870 deaths and 77,200 injuries. The total number of houses destroyed was 255,000, and 1.1 million people were left homeless. The total loss amounted to $1,100 million.

Sporadic damage was reported from northwestern El Salvador, Honduras and southeastern Mexico.

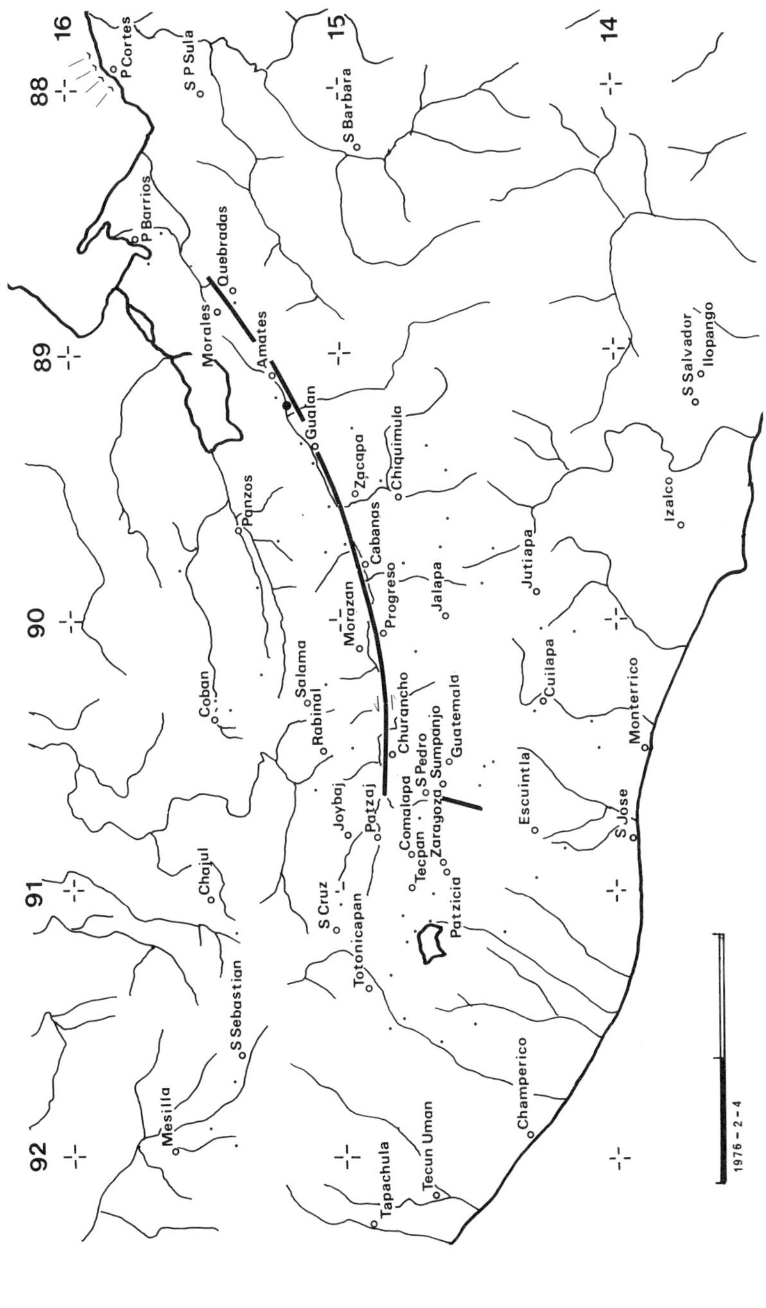

Fig. A.66 Felt effects of earthquake of 4 February 1976. Solid line shows extent of surface faulting.

The earthquake caused extensive landsliding, blocking roads and destroying communication routes in the highlands. A large slide overwhelmed two villages and dammed a river near Tecpan. In Chimaltenango two major landslides dammed rivers near Patzun and San Martin Jilotepeque, and small lakes were formed behind the dams, one of which was 60 m high.

The earthquake triggered extensive liquefaction on the delta of the Rio Villalobos on the north side of Lake Amatitlan.

The shock was strongly felt in Chiapas, and in El Salvador at Ilopango, Izalco and San Salvador, as well as in Honduras at Santa Barbara, San Pedro Sula and Puerto Cortés. In the plain of Asine near Ilopango the shock caused cracking of the ground.

The shock was felt at Tegucigalpa, Belize, Poctum and as far as Mexico City, 1,000 km from the epicentral area.

The shock caused a tsunami that was recorded on the Puerto Cortés tide gauge with a maximum amplitude of 45 cm.

The earthquake is believed to have caused fluctuations in well-water levels in Georgia, Idaho, Indiana and Wisconsin.

Thousands of aftershocks followed the main earthquake, delineating the Motagua fault and several secondary north-south faults that bound the Guatemala City graben.

[Bohnenberger and Bonis 1976; Bucknam *et al.* 1978; Davis 1976; Espinoza 1976; Kanamori and Stewart 1978; Krinitzsky and Bonis 1976; Langer and Bollinger 1979; Lisowski and Thatcher 1981; Page 1976; Plafker 1976; Sozen and Roesset 1976; USEQ; Young *et al.* 1989]

1976 Feb 6 M_S 5.6 Guatemala

An earthquake in Guatemala caused some damage but no casualties. The shock was widely felt.

1976 Mar 13 M_S 5.5 Guatemala

A small earthquake in Guatemala caused considerable damage in the area of Patzun.

At Tecpan four people were killed and about 50 injured, and more than 100 houses were ruined. Landslides added to the damage.

1977 Sep 30 M_S 5.0 Costa Rica

An intermediate-depth earthquake located off the coast of Costa Rica was strongly felt in southeastern and northwestern Costa Rica. Some slight damage was reported from Rivas and San Juan del Sur.

The shock was felt in Managua and was perceptible at San José and in some parts of El Salvador.

1978 Feb 22 M_S 6.2 Guatemala

An intermediate-depth earthquake was widely felt in western Guatemala.

It caused sporadic damage and in places, at Mirian and Coatepeque, the collapse of adobe houses and the loss of two lives.

The shock was rather strong at San Salvador.

1978 May 31 **M_S 6.0** **Nicaragua**

This intermediate-depth earthquake off the coast of Nicaragua was widely felt.

It caused some slight damage in Nicaragua and northwestern Costa Rica and triggered landslides on the Managua-Masaya road.

It was strongly felt in southern Honduras and El Salvador and was perceptible in parts of Chiapas, Guatemala and at San José.

1978 Jul 20 **M_S 5.2** **Nicaragua**

An intermediate-depth earthquake in Nicaragua was strongly felt in the Departments of Chinandega, Leon, Carazo and Managua, and was rather strong at San Salvador.

1978 Jul 29 **M_S 4.5** **Guatemala**

This was a small earthquake in Guatemala, in the region of Patzun.

In Tecpan about 400 houses were ruined, five people were killed and 12 were reported missing.

The shock triggered landslides in Patzun, and was felt at Guatemala City.

[White 1980]

1978 Aug 18 **M_S 5.9** **Guatemala**

This intermediate-depth earthquake caused some damage and injuries in southwestern Guatemala.

The shock was very strong at Guatemala City, and was felt in San Salvador.

1978 Aug 23 **M_S 7.0** **Costa Rica**

This was a large earthquake off the southwest coast of the Nicoya Peninsula, Costa Rica.

In spite of its large size the shock caused only minor damage in the region of Samara and Santa Cruz, and it was not felt further east than Guapiles and Turrialba in Costa Rica, and Rivas in Nicaragua, 210 km from the epicentre.

[Barquero and Rojas 1994b; Güendel 1986; Güendel and McNally 1982]

1978 Sep 10 **M_S 5.8** **Guatemala**

A subcrustal earthquake in Guatemala caused minor damage in unspecified localities, injuring one person.

The earthquake was felt at Huehuetenango and Guatemala City. At San Salvador a peak ground acceleration of 1.7%g was recorded.

1978 Oct 30 **M_S 6.0** **Guatemala**

A subcustal earthquake off the coast of Guatemala was strongly felt in southeastern Guatemala and western El Salvador.

It was generally felt in Guatemala City and had an acceleration of 1%g in San Salvador.

1978 Dec 6 **M_S 6.0** **El Salvador**

A subcrustal earthquake off the coast of El Salvador was very strong at San Salvador, where the recorded peak acceleration was 1.6%g. It was felt in Guatemala City.

1978 Dec 18 **M_S 5.0** **El Salvador**

A subcrustal earthquake off the coast of El Salvador was rather strong at San Salvador, where it registered 5.7%g.

1979 Jan 12 **M_S 5.8** **Guatemala**

An intermediate-depth earthquake was felt strongly in southwestern Guatemala and at Guatemala City, where it caused power failure, but no damage or injuries.

1979 Mar 15 **M_S 3.4** **El Salvador**

A small earthquake was widely felt in San Salvador, with a recorded peak acceleration of about 1%g.

1979 Jul 1 **M_S 6.4** **Figure A.67** **Panama**

This was a damaging earthquake in the Gulf of Chiriqui, Panama.

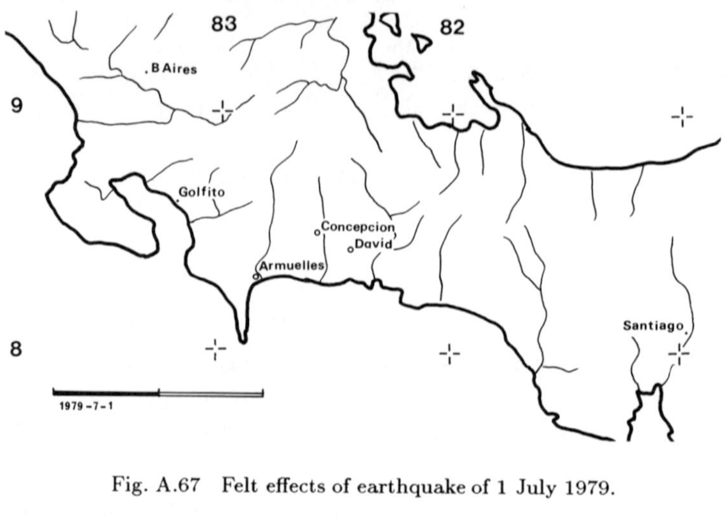

Fig. A.67 Felt effects of earthquake of 1 July 1979.

In Puerto Armuelles many houses suffered some damage, including the school and the bank, and two people were injured. In the port the shock caused cracks in the ground, and an interruption of the water and electricity supplies. Damage to the oil terminal and loading facilities in the port was estimated at $2 million.

The shock was strong at Conceptión, David and other settlements around the Bay of Charco Azul, as well as across the border in Costa Rica, where a number of houses were slightly damaged.

1979 Aug 24 M_S 6.4 Costa Rica

A large-magnitude earthquake in Costa Rica caused widespread but minor damage in the region between Puerto Cortés and Golfito, and was strongly felt at Paso Canoas and Armuelles.

The shock was felt throughout southwestern Costa Rica, causing some slight damage in places, and the injury of 10 people in the region of San José.

1979 Oct 9 M_S 5.0 Figure A.68 Guatemala

This damaging shock in the District of Santa Rosa, eastern Guatemala, was preceded and followed by earthquake swarms which caused much alarm.

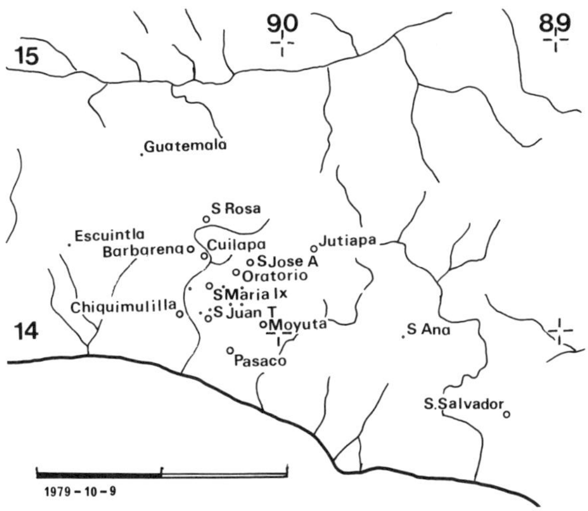

Fig. A.68 Felt effects of earthquake of 9 October 1979.

Maximum damage occurred within a small area of about 10 km radius in which 40 people were injured. At San Juan Tecuaco 50 houses were destroyed, and the school, church, municipal building, and 150 houses were severely damaged. At Santa Maria Ixhuatan 75% of the houses were cracked and four were destroyed. About 60% of the houses at Oratorio were damaged and seven were ruined. At Joya Grande 29 houses were damaged and 11 collapsed. The villages of Chuchuapa, La Consulta, El Coyolito, El Cuje, Estanzuela, La Fila, Ijorga, Nancinta, El Pital, San Antonio Cabrera, San José La Cumbre, San José, Pineda, San Luis, Santa Clara, and El Tangue were all damaged, and 164 adobe houses were destroyed.

Within this area there were landslides in the delta of the River Los Esclavos to the north of San Juan Tecuaco. North-south and east-west cracks were observed half a kilometre uphill from San Juan Tecuaco. The cracks measured 3 km long and 2 cm wide.

The earthquake occurred during the rainy season and slumping was observed along the footpath between San Juan Tecuaco and Ijorga.

The shock was felt in Guatemala City, Escuintla and at San Salvador.

The earthquake was followed by a swarm of aftershocks, which extended well into 1980.

[White and Sanches 1980]

1979 Oct 27 **M_S 6.8** **Guatemala**

An earthquake in southeastern Guatemala did considerable damage in western El Salvador, where it killed four people. Damage extended to Guatemala where a number of people were injured. Details of the damage caused are lacking.

The shock was strongly felt at San Salvador.

An aftershock of almost equal magnitude followed seven hours later.

1980 May 18 **M_S 4.9** **Guatemala**

This was an intermediate-depth earthquake in western Guatemala.

Moderate damage was reported in the southern coastal region, but details are lacking.

1980 Aug 9 **M_S 6.5** **Guatemala**

This earthquake in the Gulf of Honduras caused considerable damage in Puerto Barrios where two people were killed and many injured. Damage occurred also in Livingston and northwestern Honduras, but no details are available from there.

[BSSA 1981 p937; White 1980]

1980 Sep 2 **M_S 4.7** **Guatemala**

This earthquake was felt in the northern part of the Honduras-Guatemala area.

1980 Nov 4 **M_S 5.0** **Guatemala**

An intermediate-depth earthquake off the Pacific coast of Guatemala was perceptible at San Salvador.

1981 Jun 11 **M_S 5.4** **Honduras**

This earthquake was felt in Honduras at San Pedro Sula and Choloma.

1981 Sep 27 **M_S 4.5** **Guatemala**

This shock was felt in western Guatemala and was perceptible at San Salvador.

1982 Jan 12 **M_S 6.1** **Figure A.69** **Honduras**

An earthquake in the Gulf of Fonseca in Honduras caused moderate damage at Amapala, injuring two people. Allegedly it caused slight damage at Masaya in Nicaragua and was felt in southern Honduras and western Nicaragua.

1982 Mar 4 **M_S 4.8** **El Salvador**

This earthquake occurred off the coast of El Salvador.

It was widely felt and was rather strong at San Salvador.

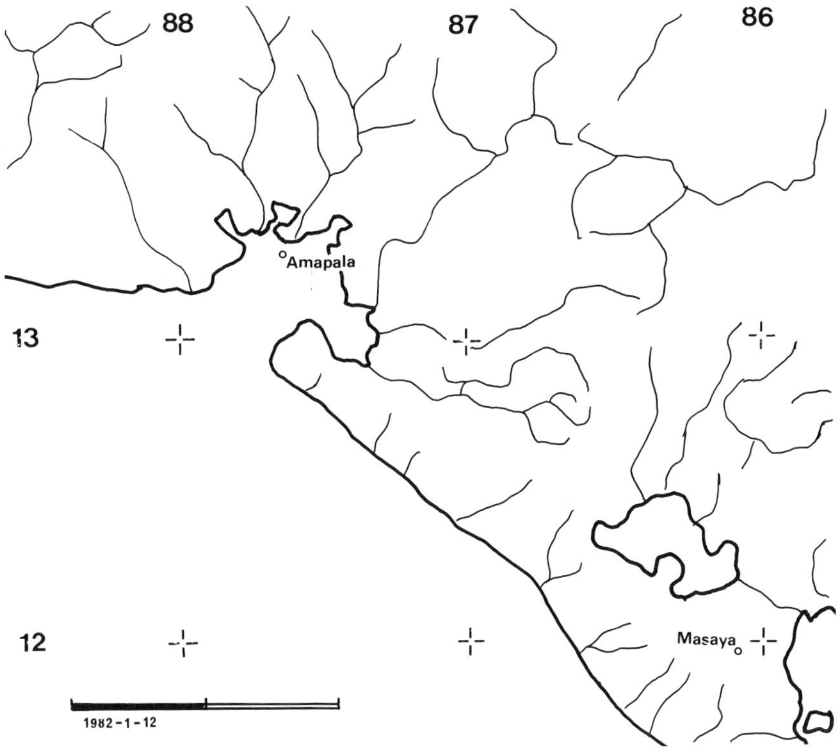

Fig. A.69 Felt effects of earthquake of 12 January 1982.

1982 Apr 6 **M_S 6.6** **Guatemala**

A subcrustal earthquake off the coast of western Guatemala was strong at Tapachula, and was felt in Chiapas, southern Guatemala and El Salvador.

1982 Jun 19 **M_S 7.3** **Figure A.70** **El Salvador**

This was a major subcrustal earthquake in the Pacific south of El Salvador.

Serious damage was reported along the border with Guatemala where virtually all roads linking the two countries, including the Pan American Highway, were blocked by landslides.

At Jalpatagua about 180 houses collapsed, killing three and injuring 40 people.

At Chinamas rural roads were blocked by landslides and fallen trees.

At Barra de Santiago and in central Sonsonate Province damage was extensive with some loss of life, but details are not available.

In Comasagua many houses collapsed, injuring 40 people, and the town was isolated because of landslides.

In San Salvador damage to houses, mostly of older type construction, was considerable and recorded peak ground accelerations reached nearly 20%g. A few modern buildings, including the US Embassy, were also damaged. Power lines and the water system were damaged in parts. In San Salvador 40 people were killed and about 300 injured.

San Miguel and the nearby village of San Juan Tepezondes were ruined; two people were killed and 73 injured.

Telephone services in El Salvador and eastern Guatemala were disrupted. Power supply in western El Salvador failed and a turbine in one of the hydro-electric dams was damaged by the shock.

The shock caused some damage in Jutiapa and was very strong in Guatemala City, but caused no damage there.

All ports along the coast were seriously shaken, but without serious damage.

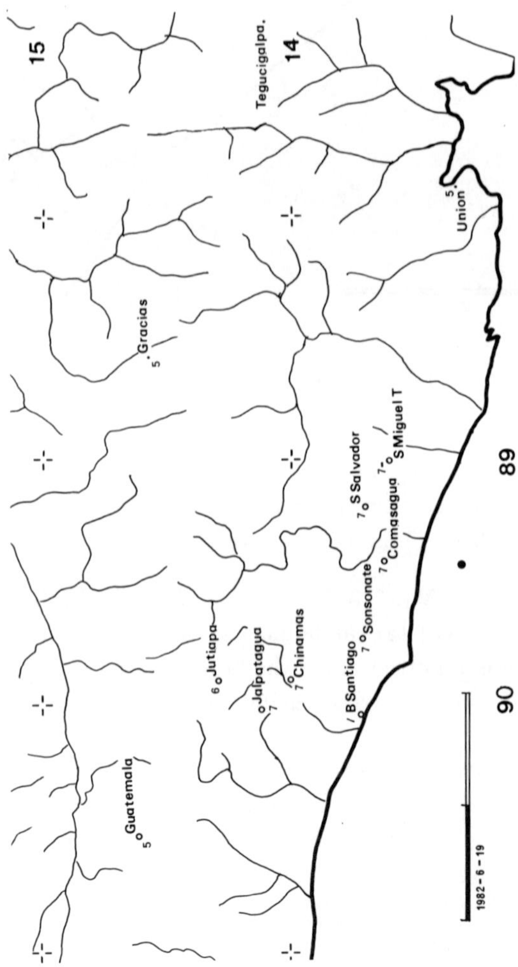

Fig. A.70 Felt effects of earthquake of 19 June 1982.

The shock was felt in Honduras and Nicaragua and was perceptible in Costa Rica.

[P]

1982 Aug 19 **M_S 6.7** **Panama**

This was a large-magnitude crustal earthquake with an epicentre off the coast of western Panama.

The shock caused sporadic, minor damage over a large area in Panama at David, and in Costa Rica in the Osa Peninsula, injuring three people. The shock was strong at San José and in other places in the Central Plateau.

The earthquake was felt throughout Costa Rica and along the southern coast of Panama.

1982 Sep 29 **M_S 5.2** **Figure A.71** **Guatemala**

This earthquake affected an area on the Honduras-Guatemala border. It was preceded two hours earlier by a strong foreshock.

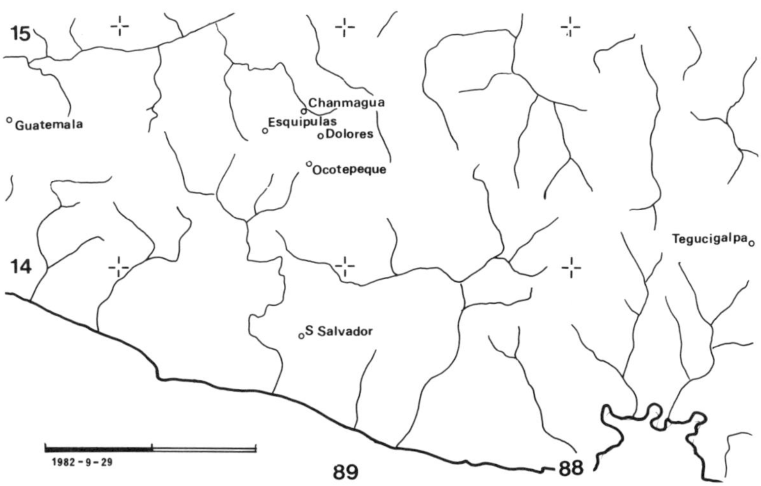

Fig. A.71 Felt effects of earthquake of 29 September 1982.

Three people were killed, several injured and many homes were damaged in eastern Chiquimula. About 50 houses were destroyed and 554 were damaged around Chanmagua where a scarp was observed that was 0.1m high and several kilometres long.

The earthquake triggered some landslides in the Dolores Merendon-Ocotepeque area in Honduras.

The shock was felt at San Salvador and was perceptible at Tegucigalpa and in Guatemala City, where it triggered an accelerograph.

[White 1980; BSSA 1983.1267]

1982 Oct 27 M_S 5.2 El Salvador

An intermediate-depth earthquake in El Salvador was widely felt and was strong at San Salvador, but caused no damage.

1983 Apr 3 M_S 7.2 Figure A.72 Costa Rica

This major earthquake near the Costa Rica-Panama border affected the same area as the earthquake of 5 December 1941.

The shock caused sporadic damage over a relatively large area, which extended from the southern part of the Valle del General to the Golfo del Dulce and into the Department of David in Panama. Within this area a few hundred houses were destroyed, one person killed and a few injured.

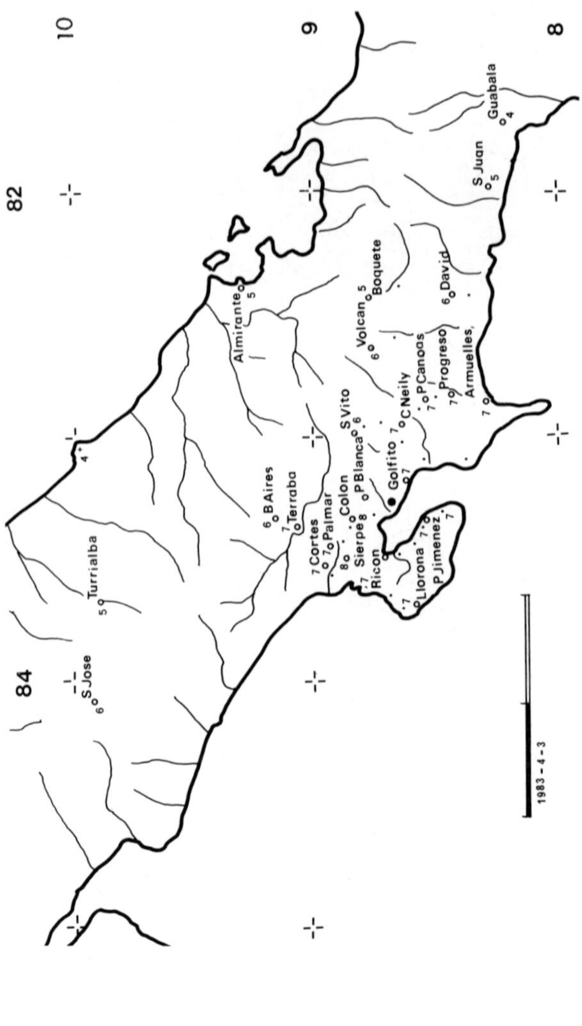

Fig. A.72 Felt effects of earthquake of 3 April 1983.

Damage, which was disproportionately small for the size of the earthquake, was associated with the liquefaction of the ground or was due to the substandard construction of masonry houses, and occurred around Palmar Norte, Palmar Sur, Ciudad Cortes and sporadically around the Golfo del Dulce and in the western part of the Gulf of David. Timber constructions suffered little or no structural damage.

Within this area spreading and slumping of the ground was responsible for the damage of approaches to bridges, causeways, retaining walls and for the partial collapse of a few engineered structures, interrupting power supplies and communications. The distribution of damage does not define an epicentral area.

Minor damage and some panic were reported from as far as San José, the Central Valley of Costa Rica, and from Almirante and David in Panama.

The shock was felt throughout Costa Rica, in southern Nicaragua and in much of Panama.

The earthquake was perceptible in Panama City.

[Barquero and Rojas 1994b; Boscini 1989; Montero and Mora 1985]

1983 Apr 7 M_S 5.6 Panama

A strong aftershock with an epicentre in the Gulf of Chiriqui caused some minor damage in the coastal area of the Department of David.

1983 Apr 14 M_S 5.6 Guatemala

A subcrustal earthquake was felt in central and southern Guatemala.

The shock was rather strong in Guatemala City, and was perceptible at San Salvador.

1983 Apr 29 M_S 4.9 El Salvador

This intermediate-depth earthquake occurred off the coast of El Salvador.

It was strongly felt at San Salvador.

1983 Jul 3 17h 13m M_S 5.5 Costa Rica

A strong foreshock of the earthquake in region of San Isidoro de El General caused no damage, but was widely felt, as far as Puntarenas and David in Panama.

1983 Jul 3 17h 14m M_S 6.3 Costa Rica

The main shock, about 40 seconds after the foreshock, caused great damage within a small area of radius of about 12 km around Buenavista. In the epicentral region which included Buenavista, Siberia, Division, Hortansia, Herradura, Canaan, Rivas and San Isidoro, 300 houses, schools and other public buildings were almost totally destroyed, two people were killed and about 60 were injured.

Damage was aggravated by landslide debris that affected an area of 25 km^2, and by landslide movements triggered by the earthquake over an area of 150 km^2, that affected roads.

At San Isidoro a few houses and buildings were damaged beyond repair.

Sporadic damage extended over a large area, as far as Pacayas.

At San José the shock was very strong, and it was felt throughout Costa Rica and in western Panama.

[Barquero and Rojas 1994b; Bengtsson and Sderberg 1991; Boschini 1989; Morales and Leandro 1985]

1983 Jul 18 M_S 6.0 Nicaragua

An intermediate-depth earthquake in Nicaragua caused some slight damage at Chinandega, and was felt strongly in the Managua area.

It was rather strong at San Salvador

The shock was perceptible as far as San José and in western Guatemala.

1983 Sep 23 M_S 5.1 Costa Rica

A strong earthquake in western Panama was reported from Conceptión and Boquete, and was widely felt in the Chiriqui Province.

1983 Sep 29 M_S 4.6 Costa Rica

A small subcrustal earthquake off the Pacific coast of Costa Rica was perceptible in the Liberia and Santa Cruz areas in Guanacaste.

1983 Oct 12 03h M_S 5.6 Panama

An earthquake in the Gulf of Chiriqui caused some damage in Costa Rica and was felt in western Panama.

1983 Oct 12 13h M_S 5.4 Guatemala

An intermediate-depth earthquake off the coast of Guatemala was perceptible in Guatemala City and at San Salvador.

1983 Oct 17 M_S 4.0 El Salvador

This earthquake occurred off the coast of El Salvador and was felt at San Salvador.

1983 Oct 19 M_S 4.1 Nicaragua

An intermediate-depth earthquake off the coast of Nicaragua was felt in northwestern Nicaragua, and was perceptible in San Salvador.

1983 Nov 30 M_S 4.7 Guatemala

An intermediate-depth earthquake with an epicentre off the coast of Guatemala was perceptible at San Salvador.

1983 Dec 2 M_S 6.9 Guatemala

This large earthquake off the southwest coast of Guatemala caused considerable concern and minor damage in southwestern Guatemala.

It was strongly felt in Huixtla, Chiapas, Guatemala City and at San Salvador.

1983 Dec 10 M_S 5.6 Guatemala

An intermediate-depth earthquake in Guatemala was widely felt and was perceptible in El Salvador.

1984 Mar 26 m_b 5.0 El Salvador

This relatively small earthquake in the Pacific was felt at San Salvador.

1984 Apr 9 **M_S 4.7** **Costa Rica**

This small subcrustal earthquake in southeastern Costa Rica was rather strong at Puerto Jiménez, and was perceptible in San José.

1984 Apr 13 **M_S 3.9** **Nicaragua**

A small intermediate-depth earthquake off the northwest coast of Nicaragua was felt at San Salvador.

1984 Aug 22 18h **M_S 5.1** **Nicaragua**

This was a deeper foreshock of the earthquake of August 23d 19h, with an epicentre off the west coast of Nicaragua.

It was rather strong in the region of Leon and Managua, and was felt at La Cruz in Costa Rica.

1984 Aug 22 23h **M_S 4.8** **Nicaragua**

Another foreshock, with an epicentre off the coast of Nicaragua, was felt at La Cruz, Costa Rica.

1984 Aug 23 00h **M_S 4.6** **Nicaragua**

An equally strong foreshock was felt at La Cruz and was perceptible at San José.

1984 Aug 23 19h **M_S 6.1** **Nicaragua**

The main shock of this sequence was strongly felt along the coast of Nicaragua and at Managua. It was also felt at La Cruz and Peñas Blancas in Costa Rica.

1984 Aug 31 **M_S 5.2** **Nicaragua**

This earthquake occurred in northwestern Nicaragua.

It was rather strong at Leon and in southern Honduras and was perceptible at San Salvador.

1984 Dec 20 **M_S 5.6** **Nicaragua**

This earthquake was located on the southwest coast of Nicaragua. It was felt at Managua, and was perceptible at Liberia in Costa Rica.

1985 Jan 5 **M_S 5.9** **Panama**

An earthquake off the north coast of central Panama was strongly felt in Colón, where it caused some minor damage.

The shock was felt in Darien, on Isla San Andrés and in the Central Valley of Costa Rica.

1985 Feb 8 **m_b 5.1** **Costa Rica**

This small earthquake located off the Pacific coast of Costa Rica was felt at Parrita and was perceptible at San José.

1985 Mar 10 **M_S 5.0** **El Salvador**

A subcrustal earthquake off the coast of El Salvador was very strongly felt at San Salvador.

1985 Apr 19 **M_S 5.9** **Nicaragua**

This intermediate-depth earthquake was located in the Pacific, and was felt along the coast of Nicaragua and at Managua. It was also felt in western Costa Rica.

1985 Apr 23 **M_S 4.0** **El Salvador**

This earthquake in El Salvador was associated with volcanism.

The shock was very strong on the east flank of Tacapa Volcano and at Santiago de Maria, where three days after the earthquake a graben 30 cm deep and 0.8 to 2.0 km wide was formed.

The shock was strong at Berlin and Chinameca, and was perceptible at San Salvador and in southern Honduras.

1985 Jul 30 **m_b 5.0** **El Salvador**

An intermediate-depth earthquake off the coast of El Salvador was felt rather strongly at San Salvador.

1985 Sep 25 **M_S 4.9** **Costa Rica**

This earthquake was located in the Pacific off the coast of Costa Rica. It was very strong at Quepos, was felt in central Costa Rica and at San José, and was perceptible at Puntarenas.

1985 Oct 11 **m_b 4.6** **Guatemala**

A local shock in northern Guatemala made many people homeless and injured a few. About a third of the houses in Uspantan were destroyed or damaged beyond repair.

The shock was felt in Guatemala City.

1985 Nov 27 **M_S 4.6** **Guatemala**

This relatively small subcrustal earthquake was located near the coast of Guatemala. It was felt over a large part of the country and was rather strong at Guatemala City.

It was perceptible in western El Salvador, at San Salvador and in Chiapas, Mexico.

1985 Dec 16 **M_S 6.1** **Figure A.73** **Nicaragua**

An earthquake in Nicaragua caused some damage in the Rivas-Masaya area. Six people were injured by landslides. The shock was widely felt along the Pacific coast of Nicaragua, and was reported from San José in Costa Rica and from San Salvador.

1986 Feb 15 **M_S 5.0** **Guatemala**

An intermediate-depth earthquake was felt in western and central Guatemala, as well as in Chiapas in Mexico. It was also felt at Tapachula and was perceptible at San Salvador.

1986 Mar 24 **M_S 4.5** **Guatemala**

This relatively small subcrustal earthquake was felt along the coast of Guatemala and at Guatemala City, and was perceptible in Chiapas at Tuxtla Gutierrez.

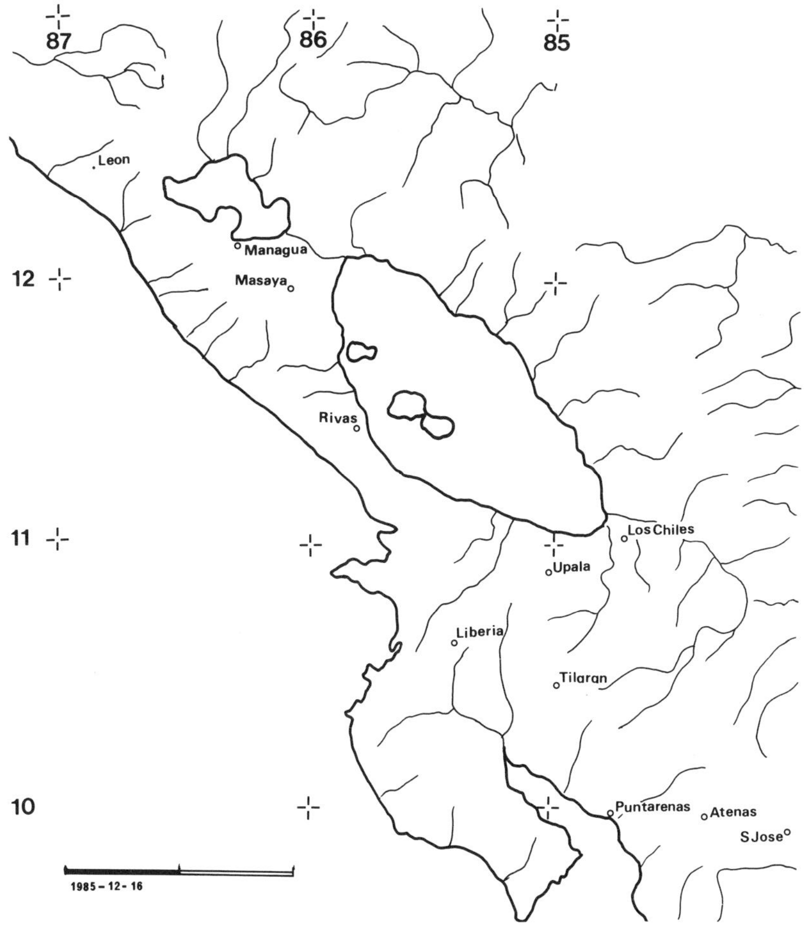

Fig. A.73 Felt effects of earthquake of 16 December 1985.

1986 Jun 30 Ms 5.7 Nicaragua

A subcrustal earthquake off the coast of Nicaragua was rather strong at Managua. It was felt in northern Costa Rica and was perceptible at San José.

1986 Jul 5 Ms 5.2 Mexico

This intermediate-depth earthquake in Mexico was felt in Guatemala, and at Tuxtla Gutierrez in the State of Chiapas.

1986 Aug 2 Ms 5.0 El Salvador

A subcrustal earthquake located off the coast of El Salvador was felt rather strongly at San Salvador.

1986 Aug 7 Ms 5.4 Panama

This earthquake was located off the west coast of the Peninsula of Azuero,

Panama.

It was strongly felt in western Panama, and was perceptible as far as Golfito and Panama City.

1986 Oct 10 **M_S 5.4** **Figure A.74** **El Salvador**

This was a locally destructive shock in El Salvador.

In the San Salvador area at least 1,000 people were killed, about 10,000 were injured, and 200,000 were made homeless.

Some slight damage was reported at Tegucigalpa, Honduras.

The shock was felt strongly in parts of Guatemala and Honduras.

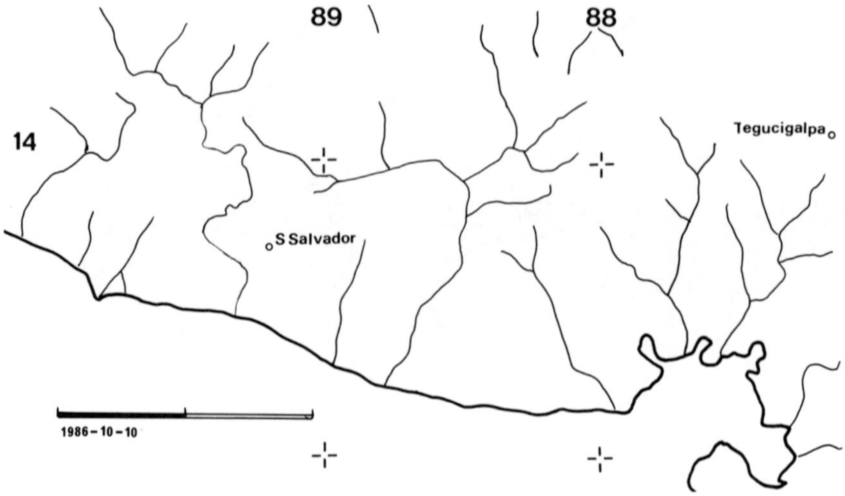

Fig. A.74 Felt effects of earthquake of 10 October 1986.

The earthquake was associated with a normal surface break that extended for 3 to 6 km northwest from the epicentre through the centre of the damaged downtown area of San Salvador. About 50 people were killed as a result of a landslide in the epicentral area.

[Anderson 1987; Chieruzzi 1987; Durkin and Hopkins 1987; Harlow *et al.* 1993; Lara 1987a,b; Ledbetter and Bommer 1987; Morgan and Swan 1987; Rios *et al.* 1986; Rymer 1987; Sauter 1987; Shakal *et al.* 1987; White *et al.* 1987]

1986 Oct 21 **M_S 5.2** **Guatemala**

An earthquake off the coast of Guatemala was felt along the Pacific coast and at Guatemala.

1987 Jan 15 **m_b 4.0** **El Salvador**

This small earthquake in El Salvador caused the collapse of some buildings that had been damaged by the earthquake of 10 October, 1986.

[NEIC]

1987 Mar 21 M_S 5.2 Figure A.75 Costa Rica

This earthquake in Costa Rica was felt in the southern part of the country and in western Panama.

The shock was strong in the region of Palmar, where, however, it caused no damage, and was felt at Golfito, Volcán, Boquete, Conceptión and Puerto Armuelles.

It was perceptible in the Central Valley, at Puntarenas and Limón.

[Barquero and Rojas 1994b; Boschini 1987]

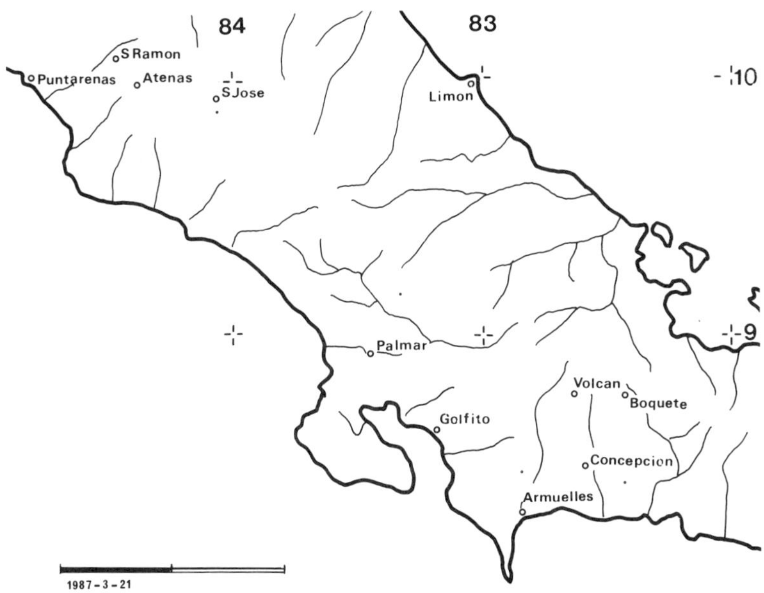

Fig. A.75 Felt effects of earthquake of 21 March 1987.

1987 Apr 8 M_S 6.3 Nicaragua

A subcrustal earthquake located off Nicaragua was strongly felt at Masachapa and in the Managua area, and was widely felt in northwestern Costa Rica at Peñas Blancas, La Cruz, and in the Playa Hermosa and Coco.

The shock was generally felt as far as Puntarenas and was perceptible in San José and in the Central Valley, as well as at San Salvador.

[Boschini 1987]

1987 Jun 2 M_S 5.1 Costa Rica

This earthquake was located at sea in the Gulf of Coronado, Costa Rica.

In spite of its relatively small magnitude, the shock was very strongly felt along the coast from Quepos to Palmar, where, however, it caused no damage.

It was strong in the Central Valley, at San Ramón, from San José to Alajuela, and as far south as Puerto Jiménez.

The shock was felt at Puntarenas, Limón, Volcán, Boquete and David in Panama, and was perceptible at Liberia and Bocas del Toro.

[Barquero and Rojas 1994b; Boschini 1987]

1987 Jun 22 M_S 5.2 Panama

This eathquake was located in the Gulf of Chiriqui.

It was felt over an area disproportionately large for its depth and size, from Laurel de Corredores in Costa Rica to Aguadulce in Panama and was perceptible as far as San José and Turrialba.

[Boschini 1987]

1987 Jul 15 M_S 4.5 Costa Rica

An eathquake in Costa Rica was very strong in the Quepos area and caused some concern at Parrita, San Marcos de Terrazu, La Lucha and in the San José area.

The shock was perceptible in Guanacaste, Limón, Palmar Norte and along the Costa Rica-Panama border.

It was followed by an equally strong aftershock about three hours later.

[Barquero and Rojas 1994b; Boschini 1987; Santana 1989; Santana *et al.* 1987]

1987 Sep 13 M_S 4.6 Guatemala

An intermediate-depth earthquake in Guatemala was felt in the southeastern part of the country, and was rather strong at San Salvador.

1987 Oct 4 M_S 6.0 Costa Rica

An earthquake with an epicentre off the Pacific coast of Costa Rica was strongly felt in the region of Guanacaste, in Playa Coco and at Liberia, where it caused no damage.

The shock was felt along the Pacific coast to Quepos, and was perceptible further inland in the Central Valley as far as Cartago.

[Boschini 1987]

1987 Oct 14 M_S 5.3 El Salvador

An intermediate-depth earthquake close to the coast of El Salvador was rather strong at San Salvador.

1987 Oct 20 M_S 3.8 Costa Rica

This was a small earthquake in Costa Rica, which was felt in the region of Joco.

It was perceptible at Atenes and Santiago de Puriscal.

[Boschini 1987]

1987 Nov 17 M_S 6.1 Nicaragua

This intermediate-depth earthquake in Nicaragua caused some damage, injuring three people, in the region of Corinto and Viejo. At Chinandega walls of several houses were cracked, and in Leon at least one house collapsed.

The shock caused some alarm in Managua. It was felt at Coco in southwestern Honduras, and at La Cruz in northern Costa Rica, and as far as Puntarenas to the south and San Salvador to the north.

[Boschini 1987]

1987 Nov 19 M_S 4.3 Costa Rica

A small earthquake in central Costa Rica was rather strong at La Suiza and Tayutic, and was felt in the region between Limón, Grecia and Quepos. It was perceptible as far as San Carlos, Sixaola and Paso Canoas.

[Barquero and Rojas 1994b; Boschini 1987]

1988 Jan 31 M_S 4.3 Costa Rica

A small earthquake in central Costa Rica was very strong at La Division, and was widely felt throughout the southern part of the country.

[Berquero and Rojas 1994b]

1988 Mar 2 M_S 5.2 Costa Rica

An earthquake in the Gulf of Nicoya, Costa Rica, was rather strong at Puntarenas, Cobano, Nicoya and in particular at Herradura. It was felt at San José and in the Central Valley, and was perceptible in northern Costa Rica.

1988 Mar 11 M_S 6.0 Figure A.76 Costa Rica

A rather large earthquake in southern Costa Rica caused considerable damage in the region of San Vito and Portero Grande. Some damage was reported from Buenos Aires, Puerto Cortés, Puerto Jiménez and Ciudad Neily.

In the region of Palmar Norte the earthquake triggered landslides.

The shock was felt as far as San Carlos and Puntarenas in Costa Rica, and in Chitré and Las Tablas in Panama. It was perceptible in Panama City.

[Barquero and Rojas 1994b; NEIC]

1988 May 6 M_S 6.3 Nicaragua

An intermediate-depth earthquake in Nicaragua was very strong in the Department of Rivas, but caused no damage there.

The earthquake caused some alarm in the Managua area and landslides were triggered at Sososca. It was rather strong at Corinto, Los Chiles, Puntarenas, San Ramón and San José.

The shock was felt at Turrialba, Quepos, and was perceptible at Bluefields, Limón, Ciudad Cortes and San Salvador.

[Barquero and Rojas 1994b]

1988 Jun 18 M_S 5.6 Guatemala

An earthquake in the Pacific Ocean was felt throughout western Guatemala and was perceptible at San Salvador.

1988 Nov 3 M_S 6.4 Guatemala

An intermediate-depth earthquake off the coast of Guatemala caused widespread minor damage in the southern part of the country, where five people were killed, some injured and about 100 houses ruined.

The shock triggered landslides at Palin, in the area of San Vicente Pacaya, and damaged roads in the District of Chimaltanengo.

The shock was strongly felt at San Salvador and at Tapachula, and was perceptible in parts of Honduras, at Managua in Nicaragua and also in parts of Oaxaca, Veracruz and as far as Mexico City.

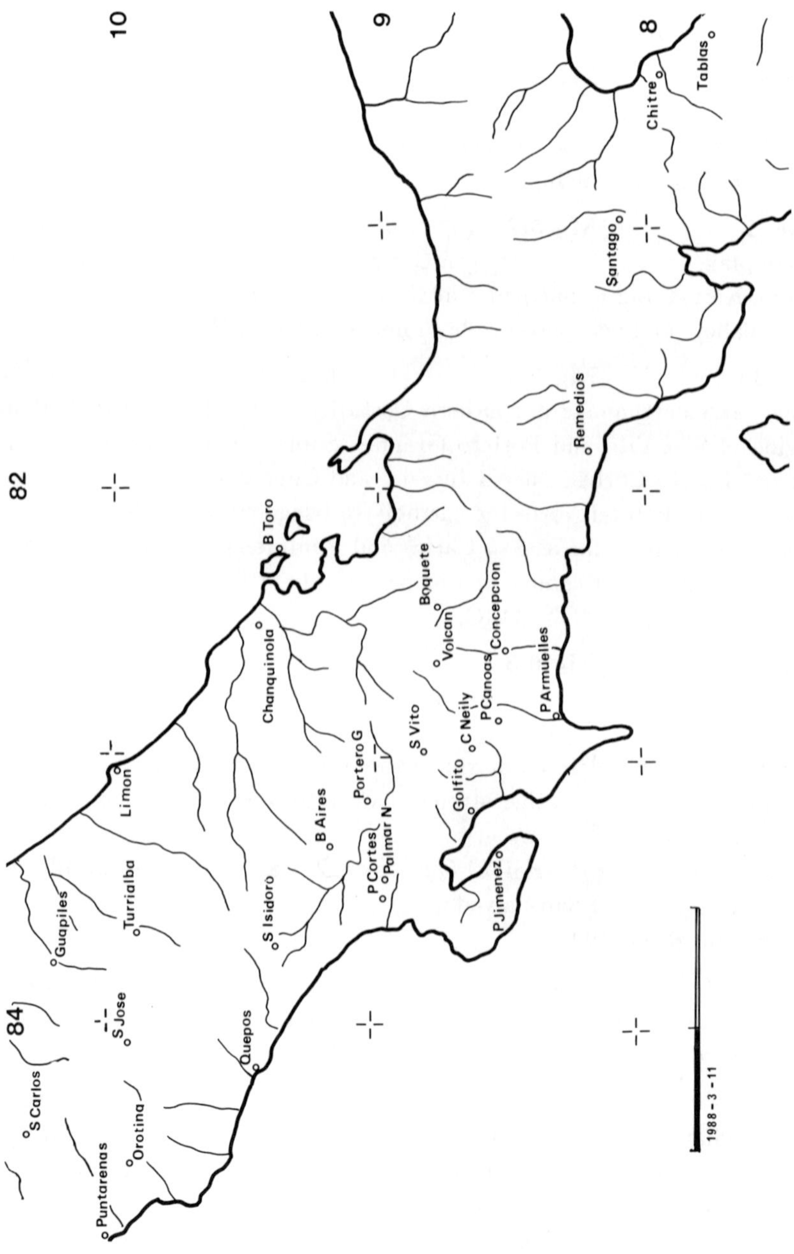

Fig. A.76 Felt effects of earthquake of 11 March 1988.

1989 Feb 26 M_S 4.8 Costa Rica

A relatively small earthquake in Costa Rica caused some slight damage and triggered a few landslides in the region of Cerro Caraigres at Guitil, San Ignacio, San Pablo and San Marcos.

The shock was felt along the coast from Herredura to Quepos and in parts of the Central Valley, and was perceptible at Limón.

[Barquero and Rojas 1994b]

1989 Apr 2 M_S 4.5 Costa Rica

An earthquake in the Gulf of Papagayo was strongly felt at Playas Coco and Hermosa.

The shock was felt throughout the District of Guanacaste and was perceptible at San José and Managua.

[Barquero and Royas 1994b]

1989 Oct 23 M_S 4.8 Nicaragua

This earthquake in Nicaragua was felt in Managua. In Costa Rica it was felt at Playa Hermosa and Peñas Blancas and was perceptible at San José. It was also perceptible in San Salvador.

1989 Dec 06 M_S 3.9 Mexico

This earthquake was felt slightly in the Mexico-Guatemala border region and in the State of Pueblo. It was also perceptible in Guatemala City and western Guatemala.

1990 Jan 12 M_S 4.4 Nicaragua

A subcrustal earthquake off the coast of Nicaragua was felt at Managua, and was perceptible at San Salvador.

1990 Feb 22 M_S 5.2 Nicaragua

An earthquake with a epicentre near the Pacific coast of Nicaragua was widely felt in the country, and was perceptible in the Liberia area in Costa Rica.

1990 Mar 25 13h 16m M_S 7.0 Costa Rica

This large earthquake in Costa Rica was followed seven minutes later by a second, somewhat larger, shock.

The shock was felt throughout Costa Rica, in western Panama and in the San Marcos area in Nicaragua, but its effects cannot be separated from those of the earthquake that followed.

[Montero *et al.* 1991a]

1990 Mar 25 13h 22m M_S 7.1 Tsunami Costa Rica

This was the main shock of Costa Rican sequence, with an epicentre in the Gulf of Nicoya.

It caused considerable damage at Cobano, Herredura, Jaco, Mata de Limon, Montezuma, Puntarenas, Tambor and around the southern part of the coast of the

Gulf of Nicoya, where a few old houses collapsed, hundreds were damaged beyond repair, and about ten people were injured.

Less serious damage was done over a large area, which extended from Nicoya to Dominical and from Quepos to San Rafael. About 50 buildings were damaged in the San José area, while at Alajuela a number of old buildings were badly cracked.

Several landslides blocked roads linking San José to the Atlantic Coast, and power supply was interrupted for a while.

Peak ground acceleration at epicentral distances of about 100 km reached 7%g.

A small seismic sea-wave was reported from Quepos and Tambor.

The shock was strong throughout Costa Rica causing some concern, and was felt in southwestern Nicaragua as far as San Marcos, and at Almirante, Puerto Armuelles and David in Panama.

[Barquero *et al.* 1971; Barquero and Rojas 1994b; Climente 1991a; Montero *et al.* 1971a,b; Montero and Boschini 1971; Wolge 1991]

1990 Apr 3 M_S 6.5 Nicaragua

A large earthquake off the coast of Nicaragua caused some slight damage in the District of Carazo at Masachapa, Montelimar and Rivas, and was strongly felt at Managua and Cuajinquil, and at Liberia in Costa Rica.

The earthquake was rather strong at Puntarenas, and was felt throughout much of Nicaragua, and in Costa Rica at San José and in the Central Valley.

It was perceptible at Limón and as far as San Salvador.

A strong aftershock on April 4 was widely felt and was perceptible as far as Atenes in Costa Rica.

[Barquero *et al.* 1991; Barquero and Rojas 1994b]

1990 Apr 28 M_S 6.4 Figure A.77 Costa Rica

A relatively large shock in southwestern Costa Rica caused only sporadic and minor damage in the region of Puerto Armuelles, Paso Canoas, Neily, Golfito, Puerto Cortéz and Puerto Jiménez.

The shock was rather strong at Bajo, David and San Isidoro and was felt at Atenes, Orotina, San José, Quedada, and as far as Barra del Colorado and Santiago in Panama.

It was followed by a large number of small aftershocks.

[Barquero *et al.* 1991; Barquero and Rojas 1994b]

1990 May 1 M_S 5.4 Guatemala

An earthquake located on the coast of Guatemala was widely felt in southern Guatemala and at Guatemala City.

1990 Jun 30 M_S 5.1 Costa Rica

This rather small earthquake in Costa Rica and its numerous foreshocks caused considerable damage in the region of Puriscal.

The earthquake and the re-activation of old slides which continued to move after the earthquake, ruined beyond repair 175 houses, damaged another 400, ruptured water mains and blocked several roads in the area between San Rafael, Santiago, Piedades, Desamperaditos and Tinamaste.

The shock was very widely felt, in the Central Valley, and was perceptible from Los Chiles in the north to Golfito in the south and from Nicoya to Limón.

[Barquero *et al.* 1991; Barquero and Rojas 1994b; Clemente 1991b; Wolge 1991]

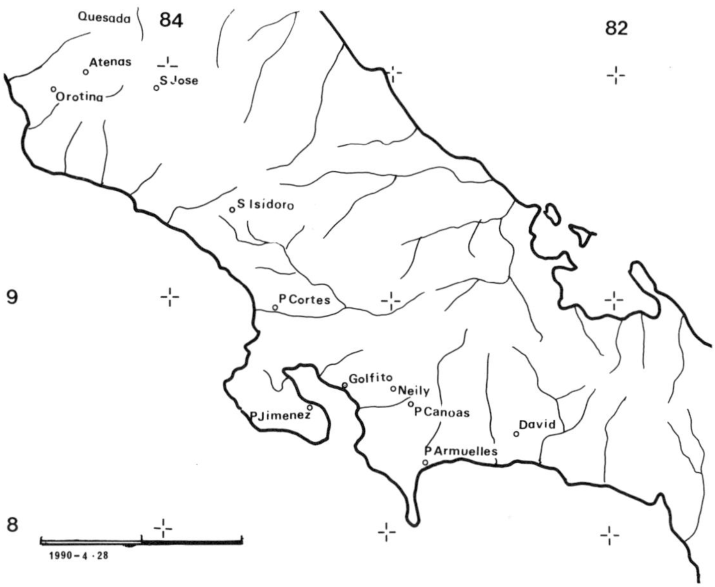

Fig. A.77 Felt effects of earthquake of 28 April 1990.

1990 Jul 13 **M_S 5.0** **El Salvador**

An earthquake in the Gulf of Fonseca caused some slight damage in the region of La Unión in El Salvador, where a few people were injured.

The shock was felt in southern Honduras.

1990 Jul 27 **M_S 4.9** **Honduras**

This small earthquake was located in the islands of La Bahia, and was felt in northern Honduras. Details are lacking.

1990 Dec 18 **m_b 5.0** **Honduras**

A small earthquake in Honduras was felt at Puerto Cortés and San Pedro Sula.

1990 Dec 22 **M_S 5.7** **Figure A.78** **Costa Rica**

This was a medium-magnitude earthquake in Puriscal, Costa Rica.

Much of the damage and the collapse of several buildings that injured 300 people, occurred within a small region between Petras Negras, Colón, Santiago and Barbacoas.

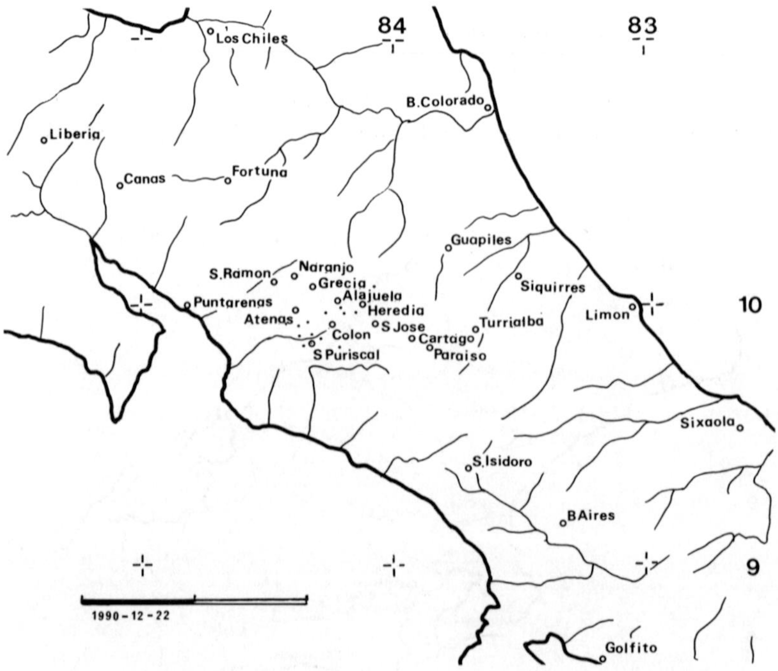

Fig. A.78 Felt effects of earthquake of 22 December 1990.

Less serious non-structural damage was done at Alajuela, Atenas, Balsa, Piedras Negras and Santiago. The peak ground acceleration recorded at Alajuela exceeded 4%g.

The shock was widely felt in Costa Rica, from Golfito to Los Chiles, and from Liberia to Sixaola.

A strong aftershock about one minute after the main shock was strongly felt at Alajuela, Atenas, Barbacoas and Pietras Negras.

[Barquero and Rojas 1994b; Montero *et al.* 1991c; Rojas and Barquero 1991; Wolge 1991]

1991 Jan 23 M_S 4.9 Nicaragua

A small earthquake in northern Nicaragua was felt in the Managua area, and was perceptible at Tegucigalpa in Honduras.

1991 Feb 16 M_S 3.5 Costa Rica

A very small earthquake in Costa Rica caused some damage to a few houses at Rio Segundo and San Joaquin de Flores.

At Alajuela the shock caused some panic, and was recorded with an acceleration of about 13%g.

1991 Mar 1 M_S 5.5 Costa Rica

An intermediate-depth earthquake in Costa Rica was perceptible in Guanacaste and in the Central Valley.

1991 Mar 16 M_S 6.3 Costa Rica

This earthquake was off the southwest coast of the Peninsula of Nicoya in Costa Rica.

Some damage was reported from Punta Guiones and Nosara.

The shock was strongly felt in the Provinces of Nicoya and Guanacaste in the region encompased by Culebra, Guayabo, Tilaran and Puntarenas, and as far as Limón and the Panama border.

[Barquero and Rojas 1994b]

1991 Apr 22 M_S 7.6 Tsunami Figure A.79 Costa Rica

This was a major earthquake near the Caribbean coast of Costa Rica and Panama.

In Costa Rica 4,450 houses were destroyed, 4,680 damaged beyond repair and 3,200 were damaged to a lesser extent, mainly in the alluvial plains between Limón and Guabito, and about another 1,000 were destroyed in the region between Guabito and Almirante in Panama.

At least 45 people were killed and over 100 were injured in Costa Rica; 28 were killed and 455 were injured in Panama.

Widespread liquefaction and spreading of the ground was responsible for much of the damage, particularly in the region of Matina, Moin, Limón, Cahuita, Sixaola, Changuinola and Almirante.

The most important losses and damage occurred to roads, railroads, bridges, ports and water pipes, particularly between Limón and Siquirres and Sixaola where 310 km of roads had to be rebuilt. In the region of Limón four bridges collapsed and six were ruined.

Slight damage extended as far as David, Puerto Armuelles, Quepos and Grecia.

The total damage was estimated at $188 million, or more than 4% of the gross national product of Costa Rica.

There is no evidence of coseismic surface faulting, but as a result of the earthquake the shoreline from Moin to Gandoca was uplifted by about 1 m.

The earthquake was associated with a seismic sea-wave which affected the whole coast from north of Limón to Panama, where it was perceptible. It was recorded at Cristobal. The maximum amplitude of the wave was about 2.0 m and in places it caused additional damage.

The shock was felt at Panama City, Colón, San Andrés Island, in southern Honduras, throughout Nicaragua, and was perceptible as far as eastern Panama and San Salvador.

Many aftershocks followed, some causing additional damage.

[Amador *et al.* 1994; Baraquero and Rojas 1994; Bermudez 1994; Boschini and Montero 1994; Camacho 1993, 1994; Camacho and Viquez 1994; Cortes *et al.* 1994;

Denyer *et al.* 1994a, 1994b, 1994c; Hafstrom and Skogsberg 1994; Jimenez and Cortes 1994; Montero *et al.* 1994a, 1994b; Mora and Mora 1994; Mora and Yasuda 1994; Morales 1994; Ponce *et al.* 1994; Ramirez and Laurent 1994; Sauter 1994; Shea 1991]

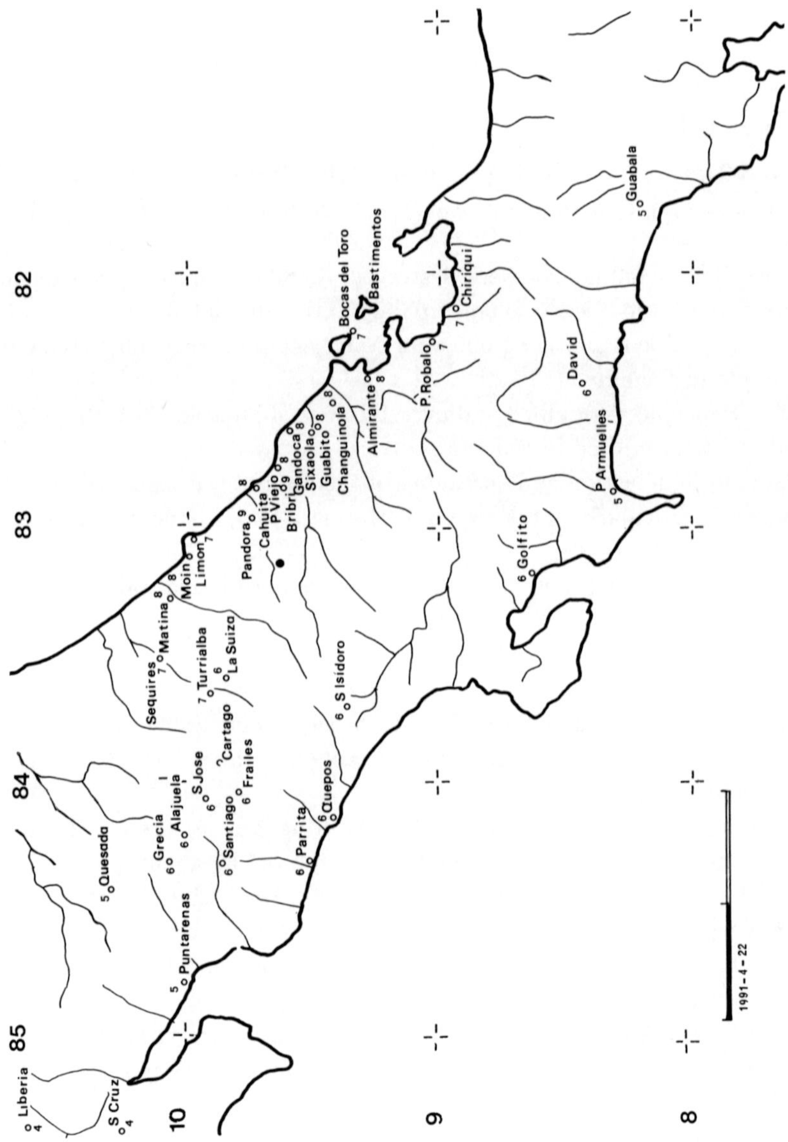

Fig. A.79 Felt effects of earthquake of 22 April 1991.

1991 Apr 23 **M_S 5.6** **Guatemala**

An intermediate-depth earthquake on the coast of Guatemala was felt at Puerto Madero, Tapachula, San Marcos and Quezaltenango, and was perceptible at San Salvador.

1991 Apr 24 **M_S 6.1** **Costa Rica**

This relatively large aftershock in central Costa Rica did some additional damage at Siberia, Hortansia and Herradura, and was very strong at San Isidoro.

The shock triggered a landslide, which blocked the Pan-American Highway, and was felt throughout almost all Costa Rica and in western Panama at Changuinola, and Chiriqui.

[Montero *et al.* 1994]

1991 May 4 **M_S 6.2** **Panama**

This was a damaging aftershock in Panama.

Much of the damage was done in the region between Changuinola, Almirante and Bocas del Toro, where two people were killed and 36 were injured. Hundreds of houses were damaged beyond repair and quite a few collapsed. The shock caused liquefaction and cracking of the ground.

The earthquake was strong at Limón and was felt from the Central Valley of Costa Rica as far as Santiago in Panama.

[Montero *et al.* 1994]

1991 May 15 **M_S 4.6** **Panama**

A small aftershock was rather strongly felt at Changuinola, Bocas del Toro, and Boquete. It caused no damage but it was associated with a small seismic sea-wave at Chiriqui Grande.

The shock was felt in David.

[Montero *et al.* 1994]

1991 May 27 **M_S 5.0** **Panama**

Another aftershock in Panama caused some slight damage in the region of Sixaola and Changuinola, where two people were injured.

The shock was rather strong on Bocas del Toro and Carenaro Islands, and was felt at Limón, Volcán, David, and in southeastern Costa Rica.

[Montero *et al.* 1994]

1991 Jun 21 **M_S 5.5** **El Salvador**

A subcrustal earthquake off the coast of El Salvador was rather strong in the San Salvador area, and was perceptible in Guatemala City.

1991 Aug 9 **M_S 4.7** **Costa Rica**

A small earthquake in central Costa Rica caused some considerable damage in the Cartago area.

At Corralillo, Frailes and La Lucha about 30 houses were damaged.

The earthquake was perceptible in much of Costa Rica and triggered small landslides.

[Barquero and Rojas 1994b]

1991 Sep 18 **M_S 6.2** **Figure A.80** **Guatemala**

A rather large earthquake in Guatemala caused extensive damage in the Pochuta-Solola area.

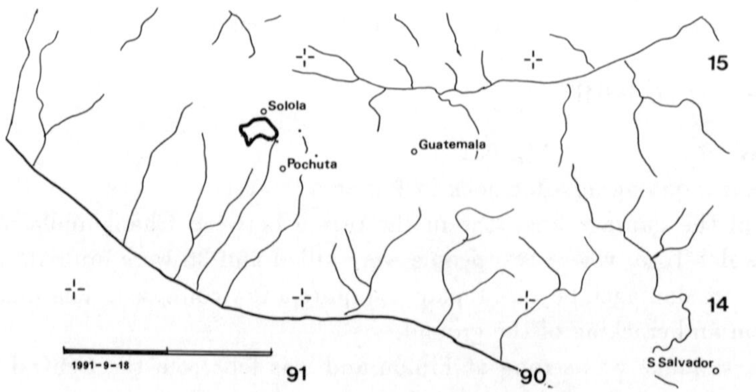

Fig. A.80 Felt effects of earthquake of 18 September 1991.

In Pochuta and in farms near the town 16 people were killed, more than 200 injured, and many hundreds of adobe houses were ruined. The shock caused ground deformation and triggered landslides in the epicentral region.

Minor damage was reported from Solola, Patzun, San Lucas and Acatenango.

The shock was strong in Guatemala City, was felt in the region of San Salvador, and was perceptible in the city.

[BSSA: 1992.1985; White and Harlow 1993]

1992 Mar 7 **M_S 5.9** **Costa Rica**

An intermediate-depth earthquake in Costa Rica caused sporadic and minor damage in the Central Valley from Puntarenas and San Ramón to Cartago and Turrialba.

The shock was felt throughout the country as well as at Changuinola and David in Panama.

[Barquero and Rojas 1994b]

1992 Mar 20 **M_S 4.5** **Guatemala**

A small subcrustal earthquake off the coast of Guatemala was perceptible at Quezaltenango, Suchitepequez, Chimaltenango, Escuintla, Chimaltenango, Guatemala City, in western El Salvador, and at San Salvador.

1992 May 18 M_S 6.1 Panama

An earthquake in the Gulf of Chiriqui, Panama was felt rather strongly at Puerto Armuelles, Corredor, and David. It was felt at Bocas del Toro and was perceptible in Panama City and in the Central Valley in Costa Rica.

1992 May 23 M_S 5.7 El Salvador

An intermediate-depth earthquake off the coast of El Salvador was felt throughout El Salvador and in large parts of southeastern Guatemala.

It was strongly felt at San Salvador.

1992 May 30 M_S 6.2 Mexico

A subcrustal earthquake off the coast of Chiapas was felt in the Mexico-Guatemala border area and along the coast of Guatemala. The shock was felt in Guatemala City and was perceptible at San Salvador.

1992 Sep 2 M_S 7.2 Tsunami Nicaragua

This was a major earthquake off the Pacific coast of Nicaragua.

Most of the damage and casualties were caused by a seismic sea-wave which affected the west coast of Nicaragua and Costa Rica, reaching heights of up to 8 m. Wave run-up to 1 km was reported at Nasachapa.

About 1,300 houses and 185 fishing boats were destroyed along a 250 km long strip of Nicaragua's west coast and at least 116 people were killed and more than 68 missing; over 13,500 were left homeless in Nicaragua.

The shock was strongly felt in the Departments of Chinandega and Leon, at Crucero, Managua and San Marcos. Some damage was also reported in Costa Rica.

Total damage in Nicaragua was estimated at about US $25 million.

The shock was felt at San José in Costa Rica.

Maximum seismic sea-wave heights were 1.11 m at Baltra Island, 0.83 m at Easter Island, 0.28 m at Socorro Island, 0.18 m at Liberia in Ecuador, 0.1 m at Valparaiso in Chile and 0.1 m at Hilo, Hawaii.

[Satake *et al.* 1993]

1993 Mar 6 M_S 5.9 Nicaragua

A subcrustal earthquake off the Pacific coast of Nicaragua was felt strongly at Corinto, and also at Chinandega, Leon, Managua and Granada. In El Salvador it was felt at La Unión, San Miguel and San Salvador.

[NEIC]

1993 Jul 8 M_S 4.6 Costa Rica

This small earthquake in Costa Rica was a foreshock of the earthquake of July 10.

It was very strong in the area of Turrialba, where, however, it caused no damage, and was felt in the Central Valley and San Isidoro.

[Barquero and Rojas 1944b]

1993 Jul 10 M_S 5.6 Costa Rica

This was a larger shock in Costa Rica.

One person was killed and 10 injured in the Turrialba area, where the shock did considerable damage.

Sporadic damage occurred within a radius of 20 km from the epicentre, where landslides blocked roads.

The shock was felt throughout Costa Rica.

[Barquero and Rojas 1994b]

1993 Sep 10 M_S 7.3 Mexico

We have not attempted to study the macroseismic effects of this event, which was located off the coast of the Mexico-Guatemala border.

The earthquake did considerable damage in southwestern Guatemala, killed a few people, blocked roads and triggered landslides. Damage extended into the District of Chiapas. The shock was felt throughout southern Mexico and as far as Mexico City.

[P]

1993 Sep 12 M_S 5.5 Guatemala

This subcrustal shock was felt in the Mexico-Guatemala border region and destroyed two old houses at Mejicanos in El Salvador. It was also felt in San Salvador.

[NEIC, TAC]

1994 Mar 14 M_S 6.2 Mexico

Slight damage was caused at Tuxtla Guitierrez in Mexico by this intermediate-depth earthquake.

It was also felt at Cobán, Guatemala City, Quezaltenango and San Marcos in Guatemala, and at Mexico City, and was perceptible at San Salvador.

[NEIC]

1995 Feb 11 M_S 5.3 Nicaragua

This shallow earthquake was centred off the Caribbean coast of Nicaragua. It caused some damage to tall buildings and power outages on Isla de San Andrés and was felt on Islas de Maiz.

[NEIC]

1995 Dec 19 M_S 4.8 Guatemala

A rather small earthquake in central Guatemala caused landslides which resulted in one death and one person being injured at Tactic. Several houses were damaged at San Miguel Tucuru, and the shock was also reported felt at Cobán and Guatemala City.

[NEIC]

Appendix B
Parametric Catalogue of Earthquakes

B.1 Notes on Parametric Table

The catalogue lists all selected earthquakes in our region from the beginning of instrumental recording in 1898 to the end of 1995. In general the focal parameters are given with a precision that reflects the quality of their determination. These notes should be read carefully before consulting the catalogue.

B.1.1 *Origin*

Following the date, the origin time of each event is given in Universal Time to an appropriate precision, in hours, minutes and seconds. For some early events, particularly those determined from felt information only, the time is given only to the nearest minute, hour or even day. For many recent earthquakes agencies report origin times to a decimal of a second. These have been rounded. An asterisk following the date signifies that the event is a foreshock or aftershock.

The epicentre is given in degrees of north latitude and west longitude. Most epicentres are given with a precision of 0.1°, but for better determined events a precision of 0.01° is given. In the middle and early periods this is usually reserved for epicentres that are exceptionally well determined by closely spaced isoseismals. At the other extreme, some poorly determined epicentres are listed only to the nearest degree. No figures are given that are not considered significant. A consequence of this is that positions given in original sources as $\frac{1}{4}^{\circ}$ or $\frac{3}{4}^{\circ}$ are rounded to .3° or .8° respectively.

Depths given as close to the standard default value of 33 km, or for which there is no specific information, are listed as *n*. If there is definite indication of upper crustal or shallower depth, the designation is *s*. Unspecified or poorly determined subcrustal and intermediate depths are given as *n+* and *i*, respectively; *s* usually signifies a depth less than 16 km, *n+* a depth between 45 and 60 km, and *i* one of more than 60 km. Standard default depths (such as .01R or 96 km used by ISS) are rounded to the nearest 5 or 10 km. Depths found by modern computer relocations to be greater than normal may be given to the nearest kilometre, as determined.

B.1.2 *Agency*

When origin determinations are available from more than one agency, the choice is generally made in the following order of preference:

1. **AM.** Well determined macroseismic location carried out by us
2. **AR** Good instrumental relocation by us
3. **Am** Less well determined macroseismic location by us
4. **Ar** Relocated by us, using combined instrumental and felt data, or poor instrumental data
5. Relocations made in special studies:
   - **AC** Acharya, 1965
   - **C** Camacho, 1991
   - **D** Dewey and Algermissen, 1974
   - **E** Engdahl, van der Hilst and Buland, 1998
   - **G** Gutenberg, 1958; Gutenberg and Richter, 1956
   - **GD** Güendel, 1986
   - **K** Kelleher, Sykes and Oliver, 1973
   - **M** Molnar and Sykes, 1969
   - **MR** Martinez, 1978
   - **S** Sykes and Ewing, 1965
   - **W** Wysession, Okal and Miller, 1991
6. Locations made by national and international agencies:
   - **B** BCIS/Strasbourg
   - **I** BAAS/ISS/ISC
   - **J** Jesuit Seismological Association
   - **OT** Ottawa, Canada
   - **SJ** San José, Costa Rica
   - **T** Tacubaya, Mexico
   - **U** USCGS/USGS/NEIC

B.1.3 *Size*

Four columns give information relating to the earthquake's size.

M_S Unbracketed values give the surface-wave magnitude determined by us, using the Prague formula as described in the text. The value is followed by the number of station readings used in the determination. Values given in brackets are determinations made by other agencies.

m_B/m_b In the early part of the catalogue this is generally the body-wave magnitude m_B, from long- or medium-period instruments, determined by us and followed by the number of stations used. When the value given is bracketed we have adopted it from another agency. For the later period most of the magnitudes from other agencies, and particularly those determined by NEIC, will be m_b from short-period instruments.

$M^*/logM_O$ For the early period up to 1917, this column gives an equivalent M_S determined from Milne seismographs, followed by the number of stations used. For the period after 1976, we list the logarithm of the seismic moment in newton·metres, usually taken from publications of Harvard University or bulletins of the US Geological Survey.

N This column lists the number of station observations reported to the main agency. Not all these are necessarily used in location. If no other information on size is available, this number may be used to give a rough estimate of comparative size of events in the same time period.

B.1.4 *Other Information*

In the *Felt* column roman numerals give the highest reported MSK intensity, where this is *V* or more. *F* denotes felt reports of unspecified intensity, or a reported intensity of *IV* or less. *D* denotes damage reported, but no specified intensity. Two-letter abbreviations in the column *Reg* indicate the region by giving the country closest to the epicentre. These are for a guide only, and minor inconsistencies may occur when earthquakes are in a border region or offshore.

BZ	Belize	CR	Costa Rica
ES	El Salvador	GU	Guatemala
HO	Honduras	MX	Mexico
NI	Nicaragua	PA	Panama

We list other agencies whose solutions we have considered, but not necessarily all other agencies that have published solutions. *Notes* signifies an entry in the detailed Case Histories in Appendix A, *ts* the report of a tsunami, and *volc* an association with volcanism. The final column gives the felt radii in kilometres where these can be determined for specific MSK intensities.

B.2 Parametric Table

Date	Time (UT) h.m.s.	Epicentre °N	°W	Depth km	Ag	M_S	$m_B/$ m_b	M^* $log M_O$	N	Felt	Reg	Other Agencies, Comments	Felt Radii
1898 Apr 29	1618	12.6	87.1	n	AM			6.8/3	>8	VIII	NI	G, Notes	3/400, 5/150, 7/35, 8/15
1899 Mar 25	1427	16.8	92.8	n	Am			6.2/3	>6	D	MX	I, Notes	
1899 Mar 25	17	13.6	88.8	s	AM			<5.3	2	VII	ES	Notes	4/14, 7/5
1899 Jun 10	0728	14.	87.	n	Am					D	HO	Notes	
1900 Jun 21	2058	10.6	85.5	n	Am			7.2/9	>20	VI	CR	G, Notes	4/200, 5/140, 6/80
1900 Nov 09	1608	13.	90.	n	G			6.9/10			ES	Notes	
1901 Oct 08	0216	11.0	86.5	n	Ar			7.0/12		V	NI	G, Ts, Notes	3/260, 5/90
1902 Jan 18	2323	14.7	91.6	n	AM			6.3/12		VII	GU	I, Notes	3/140, 6/35, 7/15
1902 Feb 27	0036										ES	Ts, Notes, non-seismic?	
1902 Mar 24	1755	12.4	86.6	n	AM			6.6/8		VI	NI	I, Ts, Notes	3/300, 5/110, 6/45
1902 Mar 25*	0322	12.4	86.7	n	Am			6.3/4		D	NI	I, Notes	
1902 Apr 19	0224	14.9	91.5	n	AM			7.5/22		IX	GU	G, Notes	3/580, 5/200, 7/90, 8/50, 9/25
1902 Sep 23	201815	16.5	92.5	n	AM			7.6/26		IX	MX	G, Notes	3/530, 4/380, 5/260, 6/190, 8/41, 9/22
1902 Oct 24	0418	14.7	91.6	s	Am			5.2/2		F	GU	Notes	
1903 Jan 14	0148	15.0	95.0	n	Ar	7.8/1	7.6/1	7.6/22		F	MX	G, Notes	3/600
1903 Feb 28	0945	10.5	84.5	n	AR	5.8/1		5.9/12		F	CR	I, Notes	
1904 Jan 20	145020	7.0	82.0	n	Ar	7.9/2	7.1/3	7.0/16		F	PA	G, Notes	
1904 Dec 02	021935	7.0	82.0	n	Ar	6.9/3		6.7/12	29	F	PA	I, Notes	
1904 Dec 20	0542	9.2	82.8	n	Am	7.5/3	7.7/4	7.3/25		VII	PA	G, Notes	2/860, 4/360,5/250 6/150, 7/80
1905 Jan 20	1808	10.0	84.3	n	AM	6.5/3	6.3/8			VII	CR	Notes	4/160, 7/20
1905 Dec 28	0035	9.9	84.0	s	Am					VI	CR	Notes	6/8
1906 Jan 03		11.97	86.17	n	Am					F	NI	Notes, volc?	
1906 Jun 20	0225	13.9	89.3	n	AM	6.1/2	6.1/3	6.0/9	50	VII	ES	I, Notes	3/190, 5/44, 7/17
1906 Jun 22	0319	15.5	90.5	n	Am	5.3/2	6.5/3	5.7/3		IV	GU	I, Notes	4/60

1907 May 31	0600	15.42	90.42	n	Am					VII	GU	Notes	3/80, 7/8
1907 Jul 01	130913	13.0	87.5	n+	AR	7.4/6	6.6/7	6.9/16	>54	F	HO	I, Notes	
1907 Sep 23	2135	14.8	90.5	n	Am	6.5/7	6.2/1	6.7/9	>33	VII	GU	B, Notes	5/70, 7/21
1907 Oct 06	0048	10.0	84.0	n	Ar	5.8/5	5.9/2	5.8/3	>16	F	CR	B, Notes	
1907 Dec 30	052652	11.1	87.3	n	AR	7.4/8	6.7/4	7.1/20	60	F	NI	I, Notes	
1908 Feb 01	2312	10.5	83.5	n	Ar	6.6/6	6.5/1	6.6/12		F	CR	I, Notes	
1908 May 19	0145	14.68	88.86	n	Am					VII	HO	Notes	7/10
1908 Jul 19	0742	14.9	92.0	n	Am					D	GU	Notes	
1909 Aug 16	0655	11.5	85.5	n	Am	6.9/11	6.4/5	6.8/20		F	NI	I, Notes	
1909 Aug 30	1318	9.0	81.0	n	AR	6.1/6	6.2/3	6.2/7		F	PA	Notes	
1910 Jan 01	1102	17.0	85.0	n	Am	6.9/10	6.9/10	7.0/23		F	HO	G, Notes	
1910 Mar 10	2109	9.6	84.0	n	Am					IV	CR	Notes	4/30
1910 Apr 13*	0613	9.8	84.0	s	Am				>2	F	CR		
1910 Apr 13	0640	9.83	84.03	n	AM	5.6/3		5.7/5	>10	VII	CR	I, Notes	3/140, 4/75, 5/30 6/15, 7/6
1910 Apr 14*	1406	9.8	84.0	s	Am				>3	F	CR		
1910 May 05	0026	9.84	83.91	s	AM	6.1/9	6.5/1	6.1/11	>17	IX	CR	I, Notes	3/220, 5/75, 7/20, 8/14, 9/5
1910 May 14	1404	9.7	83.9	n	AM					F	CR	Notes	4/30
1910 Dec 21	1024	9.0	82.0	n	Am	6.2/3		6.2/11	>20	VI	PA	Notes	3/250, 6/40
1911 Jul 11	0230	9.75	84.75	n	Am				2	F	CR	Notes	3/90
1911 Aug 25	1216	10.3	85.3	n+	Am				2	F	CR	Notes	3/150
1911 Aug 29	0343	10.25	84.27	s	AM	5.8/5		5.9/8	>15	IX	CR	Notes	5/50, 7/21, 8/9, 9/4
1911 Oct 10	1312	10.6	84.9	n	AM	6.5/17	6.3/3	6.8/21	>36	VII	CR	Notes	5/100, 7/33
1911 Nov 01	0925	11.	87.	n	Ar	6.7/12		6.7/19	>33	F	NI	Notes	
1912 Feb 21	0820	9.95	83.97	s	AM				>2	VII	CR	Notes	3/90, 5/18, 7/4
1912 Jun 06	0612	10.14	84.30	s	AM	5.1/1			>2	VII	CR	Notes	3/93, 5/30, 7/10
1912 Jun 12	124442	16.5	88.0	n	AR	6.8/18	6.6/9	6.7/22	>42		BZ	G, Notes	
1912 Jul 18	2358	13.83	89.58	s	Am					D	ES	Notes	4/50, 6/15
1912 Dec 09	083230	14.5	95.0	n	Ar	7.2/9	6.8/5	7.1/23	19	F	MX	G, Notes	
1913 Mar 08	16	14.26	90.27	s	AM	6.4/14	6.9/1	6.5/18	>42	VIII	GU	Notes	5/100, 8/15
1913 Mar 13		14.3	88.2	n	Am					D	HO	Notes	
1913 Jul 25	123806	10.0	82.5	n	AR	6.1/12		6.2/15	>42	V	CR	G, Notes	3/300, 5/100

Date	Time (UT) h.m.s.	Epicentre °N	°W	Depth km	Ag	M_S	m_B/ m_b	M^* $log M_O$	N	Felt	Reg	Other Agencies, Comments	Felt Radii
1913 Oct 02	042328	7.1	80.6	s	AM	6.7/10	6.8/1	6.4/16	>47	VII	PA	I, Notes	4/240, 6/90, 7/45
1913 Oct 04*	2206	7.5	80.0	s	Am	6.1/4		6.1/7		F	PA		
1913 Oct 23*	150032	7.5	80.0	s	Am	6.0/3		6.1/11	>25	F	PA	I, Notes	
1914 May 28	032404	11.1	78.0	n+	Ar	6.4/10	6.9/5		>47	D	PA	G, Notes	
1915 Sep 07	012052	13.92	89.65	n+	Ar	7.7/23	7.5/11	7.8/1	>55	VIII	ES	G, Notes	4/560,5/350.7/90, 8/25
1915 Sep 07*	0318	13.8	89.7	n	Am	6.4/3				F	ES		
1915 Sep 07*	0423	13.8	89.7	n	Am	6.6/1				F	ES		
1915 Sep 07*	0458	13.8	89.7	n	Am	6.0/3				F	ES		
1915 Sep 07*	0531	13.8	89.7	n	Am	5.3/1				F	ES		
1915 Sep 07*	1247	13.8	89.7	n	Am	5.9/3				F	ES		
1915 Sep 07*	2039	13.8	89.7	n	Am	5.6/2				F	ES		
1915 Sep 08*	1256	13.8	89.7	n	Am	5.2/2				F	ES		
1915 Nov 26	1912	8.0	80.0	n	Ar	5.8/2	6.3/2		>5	F	PA	Notes	
1915 Dec 01	16	14.66	90.45	n	Am					D	GU	Notes	
1915 Dec 18	12	14.83	91.51	n	Am					F	GU		
1915 Dec 20		14.5	90.7	n	AM					V	GU	Notes	5/25
1915 Dec 26*	0911	14.6	88.6	s	AM	5.7/2				F	HO		
1915 Dec 27*	0409	14.6	88.6	s	Am	5.5/1				F	HO		
1915 Dec 28	2354	14.54	88.60	s	AM	6.4/8	6.7/1	6.6/2	>10	VIII	HO	Notes	4/180, 5/100, 8/16
1916 Feb 08	1542	7.0	80.0	n	Am	5.5/1				F	PA	Notes	
1916 Feb 27	202123	10.8	86.0	n+	Ar	7.3/18	7.4/8		>52	VII	CR	G, Notes	4/330, 7/70
1916 Mar 12	0731	14.9	90.9	n	Am	6.3/4		6.5/1	23	F	GU	OT, Notes	
1916 Mar 29	1857	8.3	80.0	n	Ar	5.7/2				F	PA	Notes	
1916 Apr 24	080216	10.7	86.0	n+	AR	7.3/20	7.1/11		>46	D	CR	G, Notes	2/735, 4/340
1916 Apr 24*	0831	9.5	82.0	n	Am	6.6/1			>5	F	PA	Notes	
1916 Apr 26	022128	9.2	83.1	n	AM	6.9/16	6.8/7		>37	VII	CR	G, Ts, Notes	3/400, 4/260, 5/150, 7/45
1916 Apr 26*	062515	9.6	83.1	n	Ar	6.0/3	6.3/1		>5	F	CR		

1916 Apr 26*	071556	9.6	83.1	n	Ar	6.0/4	6.5/1		>5	F	CR		
1916 May 01*	0110	9.2	83.5	n	Am					F	CR	Notes	
1916 May 10*	213614	9.6	83.1	n	Ar	6.1/5	6.4/2		>5	F	CR	Notes	2/380
1916 Jul 28*	173732	9.6	83.1	n	Ar	6.1/3			>5	F	CR		
1916 Sep 23	054301	11.9	86.0	n	AR	6.7/6	6.9/2		>30	F	NI	I, Notes	
1917 Jun 08	005151	13.82	89.31	n	AM	6.7/14	6.3/2	6.8/3	>34	VIII	ES	I, Notes	7/36, 8/20
1917 Jun 08*	0132	13.73	89.29	n	AM					D	ES	Notes	
1917 Jun 08*	0254	13.77	89.50	n	Am	5.4/1			>4	F	ES	Notes	
1917 Jun 13	1730	14.6	91.4	n	Am					F	GU	Notes	
1917 Jun 27	122635	9.0	82.0	n	AR	6.1/3			>18		PA	I	
1917 Jun 30	175038	7.2	77.8	n	AR	6.1/4	6.3/1		>25	F	PA	I, Notes	3/270
1917 Oct 22	072012	13.	83.	n	I	5.9/3			26		NI	Notes	
1917 Nov 18*	0417	14.5	90.6	n	AM					VIII	GU	Notes	6/15, 8/4
1917 Dec 12*		14.5	90.6	n	Am					VII	GU	Notes	5/15, 7/4
1917 Dec 26*	0430	14.6	90.6	s	AM	5.6/3			>7	D	GU	Notes	
1917 Dec 26	052100	14.63	90.67	s	AM	5.6/5			>11	VII	GU	I, Notes	4/75, 5/40, 7/14
1917 Dec 26*	061810	14.6	90.6	s	Am	5.1/3			>5	F	GU	I, Notes	
1917 Dec 29*	2013	14.6	90.6	s	Am	5.2/2			>4	D	GU	Notes	
1918 Jan 04*	043005	14.6	90.6	s	Am				16	D	GU	I, Notes	
1918 Jan 04*	043241	14.56	90.55	s	AM	6.0/6	6.5/1		21	VII	GU	I, Notes	4/130, 7/30
1918 Jan 25	012030	14.6	90.5	s	AM	6.2/4	6.7/1		39	D	GU	I, Notes	
1918 Feb 01	0130	13.9	89.8	n	Am					F	ES	Notes, doubtful	
1918 Feb 22		14.5	91.1	n	Am					F	GU	Notes	
1918 Apr 05		15.1	91.6	n	Am					F	GU	Notes	
1918 Apr 20	02	14.1	90.0	n	Am					D	GU	Notes	4/50
1918 Apr 27	144345	8.7	83.0	n	I	5.3/2			19		CR	Notes	
1918 Jun 13	085858	15.3	86.7	n	AR	5.5/3			18		HO	Notes	
1918 Jun 16	122746	12.5	86.0	n+	Am	5.6/3	6.1/1		21	D	NI	I, Notes	
1918 Jun 22	220530	9.5	84.0	n	I	5.7/1			19		CR	Notes	
1918 Jul 31	143643	11.	88.	n	I	6.2/6			51		NI	Notes	
1918 Oct 19	032245	15.0	91.0	s	Am	6.2/8			46	D	GU	I, Notes	
1918 Oct 29	122600	8.	84.	n	I	6.0/1			16		CR	Notes	
1918 Nov 05	223900	13.27	88.12	n	W	5.6/2			12		ES	I	

Date	Time (UT) h.m.s.	Epicentre °N	Epicentre °W	Depth km	Ag	M_S	m_B/ m_b	M^* $logM_O$	N	Felt	Reg	Other Agencies, Comments	Felt Radii
1918 Nov 16	055630	15.0	90.9	n	Am	5.3/2			8	D	GU	I, W, Notes	
1919 Apr 17	205303	14.5	91.5	n+	Am	7.0/9	7.1/2		66	VI	GU	G, Notes	6/95
1919 Apr 27	2218	12.0	86.5	n+	Am				2	F	NI	Notes	
1919 Apr 28	064605	13.69	89.19	s	AM	5.9/3			23	VIII	ES	I, Notes	4/110, 6/37, 7/14, 8/8
1919 Jun 29	231422	13.5	87.5	n	Am	6.7/8			57	D	HO	G, Notes	
1919 Jul 01	213025	14.5	91.0	n	I	5.4/1			6		GU		
1919 Jul 06	070410	14.5	91.0	n	I	5.9/5			27		GU		
1919 Jul 17	161934	11.0	88.0	n	I	5.9/4			23		NI		
1919 Jul 22	220135	14.	90.	n+	Am	5.6/3	6.1/1		25	F	GU	G, Notes	
1919 Sep 27	033350	7.0	82.5	n	I	5.8/2			12		PA		
1919 Oct 14	165540	14.5	90.7	n	Am	5.7/2			10	F	GU	I, Notes, uncer	
1919 Oct 28	072320	13.0	83.0	n	I	5.7/1			7		NI		
1919 Dec 05	001526	13.0	85.4	n	I	5.8/2			14		NI		
1920 Mar 23	152148	14.5	91.0	n	I	6.1/5			32		GU		
1920 Jul 16	171415	10.5	85.0	n	Am	5.8/3			23	F	CR	G, Notes	
1920 Oct 08	165045	15.4	93.5	n	Am	6.1/6	6.7/1		28	F	MX	I, Notes	
1920 Oct 28	0917	13.6	89.2	n	Am					F	ES	Notes	
1920 Nov 06	104430	14.5	89.5	n	Am	5.8/3			13	F	GU	I, Notes	
1920 Dec 11	212239	14.6	90.1	n	Ar	6.3/5			32	F	GU	I, Notes	
1921 Jan 20	020224	7.0	82.5	n	I	5.8/2			18		PA		
1921 Feb 04	082244	15.0	91.0	120	G	7.2/11	7.3/3		72		GU	I, Notes	
1921 Feb 11	223936	9.5	84.0	n	I	5.4/1			9		CR		
1921 Mar 12	103024	7.0	82.5	n	I	6.0/2			21		PA		
1921 Mar 28	074939	12.9	86.8	170	AR	7.4/10	7.2/2		87	VI	NI	G, Notes	6/200
1921 May 14										F	ES	Notes	
1921 Oct 25	154525	7.0	82.5	n	I	5.8/2			11		PA		
1921 Nov 13										F	NI	Notes	
1922 Feb 16	031448	11.7	85.7	n	AM	6.4/10			53	VI	NI	I, Notes	6/50, 5/110

1922 Apr 02	170221	15.1	91.0	s	W	5.6/3		17		GU	I	
1922 Apr 20	054818	15.5	91.0	n	Am	5.8/2		16	F	GU	I, Notes	
1922 Aug 18	195026	13.0	85.4	n	I	5.6/2		9		NI	I	
1922 Dec 08	080840	10.0	83.9	n	Am	5.1/1		8	D	CR	I, Notes	
1923 Jan 10									F	ES	Notes	
1923 Feb 02									D	NI	Notes, volc.?	
1923 Feb 17									F	ES	Notes	
1923 Jul 14									F	NI	Notes	
1923 Nov 29		13.1	86.0	n	Am				D	NI	Notes	
1923 Dec 26	0756	14.5	90.7	n	Am	4.9/1			D	GU	Notes	
1924 Mar 04*	020612	9.7	85.0	n	Am	5.3/1		6	F	CR	I	
1924 Mar 04	100742	9.84	84.67	s	AM	7.0/16	6.6/3	85	IX	CR	G, Notes	3/350, 6/95, 7/55, 8/35, 9/15
1924 Mar 04*	114336	9.8	84.7	n	Am	6.5/6		32	D	CR	I,	
1924 Mar 11*	104108	9.9	84.6	n	Am	6.3/11		50	D	CR	I, Notes	
1924 Mar 11*	203406	9.8	84.5	n	Am	5.7/2		19	F	CR	I	
1924 Mar 12*	025020	9.8	84.5	n	Am			16	F	CR	I	
1924 Mar 20*	095612	9.7	84.6	n	Am			12	F	CR	I	
1924 Mar 24*	114015	9.8	84.5	n	Am			7	F	CR	I	
1924 Mar 24*	202900	9.8	84.5	n	Am	6.0/3		39	F	CR	I	
1924 Mar 25*	140020	9.6	84.3	110	AR	6.2/7		39	D	CR	I, Notes	3/280, 4/160
1924 Mar 27*	082945	9.8	84.6	n	Am			14	F	CR	I	
1924 Mar 28*	045700	9.8	84.6	n	Am			4	F	CR	I	
1924 Apr 23	1230	10.1	84.3	n	Am				F	CR	Notes	4/40
1924 May 01	195424	12.7	89.4	n	AR	6.6/14	6.6/1	80		ES	I, Notes	
1924 May 21	101250	14.5	88.7	n	I	5.3/2		18		HO		
1924 Jun 04	160930	16.1	93.8	n	Am	5.8/7		26	F	MX		
1924 Jul 12									F	ES	Notes	
1924 Oct 10	21	14.42	90.48	s	AM	4.9/1		4	V	GU	Notes	5/25
1924 Nov 01	045515	11.5	86.5	n	Am	6.0/3		25	V	NI	I, Notes	5/80
1925 Jan 15		14.0	91.0	n	Am				F	ES	Notes	
1925 Jan 18	131142	12.0	91.0	n	I	5.6/1		5		GU		
1925 Jan 26	190206	8.0	83.0	n+	Ar	6.2/13		55	F	PA	I	

Date	Time (UT) h.m.s.	Epicentre °N	°W	Depth km	Ag	M_S	m_B/ m_b	M^* $logM_O$	N	Felt	Reg	Other Agencies, Felt Radii Comments
1925 Jan 28*	105830	8.0	83.0	n	Ar	5.9/5			26	F	PA	I
1925 Feb 09	055330	14.9	92.1	n	Am	5.1/3			9	F	MX	I
1925 Feb 10	032220	17.0	95.0	n	Am	5.6/1			14	F	MX	I
1925 Feb 10		13.5	87.5	n	Am					F	ES	Notes
1925 Apr 04										F	ES	Notes
1925 Apr 15		11.5	85.5	s	Am					F	NI	Notes, volc.?
1925 May 26	082024	16.5	94.4	n+	Am	5.7/9	6.2/1		28	F	MX	I
1925 Sep 04	103600	16.0	91.5	n	Am	5.2/6			11	F	GU	I
1925 Sep 30										D	NI	Notes
1925 Oct 05	040902	12.3	85.2	135	Am	6.4/17	6.8/5		80	F	NI	G, Notes
1925 Oct 05	041055	17.5	83.0	n	I				7		HO	
1925 Nov 28	123325	16.8	92.1	n	Am	5.5/7			22	F	MX	I, Notes
1925 Dec 10	141442	15.5	94.5	n	Ar	7.1/14	6.8/2		92	F	MX	I, Notes
1925 Dec 10*	160650	15.5	92.5	n	I				8		MX	
1925 Dec 10*	203420	15.5	92.5	n	I				13		MX	
1925 Dec 11*	012748	15.5	92.5	n	I	5.9/3	5.9/1		27		MX	
1925 Dec 18		11.5	85.5	s	Am					F	NI	Notes, volc.?
1926 Feb 08	151807	12.9	88.7	100	AR	7.1/25	7.1/9		108		ES	G, I, Notes
1926 Feb 10	144850	13.0	87.0	n	Ar	5.7/6			24		NI	I, Notes
1926 Feb 15	025948	11.8	89.5	n+	Ar	6.6/22	7.1/5		94		NI	G
1926 Mar 17	115336	12.5	82.5	50	G	6.9/22	6.8/3		94		NI	
1926 Mar 24	105442	15.7	92.2	n	Am	5.9/7			12	F	MX	I, Notes
1926 Apr 12	12	12.0	86.0	n	Am					F	NI	Notes
1926 May 26	175330	14.5	88.7	n	I	5.3/6	6.1/1		20		HO	
1926 Jun 29	185621	16.92	93.26	s	W	5.4/6			21	F	MX	I, Notes
1926 Jul 21	022300	14.5	88.7	n	I	5.1/1			6		HO	
1926 Oct 19	204830	10.0	83.5	n	Am	5.8/10	5.6/1		39	F	CR	I, Notes
1926 Nov 05	075538	12.3	85.8	135	G	7.0/29	7.1/10		104	D	NI	Notes
1927 Feb 24	041352	14.5	91.0	n	I	5.7/8			24		GU	

1927 Mar 09	161315	9.5	84.0	n	I	6.0/11		41		CR	
1927 Mar 15	103818	8.0	82.0	n	U	5.1/1	5.7/1	9		PA	I
1927 May 09	200540	16.7	93.7	n	Am	6.1/19	6.2/1	57	D	MX	I, Notes
1927 Jun 22		13.7	88.3	n	Am				D	ES	Notes
1927 Aug 10	013522	7.3	81.3	n+	Am	6.7/27	6.8/5	88	F	PA	I, Notes
1927 Aug 22	025133	9.5	84.0	n	I	5.3/3		13		CR	
1927 Oct 02	044745	14.0	88.0	n	I	6.0/15	6.6/2	48		HO	
1927 Oct 02*	092915	14.0	88.0	n	I	5.1/5		19		HO	
1927 Dec 11	205702	8.0	80.5	n	Am			1	F	PA	Notes
1927 Dec 27	203145	16.5	89.5	n	I	4.9/3		6		GU	
1928 Mar 27	050928	16.2	89.2	n	Am	5.4/9		29	F	GU	I, Notes
1928 May 21	021608	13.5	92.5	n	I	5.6/1		10	F	GU	Notes
1928 Jul 01	092248	12.0	86.0	n	Am	5.5/3		26	D	NI	I, Notes
1928 Jul 30	024352	11.54	87.62	s	W	5.9/13		35		NI	I
1928 Aug 29	073137	16.6	90.9	n	T			5		MX	
1928 Oct 25	123248	12.5	85.5	s	Am	6.7/25	6.9/2	87	D	NI	I, Notes
1929 Jan 17	2240	16.0	88.5	n	Am	5.1/1			D	GU	Notes
1929 Jan 19	031754	9.5	82.0	n	Am	5.9/9		35	F	PA	G, Notes
1929 Jan 24	203628	12.8	91.0	n	I	7.1/22	6.9/5	114	F	GU	Notes
1929 Jan 24*	232900	12.8	91.0	n	I	5.5/5		13		GU	
1929 Jan 25*	012812	12.8	91.0	n	I	5.5/5		27		GU	
1929 Jan 26*	023040	12.8	91.0	n	I	5.5/5		19		GU	
1929 Jan 28*	215642	12.8	91.0	n	I	5.3/5		25		GU	
1929 Jan 31*	180517	12.8	91.0	n	I	6.1/11		42		GU	
1929 Feb 03*	180040	12.8	91.0	n	I	5.6/9		14		GU	
1929 Feb 04	101934	14.0	88.0	n	I	5.5/4		24		HO	
1929 Feb 10	033704	12.3	93.8	n	I	5.4/4		23		MX	
1929 Feb 10	153904	14.6	91.0	n	Am	6.6/21		80	F	GU	Notes
1929 Feb 13	221320	13.9	91.2	n	I	5.8/6		35		GU	
1929 Feb 15	080424	12.8	91.0	n	I	6.2/14	6.1/1	56		GU	
1929 Feb 20									F	ES	Notes
1929 Feb 28	12	11.3	85.8	n	Am				F	NI	Notes
1929 Mar 01	153940	13.9	91.2	n	I	5.6/6		22		GU	

Date	Time (UT) h.m.s.	Epicentre °N	Epicentre °W	Depth km	Ag	M_S	m_B/ m_b	M^* $log M_O$	N	Felt	Reg	Other Agencies, Comments	Felt Radii
1929 Mar 01*	205700	13.9	91.2	n	I	5.4/1			10		GU		
1929 Mar 19	205342	14.5	92.5	n	Am	6.2/17	6.1/1		72	F	MX	I, Notes	
1929 Mar 21	023706	13.8	90.5	n	AR	6.4/20	6.3/1		88	VI	GU	I, Notes	4/140, 5/70, 6/35
1929 May 12	093430	12.8	91.0	n	I	5.6/7			34		GU		
1929 Jun 09*	010245	12.8	91.0	n	I	5.2/5			13		GU		
1929 Jul 30	074334	13.9	91.2	n	I	5.6/9			29		GU		
1929 Aug 20	173654	10.5	87.0	n	I	5.5/6			24		CR		
1929 Nov 20	0126	13.5	88.3	n	Am					F	ES	Notes	
1929 Dec 20	102702	14.6	91.4	s	AM	5.3/5			17	V	GU	I, Notes	3/110, 5/25
1930 Jan 20	12	15.3	91.4	s	Am				2	F	GU	Notes	
1930 Feb 01	190400	13.0	80.0	n	I	6.1/12			>23		NI		
1930 Feb 20		13.5	87.8	n	Am					F	ES		
1930 Mar 07		11.3	85.8	n	Am					F	NI		
1930 Mar 09	1007	13.7	88.9	n	Am				1	F	ES	Notes	4/40
1930 Apr 30										F	ES	Notes	
1930 Jun 09		14.1	90.1	s	Am				>1	V	GU	Notes	5/40
1930 Jun 18*		14.2	90.2	s	Am					V	GU	Notes	4/40, 5/20
1930 Jun 19*		14.1	90.4	s	Am					VII	GU	Notes	4/60, 7/10
1930 Jun 21*		14.0	89.9	s	Am					VI	GU		3/90, 6/15
1930 Jun 24*		14.2	90.2	s	Am					D	GU	Notes	
1930 Jul 07*	133311	14.10	90.25	s	AM	6.2/22	5.9/1		55	VIII	GU	I, Notes	5/80, 7/20, 8/12
1930 Jul 07*	204252	14.1	91.2	s	Am	5.2/1			15	F	GU		
1930 Jul 14	224044	14.12	90.25	s	AM	6.9/45	6.5/6		116	VIII	GU	I, Notes	4/200, 6/90, 8/22
1930 Jul 17*	183016	14.0	90.0	s	Am	5.5/8			15	D	GU	I, Notes	
1930 Jul 27	150130	15.5	92.5	n	I	5.6/1			22		MX		
1930 Jul 27	185844	14.0	87.7	n	Am	5.8/14			43	V	HO	I	5/57
1930 Jul 29	062406	12.7	86.7	n	Am	5.8/8			>33	V	NI	I	5/60
1930 Jul 30	1652	8.5	80.5	n	Am				2	D	PA	Notes	
1930 Jul 31	0626	14.0	87.0	n	Am					F	HO	Notes	

1930 Aug 14									F	GU	Notes	
1930 Aug 29	082740	8.7	82.5	n	Ar	5.7/8	6.5/2		V	PA	I, Notes	5/47
1930 Sep 09		8.7	82.5	n	Am				F	PA	Notes	
1930 Sep 26	042214	13.9	91.2	n	I	5.3/2		11		GU		
1931 Jan 16	165203	15.75	85.0	n	G	5.1/2		11		HO		
1931 Jan 24		14.8	90.5	n	Am				F	GU		
1931 Feb 07	033035	13.	87.	100	G	5.6/4		>15		NI		
1931 Feb 12	080636	15.5	92.5	n	I			17		MX		
1931 Mar 07	004156	11.3	85.7	80	I	6.2/7	6.0/1	>39		NI	G	
1931 Mar 14	04	14.5	90.6	n	Am				V	GU	Notes	5/18
1931 Mar 31	160221	12.15	86.28	n	AM	6.0/17		>51	VI	NI	G, Notes	4/130, 6/30
1931 Apr 01*	131340	12.15	86.29	n	Am	5.4/6		>20	F	NI	I, Notes	
1931 Apr 07*	2115	12.15	86.30	n	Am				F	NI		
1931 Apr 12	0058	8.1	81.5	n	Am			1	D	PA	Notes	
1931 Apr 20	1220	12.15	86.30	n	Am				F	NI		
1931 Apr 25	110331	15.7	97.1	n	Am	4.9/3		7	F	ES	I	
1931 Jul 27	071615	17.0	85.0	n	I	5.6/6		45		HO	G	
1931 Aug 25	222055	13.2	89.7	n	AR	5.6/3		>17		ES	G, Notes	
1931 Sep 26*	195033	15.0	91.5	n	I	6.1/10		35		GU	G	
1931 Sep 26	200314	15.1	92.3	n	Am	6.3/19	6.1/2	60	F	MX	G	
1931 Oct 12	035732	7.5	82.3	n	I	6.0/3	6.0/1	24	F	PA	G, Notes	
1931 Dec 20	145955	11.0	84.2	350	I	5.4/1		15		CR	G	
1932 Jan 21	2055	14.45	90.8	n	Am	4.8/1		>1	V	GU	Notes	5/18
1932 May 21	101018	12.8	88.0	150	AR	7.1/37	6.6/3	129	VI	ES	G, Notes	4/300, 4/160, 6/25
1932 May 22	224007	14.7	89.8	n	I	5.9/12	5.5/1	>39		GU	G	
1932 Jun 20	090147	12.6	89.3	n	AR	5.8/6	5.9/2	44		ES	G, Notes	
1932 Jun 24	094340	12.5	89.2	n	I			13		ES		
1932 Aug 17	084653	16.2	92.0	n	Am	5.8/8		>31	F	MX	G	
1932 Aug 19	180740	15.5	91.7	n	I	4.9/2		8		GU		
1932 Sep 10		13.3	87.8	n	Am				F	ES	Notes	
1932 Oct 02	025915	11.4	86.4	n	I	6.9/21	6.4/2	>102		NI	G	
1932 Dec 19	062852	16.2	92.3	n	Am	5.9/2	6.1/1	40	F	GU	I, G, Notes	
1933 Jan 12	011742	11.0	87.0	n	G	5.5/1		>12		NI		

Date	Time (UT) h.m.s.	Epicentre °N	°W	Depth km	Ag	M_S	m_B/ m_b	M^* $log M_O$	N	Felt	Reg	Other Agencies, Comments	Felt Radii
1933 Feb 18	194543	16.5	86.0	60	G				15		HO		
1933 Mar 12	042552	16.5	86.5	n	G				8		HO		
1933 Mar 18	233257	16.5	86.0	n	I	5.8/2			23		HO	G	
1933 Mar 22		9.9	83.8	n	Am					V	CR	Notes	5/15
1933 Apr 03		13.7	89.2	n	Am					F	ES	Notes	
1933 May 30	114342	9.0	83.3	n	Am	5.5/4			>40	F	PA	G, Notes	
1933 Jun 04	131826	19.0	05.0	n	Am				11	F	MX	I, outside area	
1933 Jun 08	181143	7.5	82.3	n	I				6		PA		
1933 Jun 10	112659	17.3	86.3	n	G				>21		BE		
1933 Jul 04	2150	14.9	88.5	n	Am					F	HO	Notes	4/120
1933 Jul 11	1245	12.2	86.5	n	Am					V	NI	Notes	5/25
1933 Aug 24		12.2	86.5	n	J						NI		
1933 Sep 30	18	14.7	88.9	n	Am					VI	HO	Notes	6/25
1933 Nov 21	234838	8.0	83.0	n	G	6.1/8			>58	D	PA	Notes	
1933 Nov 22*	045206	8.0	83.0	n	I	5.9/10			54		PA		
1933 Nov 22*	081029	8.0	83.0	n	I	5.3/2			>15		PA		
1933 Nov 23*	185744	8.0	83.0	n	G	6.1/8	5.7/1		>61	F	PA	Notes	
1933 Nov 29*	050328	7.4	83.1	n	I	5.9/10			>40		PA	G	
1933 Dec 02*	030030	7.4	83.1	n	I				9		PA		
1934 Feb 12	064317	10.0	83.8	n	Am	5.6/2	5.9/2		>30	F	CR	I	
1934 Feb 24	053330	12.8	86.8	200	G	5.6/1			21		NI		
1934 Mar 07	224153	13.4	87.8	n	AR	6.3/14	6.1/2		>52		ES	G, Notes	
1934 May 19	104737	16.2	92.3	n+	Am	5.8/4	6.3/1		46	F	MX	G	
1934 Jul 13	112936	7.4	83.1	n	I				5		CR		
1934 Jul 18	013628	8.1	82.6	n	C	7.5/31	7.4/7		>170	VIII	PA	G, Ts, Notes	4/360, 5/210, 6/140, 8/50
1934 Jul 18*	040040	8.0	83.0	n	C	6.5/12	6.4/3		92		PA	G	
1934 Jul 18*	063539	8.4	82.5	n	C	5.4/1	5.7/1		30		PA	G	
1934 Jul 18*	160954	8.1	82.4	n	C	5.7/3	5.8/1		37		PA	G	

1934 Jul 18*	165943	7.9	82.3	n	C	6.7/28	6.9/2	>138		PA	G	
1934 Jul 19*	022104	8.0	82.9	n	K			12		CR	I	
1934 Jul 19*	032539	7.4	83.1	n	I			4		CR		
1934 Jul 19*	081719	7.4	83.1	n	I			4		CR		
1934 Jul 19*	195441	7.4	83.1	n	I			3		CR		
1934 Jul 21*	103915	8.6	82.7	n	C	6.8/28	6.8/1	>126	D	PA	G, Notes	
1934 Jul 21*	131924	8.2	82.9	n	C	5.5/1		22		PA	I	
1934 Jul 23*	140139	8.0	81.5	n	I			9		PA		
1934 Jul 24	024724	8.2	87.9	n	I	5.3/2		21		CR		
1934 Jul 27	022545	15.4	92.2	n+	Am	5.8/4	6.3/1	>29	F	GU	G	
1934 Aug 15	110414	12.0	89.0	n	I	5.5/1		>32		NI		
1934 Nov 09	161139	7.2	79.5	n	W	5.0/1		10		PA	U	
1934 Dec 03*	013537	14.7	89.0	n	Am	5.8/6		24	F	HO	I	
1934 Dec 03	023829	14.65	89.00	s	AM	6.3/19	6.3/4	>82	VIII	HO	G, Notes	3/300, 4/130, 6/40, 7/16, 8/7
1934 Dec 22	142931	11.1	86.9	n	I	6.6/16	6.3/2	>93		NI		
1934 Dec 24	143541	15.9	92.3	n	Am	5.8/11		>29	F	MX	I	
1935 Feb 27	152716	7.4	83.1	n	I			9		PA		
1935 Mar 17	213318	16.3	92.1	n+	Am	5.9/7		>43	F	MX	G	
1935 Mar 26	213224	16.9	94.0	n	I	5.3/2		24		HO	G	
1935 Apr 24	185145	15.4	92.9	n	Am	5.8/9		40	F	MX	I, G, Notes	
1935 Jul 26	044332	9.3	82.2	n	AC	5.7/5	5.8/1	47	D	PA	I, Notes	
1935 Aug 01	160819	10.58	85.25	s	AM	6.1/17	6.0/2	69	VII	CR	I, AC, Notes	3/200, 5/85, 6/25, 7/9
1935 Sep 01	004847	13.1	87.3	n	I	5.3/2		>18		HO		
1935 Oct 07	045832	13.5	92.5	n	I	5.5/3		>41		GU		
1935 Oct 31		15.2	87.5	n	Am				D	HO	Notes	
1935 Nov 30	033950	9.9	79.7	n	I	6.2/16		92	D	PA	G, Notes	
1935 Dec 14	220519	14.8	92.5	n	K	7.3/24	6.8/7	154	D	MX	G, Notes	
1935 Dec 21*	052825	15.4	92.2	n	Am	5.6/1		23	F	MX	I, AC, Notes	
1935 Dec 21*	072350	15.3	92.7	n	Am	5.4/1		>27	F	MX	I	
1935 Dec 21*	115107	15.3	92.7	n	Am	5.8/8		56	F	MX	I, Notes	
1936 Mar 07	203542	9.6	86.7	n	U	5.3/4		12	F	CR	Notes	

Date	Time (UT) h.m.s.	Epicentre °N	°W	Depth km	Ag	M_S	m_B/ m_b	M^* $logM_O$	N	Felt	Reg	Other Agencies, Comments	Felt Radii
1936 Mar 20	184628	11.	84.	n	G	5.6/4			>39		NI		
1936 Apr 02										F	NI	Notes	
1936 Apr 24	143532	14.3	92.5	n	AC	5.6/1			16		GU		
1936 Apr 27	063111	17.	87.	n	G	5.8/6			53		HO		
1936 May 06	0819	8.5	80.2	n	Am				2	D	PA	Notes	
1936 Jun 11	032634	7.4	83.1	n	I	5.3/1			14		CR		
1936 Aug 04	035048	16.0	93.0	n	U				10		MX		
1936 Sep 03	050557	16.5	92.8	n	Am	5.6/1			>24	D	MX	I, AC, Notes	
1936 Oct 29	055307	7.0	82.2	n	AC	6.1/11	5.9/1		64	F	PA	I, AC, Notes	
1936 Nov 01	2045	13.4	88.7	n	Am				1	F	ES	Notes	
1936 Nov 10	0931	13.4	88.7	n	Am				1	F	ES	Notes	
1936 Nov 14	013708	13.5	92.5	n	I	5.5/1			16		MX		
1936 Nov 15	0744	13.4	88.7	n	Am				1	F	ES	Notes	
1936 Nov 19	211027	14.2	90.9	80	AR	6.9/21	6.6/5		132	F	GU	I, AC, Notes	
1936 Nov 22*	181917	14.2	90.6	n+	AC	6.4/16	6.3/2		100	D	GU	I, Notes	
1936 Nov 23	003624	7.8	82.3	n	AC	5.1/1			10	F	PA	I	
1936 Nov 23	0946	14.3	88.2	n	Am				1	F	HO	Notes	3/120
1936 Nov 26	021206	10.7	84.1	n	I	6.0/12	5.8/1		62	D	CR	Notes	
1936 Nov 27	0157	13.6	88.8	n	Am				1	F	ES	Notes	
1936 Nov 27	020837	10.7	84.1	n	I	5.6/1			39		CR		
1936 Nov 27*	1413	13.6	88.8	n	Am				1	F	ES		
1936 Dec 18*	0455	13.6	88.8	n	Am				1	F	ES		
1936 Dec 19*	0843	13.6	88.7	n	Am				1	F	ES		
1936 Dec 20	024327	13.64	88.80	n+	AM	6.1/11			>63	VIII	ES	I, AC, Notes	4/110, 5/47, 6/20, 7/13, 8/6
1936 Dec 20*	132318	13.6	88.7	n	Am	5.3/1			>13	F	ES	I, AC	
1936 Dec 29*	135958	13.6	88.6	n	Am	4.9/1			28	F	ES	I, AC, Notes	
1937 Jan 11	132110	16.8	93.0	n	Am	5.6/2	5.9/1		50	F	MX	I, Notes	
1937 Jan 12		12.6	87.0	n	Am					D	NI	Notes	

1937 Mar 09	154020	9.0	83.4	n	I	6.4/22	6.2/2	131	F	CR	G, Notes	
1937 Mar 17*	135951	9.0	84.1	n	AC	5.7/5		50		CR	I	
1937 Mar 29*	061847	9.0	84.1	n	AC	5.6/3		>29	F	CR	I, Notes	
1937 Apr 03	221538	15.4	92.1	n	Am	5.5/1		17	D	GU	I, Notes	
1937 Apr 13	050833	8.8	82.8	n	AC	5.5/3		25		PA	I	
1937 May 28	153553	17.0	93.0	150	G	5.6/2	6.3/3	64	D	MX	Notes	
1937 Jun 08	222940	17.3	93.2	200	Am	5.8/9	6.2/1	>99	F	MX	G, Notes	
1937 Jun 24	131325	8.4	83.9	n	AC	6.0/9	6.3/3	86	F	CR	I, Notes	
1937 Jun 30	22	13.7	89.2	n	Am			1	F	ES	Notes	
1937 Aug 10	160612	10.	86.	n	I	5.4/3		26		CR		
1937 Sep 15	234900	16.2	92.3	n	Am	6.5/14	6.3/4	118	F	GU	Notes	
1937 Sep 25	072800	11.	85.	n	U	5.1/1		10		CR		
1937 Sep 28	062050	15.0	92.0	n	Am	6.0/11	5.7/1	73	D	GU	I, Notes	
1937 Sep 28*	181916	15.0	92.0	n	Am	5.8/6		53	F	GU	I	
1937 Sep 29		13.8	89.7	n	Am				F	ES	Notes, volc?	3/50
1937 Oct 12*	155947	15.0	92.0	n	Am	5.7/6	5.7/1	>35	F	GU	I	
1937 Oct 20	054733	13.0	88.5	n	Ar	5.8/1		19		ES	I	
1937 Dec 05	054205	13.9	89.5	80	Ar			20		ES	G	
1937 Dec 06*	210725	14.	82.	n	I			27		NI	I	
1937 Dec 06	214307	13.9	81.8	n	M	5.6/5	6.2/1	34		NI	G	
1937 Dec 07*	180042	13.7	82.2	n	U	5.3/1		29		NI	I	
1937 Dec 08*	022515	13.9	81.4	n	M	5.5/6		39		NI	I	
1937 Dec 18	204905	7.8	81.0	n	AC	5.4/3	5.9/2	25	F	PA	I, Notes	
1937 Dec 26	234342	13.95	89.82	s	AM	5.8/5		>21	VIII	ES	I, Notes	4/85, 7/12, 8/8
1938 Mar 05		13.7	89.2	n	Am				D	ES	Notes	
1938 Mar 25	082239	16.	86.	n	I	5.5/4	5.8/1	28		HO		
1938 Apr 10	193015	14.5	92.3	n	I			11		GU		
1938 Apr 12	110221	12.0	91.9	n	I	5.4/2		38		GU		
1938 Apr 25	170746	12.6	87.0	n	Am	6.0/12		72	D	NI	I, Notes	
1938 May 06*	181725	12.5	86.9	n	Am	6.0/12		77	D	NI	I, Notes	
1938 May 14	234148	12.0	88.3	n	U	5.0/1		12		NI		
1938 Sep 01	224821	12.6	89.3	n	I	6.1/13	6.2/2	107	F	ES	Notes	
1938 Sep 20	030508	7.0	82.7	n	I	4.9/1		16		PA		

Date	Time (UT) h.m.s.	Epicentre °N	°W	Depth km	Ag	M_S	m_B/ m_b	M^* $logM_O$	N	Felt	Reg	Other Agencies, Felt Radii Comments
1939 Jan 20	204027	13.5	91.5	70	G	6.2/11	6.2/3		101	F	GU	Notes
1939 Jan 27*	141148	13.0	89.8	120	AR	5.7/5			>47	F	ES	I, Notes
1939 Jan 29*	185022	13.3	89.7	120	AR	5.7/6			>50	D	ES	I, Notes
1939 Feb 09	153035	11.2	86.1	n	I	5.9/9			74		NI	
1939 Jun 18	164558	9.7	83.7	n+	I	5.8/9	5.8/2		61		CR	G, Notes
1939 Jul 08	213144	12.5	88.0	90	G				14		NI	
1939 Aug 05	092905	14.3	91.7	n	I				19		GU	
1939 Aug 16	170706	13.0	90.0	n	U	6.0/11			70	F	ES	I, Notes
1939 Sep 28	145827	14.9	92.0	I	Am				40	F	GU	G, Notes
1939 Oct 20	200604	8.2	82.8	n	AC	5.6/1	5.9/1		58		PA	G, Notes
1939 Nov 28	020956	8.8	78.5	80	G	5.2/1			18	F	PA	I, Notes
1939 Dec 05	083607	14.0	92.5	n	Ar	6.9/16	6.4/2		133	D	GU	G, Notes
1939 Dec 05	175148	12.7	88.8	n	U				12		ES	I
1939 Dec 12	025012	16.7	92.8	240	Am				17	F	MX	G, Notes
1939 Dec 13	18	13.7	88.8	n	Am				10	D	ES	Notes
1939 Dec 19	194330	12.4	87.8	n	U				19		NI	I
1939 Dec 21*	205447	10.1	84.6	n+	K	6.9/1	6.6/3		97	D	CR	G, Notes
1939 Dec 22	044358	9.8	84.6	70	K	6.9/14	6.5/3		128	VI	CR	G, Notes 3/560, 6/100
1939 Dec 23*	171106	9.6	85.0	n	U	5.2/1			46		CR	I
1939 Dec 26	115512	13.3	88.7	60	AR	5.7/1	5.9/1		56	F	ES	G, Notes
1939 Dec 26*	214947	9.8	85.0	n	AC				34		CR	I, K
1939 Dec 29	172724	13.	86.	n	U				11		NI	
1940 Jan 28	073034	10.7	85.1	s	W				6		CR	U
1940 Feb 06	12	13.8	89.3	n	Am					F	ES	Notes
1940 Feb 25	075806	7.0	82.5	n	U				7		PA	
1940 Mar 28	174836	12.2	89.5	75	U	5.4/3			24	F	ES	I, Notes
1940 Mar 30	044628	14.2	92.1	n	U						GU	
1940 Apr 04		13.6	88.8	n	Am					F	ES	Notes
1940 Apr 07	0348	13.3	88.8	n	Am					D	ES	Notes

1940 May 17	015940	7.5	81.8	n	I	5.8/8	5.7/2	74	F	PA	Notes
1940 Jul 13	164726	7.7	82.4	n	AC	6.2/13	6.7/2	102		PA	I, Notes
1940 Jul 27	133230	14.0	91.5	90	G	6.8/21	6.4/4	111		GU	
1940 Aug 16	025142	8.5	82.5	n	U			6		PA	
1940 Aug 22	110330	7.9	82.3	n	U			9		PA	
1940 Sep 20	055000	9.9	84.4	n	U					CR	
1940 Oct 02*	031544	9.5	83.9	50	I	5.6/3	6.0/2	>47		CR	
1940 Oct 05	143843	9.5	84.3	n	G	6.1/5	6.1/2	76		CR	
1940 Oct 05*	171713	9.5	83.9	n+	I			10		CR	
1940 Oct 27	053537	9.8	84.5	n	G	6.8/23	6.2/3	119	D	CR	Notes
1940 Nov 10	204025	17.4	83.3	n	I	5.8/1	5.8/2	42		HO	G
1940 Nov 23	034851	9.1	84.2	n	AC	5.6/2	6.0/1	59		CR	I
1940 Dec 01	211020	7.2	82.3	n	AC	5.4/1	5.8/2	31		PA	I
1940 Dec 07	031954	9.	84.	n	U					CR	
1940 Dec 15	235306	13.6	90.1	n	U			8		ES	
1940 Dec 29	163800	8.8	82.7	n	I	5.2/1		13		PA	
1941 Jan 03	091541	12.9	91.4	n	I	5.3/1	5.5/1	34		GU	
1941 Jan 06	094836	11.8	86.5	60	G	6.2/2	5.9/1	62		NI	
1941 Feb 04	122951	9.0	83.4	n	I	5.0/1		9		CR	
1941 Feb 15	085608	15.3	89.4	n	I			>23		GU	
1941 Mar 10*	034730	8.0	80.5	n	Am			6	F	PA	I, Notes
1941 Mar 10	040542	7.5	80.5	n	Am	5.2/1	5.7/1	>31	F	PA	G, Notes
1941 Mar 19	212346	12.0	91.9	n	I			13		GU	
1941 Mar 23	090027	17.0	83.9	n	I			20	F	HO	G, Notes
1941 Apr 06	225516	13.3	88.8	n	Am	5.2/1		>32	D	ES	I, Notes
1941 May 11	132900	14.5	91.5	n	U			>17		GU	
1941 Jul 24	051936	7.5	83.0	n	U					PA	
1941 Jul 24	103030	13.5	91.5	n	U			20		GU	
1941 Oct 08	042213	10.5	86.3	n	G	5.6/3	5.6/1	38		CR	
1941 Nov 16	093942	13.3	88.5	80	G			30		ES	
1941 Nov 22	052337	7.5	80.1	n	I	4.8/1		11		PA	
1941 Dec 05	204658	8.7	83.2	n	K	7.6/21	6.8/4	143	D	CR	G, Ts, Notes
1941 Dec 06	012501	10.5	85.3	n	Am	5.9/3	5.8/1	38	D	CR	G, Notes

Date	Time (UT) h.m.s.	Epicentre °N	°W	Depth km	Ag	M_S	m_B/ m_b	M^* $logM_O$	N	Felt	Reg	Other Agencies, Felt Radii Comments
1941 Dec 06*	212442	8.8	84.3	n	K	7.0/14	6.6/5		117	D	CR	G, Ts, Notes
1941 Dec 06*	235439	8.8	83.5	n	K	5.5/2	5.6/1		27		CR	I
1941 Dec 07*	002514	8.7	83.2	n	I				16		CR	
1941 Dec 07*	042303	8.6	83.1	n	K	5.2/2			14		CR	I
1941 Dec 07*	114154	8.3	83.0	n	K				23		CR	I
1941 Dec 08*	074114	8.7	83.2	n	K	5.7/2	5.8/1		>34		CR	I
1941 Dec 08*	074223	9.1	83.3	n	K				15		CR	I
1941 Dec 08*	104534	8.7	83.2	n	I				6		CR	
1941 Dec 08*	110047	8.7	83.2	n	I				6		CR	
1941 Dec 08*	162458	8.7	83.2	n	I				8		CR	
1941 Dec 13*	110325	8.9	84.4	n	K	5.4/2	5.9/3		13		CR	I
1942 Jan 05*	223506	8.5	83.0	n	U				8		CR	
1942 Feb 23	024400	15.5	92.0	n	U	5.3/1			23		GU	I
1942 Feb 23	062708	14.0	89.4	n	I				18		ES	
1942 Mar 01	095207	14.3	91.2	n	I	6.2/4	6.0/1		>58		GU	
1942 Apr 11	012511	14.7	91.2	100	I	5.7/1	6.0/3		50		GU	G
1942 Apr 22*	114924	8.0	80.5	n	U	5.2/1			10		PA	I
1942 Apr 22	232012	7.3	82.2	n	U	5.9/2			>28		PA	I
1942 Jun 28	000701	14.7	88.8	n+	I				22		HO	
1942 Aug 06	233654	13.9	90.9	n	K	7.9/17	7.4/3		130	D	GU	G, Notes
1942 Aug 07*	060430	12.5	90.6	n	K	5.2/1			31		GU	I
1942 Aug 08*	071924	14.3	91.2	n	I				34		GU	
1942 Aug 08*	223630	13.9	91.2	n	K	5.8/4	6.0/2		71		GU	G
1942 Aug 11*	044810	13.4	91.0	n	K	5.3/1			43		GU	I
1942 Aug 11*	071116	12.5	90.5	n	K				>32		GU	I
1942 Aug 15*	063532	13.0	90.8	n	I	5.4/1			>22		GU	I
1942 Aug 16*	200745	13.7	91.0	n	K	5.6/2	5.8/1		52		GU	I
1942 Aug 20*	214015	13.9	90.8	n	I				33		GU	
1942 Aug 20*	223700	13.9	90.8	n	I				38		GU	

1942 Aug 22*	195210	13.9	90.8	n	I			17		GU	
1942 Aug 29*	122350	13.7	90.9	n	K			34		GU	I
1942 Aug 29*	214017	14.0	90.7	n	K			33		GU	I
1942 Sep 04*	025353	14.5	91.1	n	K	6.0/1		46		GU	I
1942 Sep 07	045142	13.9	90.8	n	I			16		GU	
1942 Sep 26	040016	13.0	87.8	n	I	6.3/7	6.0/2	73		HO	
1942 Sep 27	170212	14.5	91.0	n	U			35		GU	I
1942 Nov 05	114736	15.5	91.5	n	U			18		GU	I
1942 Nov 11	130600	9.6	83.9	n	U	5.7/1		28	F	CR	I, Notes
1942 Dec 22	062454	7.5	79.5	n	U	5.4/2		23	F	PA	I, Notes
1942 Dec 23*	011248	7.5	79.5	n	U	5.3/2		>18		PA	I
1943 Jan 24	092642	14.9	91.0	n	U	5.0/1		28		GU	I
1943 Jan 24	204148	14.5	91.0	n	U	6.1/6	6.0/2	>65		GU	I
1943 Apr 03	221535	11.1	90.7	n	AC					GU	
1943 Apr 19	011915	17.2	81.3	n	I			23	F	HO	G, Notes
1943 May 02	171807	6.9	80.4	n+	I	7.0/16	7.2/7	138	D	PA	G, Notes
1943 Jun 02	052405	16.9	86.0	n	I	5.2/1		30		HO	
1943 Jun 15	182142	14.6	93.0	n	I	6.5/12	6.2/1	103	F	GU	Notes
1943 Jun 15*	183207	14.6	93.0	n	I			11		GU	
1943 Jun 15*	194522	14.6	93.0	n	I	5.6/2		46		GU	
1943 Jun 15*	202539	14.6	93.0	n	I	5.8/1		>41		GU	
1943 Jul 04	095158	9.4	84.7	n	AC	5.7/6	6.2/2	67		CR	I
1943 Aug 31	161040	14.3	91.5	80	G	6.0/3	5.8/2	72		GU	I
1943 Sep 23	150044	15.0	91.5	110	G	6.1/4	6.3/4	94	D	GU	I, Notes
1944 Jan 16	142742	11.0	89.0	n+	U			14		NI	
1944 Mar 24	170151	8.1	83.2	n	I			10		PA	
1944 Apr 07	133258	12.0	85.5	200	G			38		NI	
1944 Apr 23	091010	8.0	82.9	n	AC			8		CR	I
1944 Apr 28	055020	8.0	82.9	n	AC	5.3/1	5.5/1	>27		CR	I
1944 May 27	23	13.5	88.2	n	Am				F	ES	Notes
1944 Jun 25	010818	14.5	93.0	n	U	5.2/1				MX	
1944 Jun 28*	053149	14.7	92.8	n	AC	5.7/1		29		GU	I
1944 Jun 28	075849	14.3	92.7	n	K	7.2/19	6.5/3	108		GU	G, Notes

Date	Time (UT) h.m.s.	Epicentre °N	°W	Depth km	Ag	M_S	m_B/ m_b	M^* $log M_O$	N	Felt	Reg	Other Agencies, Felt Radii Comments
1944 Jul 02*	221224	14.5	93.0	n	U	5.4/3			38		GU	I
1944 Aug 02	122806	9.5	84.0	n	U	5.2/1			15		CR	I
1944 Aug 02	130653	13.0	87.4	n	I	5.6/2			>20		HO	
1944 Aug 24	233748	16.2	93.8	n	Am	5.9/6			63	F	MX	I, Notes
1944 Oct 02	172200	14.5	89.8	160	G	5.2/1	6.3/1		50		GU	
1944 Dec 31	215527	12.0	92.0	n	I				13		NI	
1945 Jan 02	194017	13.7	90.2	n	I	5.4/2			20		GU	
1945 Jan 12	215925	16.2	93.5	110	Am	6.0/3	5.7)		29	F	MX	G, Notes
1945 Feb 18	064623	7.1	82.4	n	I	5.8/2			35		PA	
1945 Mar 24	142423	13.	92.	n	I	5.4/1			18		GU	
1945 Jun 03	130533	8.4	82.7	n	I	6.3/10	6.6/2		90	D	PA	Notes
1945 Jun 04*	155329	8.6	82.6	n	U	5.0/1			21		PA	
1945 Jul 05	120230	8.5	81.1	n	Am	4.9/1			17	D	PA	Notes
1945 Jul 17	064736	15.5	92.5	n	U						MX	
1945 Aug 10	112015	15.2	89.1	n	Am	6.0/5	5.9/1		61	D	GU	I, Notes
1945 Aug 10*	140905	15.2	89.1	n	Am				17	F	GU	I
1945 Aug 11	003352	7.1	81.6	n	B	6.0/3			64		PA	I
1945 Oct 03	061925	12.0	91.0	n	U				16		NI	I
1945 Oct 05	231728	13.0	87.0	100	U				11		NI	
1945 Oct 07	132325	12.5	89.1	n	I	6.1/12	6.4/3		89		ES	G
1945 Oct 27	112438	15.1	91.2	190	I	6.0/6	6.1/1		97	D	GU	G, Notes
1945 Nov 27	112428	15.5	91.0	100	U						GU	
1945 Dec 09	204546	14.0	92.2	n	I	5.9/5	6.0/1		44		GU	
1946 Jan 04	194353	10.1	86.5	n	I	5.9/2			>29		CR	
1946 Jan 05	011511	15.0	91.0	210	G	5.7/1	5.8/1		60	F	GU	I, Notes
1946 Feb 22	172452	16.3	92.5	170	G				29	F	MX	Notes
1946 Mar 25	221633	14.5	92.3	n	I				>30		GU	
1946 Jun 24	154817	14.4	89.6	220	I				38		GU	G
1946 Jun 26	075333	14.8	90.8	90	G	5.8/2			72		GU	

1946 Aug 16	171520	12.5	86.6	n	I	5.2/1		>17		NI	
1946 Aug 24	024204	11.5	86.3	n	I	5.3/2		>26		NI	
1946 Oct 21	134509	8.4	82.6	n	AC	5.4/1		41		CR	I
1946 Nov 30	004921	15.5	91.7	280	I			21		GU	
1947 Jan 25*	031128	13.0	87.8	n	I	5.3/1		17		ES	
1947 Jan 25*	035048	13.0	87.8	n	I			16		ES	
1947 Jan 25	035240	13.5	87.5	n	Am	6.1/8	6.4/1	58	D	HO	I, Notes
1947 Jan 25*	044657	13.3	87.3	n	Am			21	F	HO	I, Notes
1947 Jan 26	100606	12.2	86.3	160	AR	6.7/17	6.9/6	130	F	NI	G, Notes
1947 Feb 26	015028	10.5	88.2	n	I	5.4/1		34		NI	
1947 May 09	133232	15.3	89.4	n	I			22		GU	
1947 May 26	000640	15.9	92.7	n	Am			28	F	GU	I, Notes
1947 Jun 05	225811	14.0	90.1	n	I	5.4/1		51	F	GU	Notes
1947 Jun 25	225004	15.5	91.7	n	I			22		GU	
1947 Jul 15									F	NI	Notes
1947 Sep 05	054130	12.0	90.7	n	I			18		ES	
1947 Sep 12	114345	12.0	90.7	n	I			20		ES	
1947 Sep 15	145635	7.1	82.4	n	I	5.2/1		24		PA	
1947 Oct 09	003019	7.5	80.1	n	I			9		PA	
1947 Dec 23	015604	15.3	92.3	n	Am			26	F	GU	I, Notes
1947 Dec 30	015518	9.5	84.5	n	U	5.8/5		55	F	CR	I, Notes
1948 Mar 22	213427	11.3	85.7	n	J	6.0/5		>51		NI	I
1948 Mar 22*	234115	11.5	86.3	n	I	5.3/1		32		NI	
1948 Apr 12	061515	14.8	90.1	65	I	5.9/1	6.0/1	42		GU	
1948 Apr 20	021058	14.0	92.2	n	I			31		GU	
1948 Jun 27	124818	17.0	85.0	n	U	6.0/7		94		HO	
1948 Jul 16*	071228	14.5	91.1	100	J	6.0/6		70	F	GU	Notes
1948 Jul 16	071934	14.3	91.2	100	J	6.2/14	5.7/1	83	F	GU	Notes
1948 Jul 17*	093043	14.3	90.7	70	J			26		GU	
1948 Jul 18*	0315	15.0	91.0	n	U			12		GU	
1948 Jul 19*	222612	15.0	91.4	n	J			>36		GU	
1948 Jul 28*	142152	7.5	81.5	n	I	5.3/2		>34	D	PA	Notes
1948 Jul 28	150545	7.5	81.5	n	I	5.5/1		>34	D	PA	Notes

Date	Time (UT) h.m.s.	Epicentre °N	°W	Depth km	Ag	M_S	m_B/ m_b	M^* $log M_O$	N	Felt	Reg	Other Agencies, Felt Radii Comments
1948 Jul 29*	185758	7.5	81.5	n	I	5.2/2			13		PA	
1948 Jul 29*	195802	7.5	81.5	n	I				7		PA	
1948 Jul 29*	204953	7.5	81.5	n	I				6		PA	
1948 Jul 31*	190414	7.5	81.5	n	I	5.5/2			52		PA	
1948 Aug 03	093032	12.5	86.8	n	I	5.0/1			>21		NI	
1948 Aug 13	111832	16.8	93.8	n	Am				31	F	MX	I, Notes
1948 Sep 16		8.	84.	n	U						CR	Notes
1948 Sep 18	083612	9.5	84.7	n	J	5.2/1			30		CR	
1948 Oct 21	045015	12.5	88.0	n	U	5.2/1			54		ES	
1948 Nov 19	010414	9.9	83.8	80	Am	6.3/13	6.7/1		114	V	CR	G, AC, Notes 5/85
1949 Jan 25	075302	11.	86.	n	U	5.0/1			22		NI	I
1949 Jun 09	104910	16.3	92.0	100	Am				>20	F	MX	I, U, Notes
1949 Jun 11	073444	12.1	87.9	n	I	5.2/1			>41		NI	
1949 Jul 08	124047	14.0	91.0	100	U	5.9/2			67		GU	I
1949 Jul 14	030129	13.	88.	n	Am				17	F	ES	
1949 Jul 16	095720	13.5	92.5	n	U	5.3/1			40		GU	I
1949 Aug 18	133325	8.5	83.0	n	U	6.1/11			89	D	PA	I, Notes
1949 Nov 13	044240	11.0	85.5	60	U	5.8/5			80		CR	I
1949 Nov 26	062409	7.	83.	n	U				17		PA	
1949 Nov 28	165321	8.	83.	n	U	5.0/1			16		CR	I
1949 Dec 05*	112157	9.3	84.7	n	AC	4.8/1			23		CR	I
1949 Dec 05	124240	9.3	84.7	n	AC	5.4/1			45		CR	I
1949 Dec 22	093047	16.4	93.1	100	Am	6.2/10	6.4/4		137	F	MX	I, U, Notes
1950 Feb 10	202324	12.	90.	n	B	4.9/1			16		GU	
1950 Feb 17	034721	13.5	91.0	n	U	5.5/1			>61		ES	I
1950 Mar 24	120729	8.5	80.5	n	Am				9	F	PA	B, Notes
1950 Apr 15	145125	14.	91.	100	U	5.8/4			>86		ES	I
1950 May 09	023526	11.	89.	n	U				10		NI	
1950 Jun 23	034825	14.5	93.0	60	U				45		GU	I

1950 Sep 20	030407	16.	93.	100	U			26		MX	I	
1950 Oct 05	160927	10.0	85.7	60	Am	7.9/34	7.2/10	220	VII	CR	G, GD, Ts, Note	7/130
1950 Oct 05*	164819	10.4	84.7	n	K			26		CR	I	
1950 Oct 05*	200913	10.4	85.4	n	K	5.4/1		46		CR	I	
1950 Oct 05*	205231	10.5	85.0	100	U			10		CR		
1950 Oct 10*	162020	10.5	85.0	100	U	5.0/2		24		CR		
1950 Oct 11	040100	9.	85.	n	U			15		CR		
1950 Oct 14*	132707	10.5	85.0	n	U	5.0/1		27		CR	I	
1950 Oct 14*	174129	10.4	85.7	n	I			26		CR		
1950 Oct 17*	103146	10.5	85.0	n	U			13		CR		
1950 Oct 17*	145610	10.1	84.8	n	K	5.3/1		33		CR	I	
1950 Oct 17	150633	10.1	88.3	n	M	5.1/1		30		CR	I	
1950 Oct 17*	220721	9.6	85.2	n	K	5.4/1		45		CR	I, M	
1950 Oct 23*	145536	14.5	92.0	100	U			4		GU		
1950 Oct 23*	150340	14.5	92.0	100	U					GU		
1950 Oct 23	161317	14.3	91.8	n	K	7.3/32	6.9/5	185	D	GU	G, Ts, Notes	
1950 Oct 23*	170526	15.0	91.5	n	K			51		GU	I	
1950 Oct 23*	174801	14.5	91.7	n	K	5.6/1		70		GU	I	
1950 Oct 23*	175938	15.0	91.7	n	K			38		GU	I	
1950 Oct 23*	195047	14.3	91.5	n	K			35		GU	I	
1950 Oct 23*	213200	14.7	92.0	n	K	5.1/1		30		GU	I	
1950 Oct 23*	233843	14.3	91.9	n	K	5.5/2		74		GU	I	
1950 Oct 23*	234910	14.5	92.0	n	U			18		GU	I	
1950 Oct 24*	002328	14.5	92.0	100	U			21		GU	I	
1950 Oct 24*	005201	14.5	92.0	n	K	5.8/5		73	F	GU	I, Notes	
1950 Oct 24*	055015	14.2	91.9	n	K	5.1/1		>35		GU	I	
1950 Oct 24*	061648	14.3	91.7	65	I			15		GU		
1950 Oct 24*	092857	14.3	91.7	65	I			31		GU		
1950 Oct 24*	155504	14.4	91.2	n	K			31		GU	I	
1950 Oct 25*	050503	14.7	91.9	n	K			19		GU	I	
1950 Oct 28*	221539	15.4	92.2	n	Am	5.4/2	6.0/1	>61	F	GU	I, K, Notes	
1950 Nov 01*	012257	14.8	92.3	n	Am			26	F	GU	U, Notes	
1950 Nov 01	124525	10.2	85.1	n	K	5.4/1		>49		CR	I	

Date	Time (UT) h.m.s.	Epicentre °N	°W	Depth km	Ag	M_S	m_B/ m_b	M^* $logM_O$	N	Felt	Reg	Other Agencies, Comments	Felt Radii
1950 Nov 05*	163518	14.8	92.3	n	Am	5.8/6	6.1/1		>72	F	GU	I, K, Notes	
1950 Nov 11	135109	9.8	85.9	n	K	5.5/1			>38		CR	I	
1950 Dec 18	080445	14.	90.	200	U	5.2/1	6.0/4		69		GU	I	
1951 Jan 05*	005240	7.	81.	90	U	5.9/4	6.0/5		95	F	PA	G, Notes	
1951 Jan 06	075131	7.5	80.4	100	U	6.6/20	6.4/5		153	F	PA	G, Notes	
1951 Jan 06*	145848	7.5	81.0	n	B	5.0/1					PA		
1951 Jan 19	014048	13.	88.	100	U	4.8/1			34		ES	I	
1951 Jan 20	131218	14.6	90.5	n	B	5.1/1			29		GU	I	
1951 Jan 22	042508	14.3	88.6	100	T				7		HO		
1951 Jan 29	102759	15.4	92.2	100	Am				34	F	GU	I, U, Notes	
1951 Feb 13	085015	15.4	92.0	n	Am				36	F	GU	I, K, Notes	
1951 Feb 13	162800	14.	91.	n	U	4.8/1			27		GU	I	
1951 Mar 02	003424	15.8	91.5	100	B				22		GU	I	
1951 Mar 05	005036	14.5	91.0	n	B				13		GU		
1951 Mar 24	205230	13.1	88.4	100	U	5.2/1			>60		ES		
1951 Mar 25	153315	13.1	88.4	100	U				19		ES		
1951 Apr 02	001334	13.0	90.0	n	I	6.0/9	6.1/1		88		ES		
1951 Apr 02	180034	12.	87.	n	U				11		NI		
1951 Apr 12	041125	11.6	78.3	n	W				7		PA	B	
1951 May 06	214219	10.5	85.6	n	D	5.3/2			>33		CR	I	
1951 May 06*	230333	13.5	88.4	90	AM	5.9/10	6.0/3		84	D	ES	I, Notes	
1951 May 06	230801	13.5	88.4	70	AM	6.0/7			66	VIII	ES	I, Notes	7/9, 8/5
1951 May 07*	202221	13.5	88.5	100	AḾ	5.5/6	5.7/1		57	D	ES	I, Notes	
1951 May 11*	021557	13.0	87.8	100	I				40	F	ES	Notes	
1951 May 14	130234	10.0	85.6	n	AC	5.4/1			41		CR	I	
1951 Jun 05	013415	10.0	85.6	n	AC	5.3/3			57		CR	I	
1951 Jun 18	174425	10.2	84.8	100	B	5.3/1	6.0/1		56	F	CR	I, Notes	
1951 Jun 25		13.5	88.8	n	Am				1	F	ES	Notes	4/45
1951 Jul 08	110012	12.	86.	n	U				16		NI		

1951 Jul 09	113732	17.1	84.8	75	M					HO	U	
1951 Jul 09	211700	11.	85.	n	U			25		CR		
1951 Jul 25	184214	14.0	90.5	100	U			37		GU	I	
1951 Jul 30	165130	10.2	85.6	n	D			26		CR	I	
1951 Aug 02*	203017	13.0	87.5	n	Am	5.6/6		76	D	NI	I, Notes	
1951 Aug 03	002358	13.0	87.5	n	Am	5.9/11	6.0/4	92	D	NI	I, Notes	
1951 Aug 03*	003510	13.0	87.5	n	U					NI		
1951 Aug 03*	052544	13.0	87.5	n	Am	5.5/5		58	F	NI	I	
1951 Aug 06*	080856	13.0	87.6	n	Am	5.1/2	5.2/1	>48	F	NI	I, Notes	
1951 Aug 22	054130	9.82	83.86	s	AM	5.4/3		47	VII	CR	I, Notes	4/65, 5/28, 7/8
1951 Sep 28	120716	11.4	86.6	55	D	5.8/6	5.9/1	>61		NI	I	
1951 Sep 28*	145117	11.2	86.7	s	D	5.7/1		>44		NI	I	
1951 Oct 25	050622	16.5	92.7	n	AM	5.3/1		48	D	MX	I, Notes	
1951 Nov 11*	120314	16.5	92.7	n	Am			7	F	MX	T	
1951 Dec 17	175153	12.1	86.4	80	D	5.4/1		28		NI	I	
1951 Dec 31	200540	13.0	90.5	n	U	5.1/1		36		GU	I	
1952 Jan 15		14.2	90.3	s	AM				D	GU	Notes	
1952 Jan 18	114219	13.5	92.0	150	B			20		GU	I	
1952 Feb 04	053040	15.3	92.7	n	Am				F	MX	T, Notes	
1952 Feb 06	163849	13.0	88.1	100	U			25		ES	I	
1952 Feb 26	153927	11.5	86.4	60	D	6.4/12	6.2/5	105	F	NI	I, Notes	
1952 Mar 02*	185300	11.4	86.2	65	D	5.9/4		62		NI	I	
1952 Mar 03*	174205	11.3	86.3	s	D	5.7/4		57		NI	I	
1952 Mar 08	192859	11.8	88.2	n	I	5.1/1		27		NI		
1952 Mar 12*	134018	11.	86.	n	B	4.9/1				NI		
1952 Apr 21	231507	7.5	83.0	n	U	4.6/1		22		PA	I	
1952 Apr 22	062600	15.0	92.5	n	B					MX		
1952 Apr 25	060200	8.2	82.7	n	AC	5.2/3	6.2/2	50		PA	I	
1952 May 10	092503	16.6	92.6	n	Am	5.1/1			F	MX	T, Notes	
1952 May 13	193145	10.3	85.3	65	I	6.4/15	5.9/3	155	F	CR	Ts?, Notes	
1952 May 14	211141	16.7	86.2	40	M	5.6/3	5.9/4	77		HO		
1952 May 15*	184357	14.5	92.3	100	T	5.5/2	5.8/1	50	F	GU	I, B	
1952 May 16	054211	14.5	92.3	100	T	5.6/4	5.7/2	60	F	GU	I, B, Notes	

Date	Time (UT) h.m.s.	Epicentre °N	°W	Depth km	Ag	M_S	m_B/ m_b	M^* $logM_O$	N	Felt	Reg	Other Agencies, Felt Radii Comments	
1952 May 29	155112	14.5	91.8	n	B	5.7/1					GU		
1952 Jul 09	181518	7.5	82.0	n	U	5.8/6	6.1/3		131	F	PA	I, Notes	
1952 Jul 09*	203648	7.5	82.0	n	U	5.5/4	5.9/3		80	F	PA	I	
1952 Jul 15	060620	14.5	92.5	n	U	5.7/5	6.0/1		74	F	MX	I, Notes	
1952 Jul 23	005917	14.0	91.5	n	U	6.1/1	6.0/2		55		GU	I	
1952 Jul 28	173122	7.7	82.5	n	AC				14		PA	I	
1952 Aug 02	090031	12.1	88.9	100	B						NI		
1952 Aug 20	054928	16.	92.	200	U		6.0/1		44	F	MX	I, Notes	
1952 Aug 28	152315	16.0	91.5	150	U				35		GU	I	
1952 Sep 09	125442	8.7	84.3	n	GD	6.8/18	6.7/10		194	V	CR	I, U, Notes	3/320, 4/200, 5/125
1952 Sep 10*	064102	8.5	84.5	n	B	5.6/2					CR		
1952 Sep 11*	052822	9.2	84.2	n	U	5.0/1			41		CR	I	
1952 Sep 24	071905	14.3	92.1	n	I	5.0/1			25		GU		
1952 Oct 14	235603	8.5	83.0	n	U	5.2/1			52		CR	I	
1952 Oct 21*	021125	9.5	84.0	n	U	5.2/2			24		CR	I	
1952 Oct 21*	023046	9.5	84.5	n	U	5.3/2	6.1/1		45	F	CR	I, Notes	
1952 Oct 21*	060905	9.5	84.5	n	U				25		CR		
1952 Oct 21*	063544	9.5	84.5	n	U	5.2/1	6.1/1		29		CR	I	
1952 Nov 20	153719	12.3	87.7	n	D	6.2/10	5.9/1		132		NI	I	
1952 Dec 01	120203	15.3	92.6	n+	Am				10	F	MX	I, Notes	
1952 Dec 03	101708	12.1	90.9	n	T						GU		
1952 Dec 05	065949	11.5	86.3	n	I	5.2/1			17		NI		
1952 Dec 10	084129	13.8	91.8	100	T						GU		
1952 Dec 14	105532	9.	83.	n	U						CR		
1952 Dec 30	120703	0.05	3.90	s	AM	5.7/6	5.8/1		65	VIII	CR	I, Notes	
1953 Jan		13.5	86.6	n	Am					F	NI	Notes	
1953 Jan 05	224942	8.	85.	n	B						CR		
1953 Jan 06	232343	9.5	83.0	n	U	5.3/1				F	CR	Notes	
1953 Jan 07	120035	10.00	83.05	s	AM	5.6/2	6.2/1		65	VII	CR	I, Notes	

1953 Jan 07*	232343	9.5	83.0	n	U					CR		
1953 Feb 07	091537	8.9	82.6	n	Am			1	F	PA	otes	
1953 Feb 16	101022	8.5	83.0	n	U	5.3/1	6.1/1	59		CR	I	
1953 Feb 25		13.7	89.2	n	Am				F	ES	Notes	
1953 May 21	015629	14.8	92.3	i	Am		5.6/1		F	GU	U, Notes	
1953 May 24	055840	10.5	85.5	n	U		5.8/1			CR		
1953 Jun 29	154719	12.3	90.2	100	T					GU		
1953 Jul 28	054250	13.1	88.5	100	T					ES		
1953 Jul 29	181534	13.0	90.3	n	U	5.8/9	5.8/2	104		ES	I	
1953 Aug 13	100734	12.1	90.7	n	T					ES		
1953 Aug 24	132102	14.1	91.4	100	I	5.6/2	5.8/2	57	D	GU	Notes	
1953 Aug 28	114929	7.8	83.0	n	B					CR		
1953 Sep 05	152446	14.8	92.3	100	Am				F	GU	T, Notes	
1953 Oct 22	015015	16.7	92.6	100	Am				F	MX	T, Notes	
1953 Oct 25	005805	9.0	85.5	n	B	5.0/1				CR		
1953 Nov 17	132951	12.6	92.2	n	K	6.9/18	6.9/9	210		GU	I, Notes	
1953 Nov 21	082736	17.0	92.5	n	B		6.0/1			MX		
1953 Nov 24	090818	14.	92.	100	U					GU		
1953 Dec 01	203536	12.	89.	n	B		5.1/1			ES		
1953 Dec 17	062634	12.1	92.2	n	T					GU		
1954 Jan 04	111636	9.0	84.4	42	M	4.9/1				CR	U	
1954 Jan 13	002630	13.0	88.5	n	Am				F	ES	T, Notes	
1954 Feb 19	004042	11.7	86.9	97	M	6.8/13	6.2/4	132	V	NI	I, Notes	3/350, 5/80
1954 Feb 19*	213443	12.1	86.8	99	M	6.7/13	6.3/4	130		NI	I	
1954 Feb 20*	020054	11.9	87.0	32	D	5.8/3		70	F	NI	I, Notes	
1954 Feb 21*	012915	11.6	86.6	56	D	4.9/1		50		NI	I	
1954 Feb 21*	233939	12.3	86.4	153	D					NI	U	
1954 Mar 08	180557	12.4	88.8	70	M	4.9/1			F	ES	U, Notes	
1954 Mar 11	103013	14.0	90.9	96	M	5.3/1	6.1/1	79	F	GU	I, Notes	
1954 Mar 16	010257	7.3	82.3	s	M	5.0/1				PA	U	
1954 Mar 21	204328	13.9	92.2	52	M					GU	T	
1954 Mar 27	184101	9.6	83.8	90	M	4.7/1	5.5/1		F	CR	U, Notes	
1954 Apr 03	214006	16.3	92.0	n	Am				F	MX	T, Notes	

Date	Time (UT) h.m.s.	Epicentre °N	°W	Depth km	Ag	M_S	m_B/ m_b	M^* $logM_O$	N	Felt	Reg	Other Agencies, Felt Radii Comments
1954 Apr 15	040230	7.4	81.0	s	M						PA	U
1954 Apr 15	044005	7.2	81.0	s	M						PA	
1954 Apr 23	012119	8.3	83.5	47	M					F	CR	U, Notes
1954 May 02	181707	13.5	90.6	n	T						GU	
1954 May 03	171336	11.9	86.3	149	D	5.4/1	6.0/3				NI	I
1954 May 12	234140	16.6	90.9	n	T						GU	
1954 Jul 06	155630	13.7	92.3	168	M					F	GU	B, Notes
1954 Aug 17	045611	12.5	90.0	n	B						GU	
1954 Aug 23	112400	8.3	82.3	n	B						PA	
1954 Aug 23	120436	8.5	82.9	n	M					F	PA	U, Notes
1954 Sep 09	034356	12.1	85.1	s	M						NI	
1954 Sep 19	212427	13.	88.	n	B						ES	
1954 Oct 11	132703	16.6	90.9	n	T						GU	
1954 Oct 21	065144	13.9	90.7	58	M	6.0/2	6.2/3		92	F	GU	I, Notes
1954 Nov 01	205623	13.7	92.2	37	M	5.3/2	6.1/1		66		GU	I
1954 Nov 16	120358	7.	83.	n	U						PA	
1954 Nov 27	160230	11.8	86.9	54	D	5.4/1			45		NI	I
1954 Dec 02	231335	11.9	86.9	107	M						NI	U
1954 Dec 21	100624	13.8	91.8	100	T					F	GU	Notes
1954 Dec 28	071047	11.5	87.5	n	B						NI	
1955 Jan 07	053059	14.3	90.0	151	M						GU	U
1955 Jan 10	193153	14.7	92.1	87	M						MX	U
1955 Jan 11	230338	12.7	87.2	128	M	4.9/1					NI	U
1955 Mar 05	032955	13.8	90.9	101	M	5.1/1				F	ES	U, Notes
1955 Mar 07	193250	13.	87.	n	U						HO	
1955 Mar 10	033842	12.0	86.8	s	D	4.9/1			42		NI	I
1955 Mar 10	045521	11.3	90.1	s	W						CR	U
1955 Mar 19	171535	13.3	90.6	53	M						ES	U
1955 Mar 20	201334	14.5	91.7	77	M				56	F	GU	I, Notes

1955 Mar 22	135412	14.5	91.1	136	M					GU	T	
1955 Mar 28	065216	12.8	88.8	59	M				F	ES	Notes	
1955 Apr 04	192407	12.8	87.2	n	AM	6.0/12	6.0/5	117	D	NI	I, D, Notes	
1955 Apr 11*	202116	12.8	87.2	n	Am	4.9/2			F	NI	U	
1955 Apr 14	082506	15.9	91.4	180	M				F	GU	Notes	
1955 Apr 23	134014	13.7	89.2	n	Am			1	F	ES	Notes	
1955 Apr 26	030328	13.1	89.4	66	M	5.8/2	6.3/1	93	F	ES	I, Notes	
1955 Apr 26*	214333	13.5	89.5	60	T					ES		
1955 Apr 27	220936	13.7	92.8	n	T					GU		
1955 Apr 28	1805	13.7	89.2	n	Am				F	ES	Notes	
1955 Apr 30*	013227	12.4	86.5	n	Am	5.5/1	6.3/1	88	D	NI	I, D, Notes	
1955 Apr 30	014355	12.4	86.5	n	Am	5.9/8	6.1/4	72	D	NI	I, D, Notes	
1955 Apr 30*	091507	12.2	86.6	43	D					NI	U	
1955 Apr 30*	091958	12.5	86.5	s	M					NI	U	
1955 May 05	081639	17.5	92.5	53	M					MX	T	
1955 May 25	030858	13.3	92.1	n	M	5.1/3	5.7/3	80		GU	I	
1955 May 26	220155	13.5	92.3	n	T					GU		
1955 Jun 16	234213	14.3	91.0	n	B					GU		
1955 Jul 04	024534	13.6	92.0	59	M					GU	B	
1955 Jul 07	045731	13.1	87.5	s	M	5.1/1				NI	U	
1955 Aug 12	054548	16.2	91.8	200	T				F	MX	Notes	
1955 Aug 16	041850	12.4	88.0	s	M	5.2/3				NI	U	
1955 Aug 16	191010	11.3	86.7	s	M	5.2/1		55		NI	I	
1955 Aug 28*	200319	13.3	91.5	s	M				F	GU	T	
1955 Aug 28	201336	13.8	90.7	77	M	6.6/16	6.4/7	162	F	ES	I, Notes	
1955 Aug 29*	011433	13.4	91.3	73	M					GU	T	
1955 Aug 29	075339	11.8	86.8	42	M					NI	T	
1955 Sep 01	173303	10.18	84.30	s	AM	5.5/4		60	VIII	CR	I, M, Notes	4/60, 7/10, 8/5
1955 Sep 03	123622	13.8	90.8	78	M	6.5/15	6.6/3	165	F	ES	I, Notes	
1955 Sep 04	025055	14.	91.	n	B					GU		
1955 Sep 26	082830	15.9	92.2	213	M	6.4/7	7.0/11	194	D	MX	I, Notes	
1955 Oct 13	215102	11.9	86.7	n	M	5.2/1		60		NI	I	
1955 Nov 12	225437	7.0	82.5	n	U					PA		

Date	Time (UT) h.m.s.	Epicentre °N	°W	Depth km	Ag	M_S	m_B/ m_b	M^* $logM_O$	N	Felt	Reg	Other Agencies, Comments	Felt Radii
1955 Nov 23	031433	13.5	90.1	66	M						GU	U	
1955 Dec 24	033445	9.0	84.6	s	M						CR	U	
1955 Dec 28	0120	13.7	89.2	n	Am					F	ES	Notes	
1956 Jan 09	131138	14.9	92.8	58	M					F	MX	Notes	
1956 Jan 09	170125	16.1	92.1	169	M				46	F	MX	I, Notes	
1956 Jan 21	080933	15.	93.	150	U						MX		
1956 Mar 01	231150	9.5	85.5	n	U						CR		
1956 Mar 07*	215148	9.5	85.5	n	U						CR		
1956 Mar 13	131315	7.4	82.5	s	M	6.0/10	6.4/5		138	F	PA	I, Notes	
1956 Mar 15*	152838	7.8	82.3	s	M						PA	U	
1956 Mar 15*	154455	7.5	82.5	n	U						PA		
1956 Mar 24	204851	13.8	90.0	n	Am				1	F	ES	Notes	
1956 Mar 25	054431	13.7	90.7	82	M						GU		
1956 May 07	081703	14.0	90.6	189	M				60	F	ES	I, Notes	
1956 Jun 05	053223	10.	84.	n	U						CR		
1956 Jun 16	165741	15.6	92.8	245	M					F	MX	Notes	
1956 Jun 23	175943	7.3	82.8	74	M						PA	U	
1956 Jul 03	154648	13.5	90.7	72	M	5.3/5			62		GU	I	
1956 Jul 05	201738	13.0	89.5	n	Am				1	F	ES	Notes	
1956 Jul 19	232637	9.6	84.1	93	M	5.7/8	5.9/3		100	V	CR	I, Notes	3/150, 4/90, 5/40
1956 Jul 19*	233814	9.5	84.0	81	M	5.6/5				F	CR	B, Notes	
1956 Jul 29	132852	13.9	90.0	61	M						ES	U	
1956 Aug 09	170104	10.5	86.1	124	D				42		CR	I	
1956 Aug 10	021711	9.5	83.5	n+	Am					F	CR	U, Notes	4/80
1956 Aug 20	053346	7.2	79.8	n	M	5.5/6	6.0/5		96	F	PA	I, Notes	3/190
1956 Aug 20*	070620	7.5	79.6	n	M					F	PA	U	
1956 Aug 20*	071956	7.0	79.9	s	M					F	PA	U	
1956 Aug 20	094350	13.5	91.5	100	U						GU		
1956 Sep 04	073138	14.0	89.8	110	M						ES		

1956 Sep 10	112528	14.5	89.5	n	U				F	ES	Notes	
1956 Sep 10	235935	8.4	82.4	66	M					PA	U	
1956 Sep 11	095436	13.5	91.2	61	M	5.8/3		8		GU	I	
1956 Sep 25	182730	8.3	83.0	42	M				F	PA	U, Notes	
1956 Sep 25*	213132	8.5	82.7	114	M					PA	U	
1956 Oct 23	042452	13.0	88.7	n	MR			1	F	ES	Notes	
1956 Oct 24	144216	11.5	86.6	n	D	7.2/17		200	D	NI	I, Notes	
1956 Oct 25*	052147	11.5	86.4	44	D	6.2/9		91	F	NI	I, Notes	
1956 Oct 27*	153300	11.7	86.4	52	D	5.5/2		45		NI	I	
1956 Nov 01	154212	10.	80.	n	U				F	PA	Notes	
1956 Nov 10	000825	10.7	86.0	n	D	5.0/1		46		CR	I	
1956 Nov 22	083249	7.8	82.1	63	M	5.0/1				PA	U	
1956 Dec 04	230132	14.9	92.2	84	M	5.8/8		108	F	GU	I, Notes	
1956 Dec 24	043429	10.4	84.7	126	M					CR		
1957 Jan 15	214026	11.0	86.5	n	U					CR		
1957 Jan 28	235951	11.1	86.3	76	M	5.1/1				NI	U	
1957 Feb 01	114033	13.5	90.0	n	MR			1	F	ES	Notes	
1957 Feb 04	090154	9.5	83.8	67	M	4.7/1		61	F	CR	I, Notes	3/100
1957 Feb 09	072318	7.5	83.0	n	U	5.2/1				PA		
1957 Mar 21	084454	14.2	93.0	77	M	5.6/5	6.0/3	72		MX	I	
1957 Mar 25	022839	13.0	91.0	60	M	4.9/1		62		GU	I	
1957 Apr 05	161223	12.6	87.8	76	M			76	F	NI	I, Notes	
1957 Apr 08	201815	8.4	82.7	n	M	5.8/8	5.9/2	107	F	PA	I, Notes	2/290
1957 Apr 10	072811	13.8	91.5	95	M	5.3/2				GU	U	
1957 Apr 17	180933	14.4	92.3	77	M	5.6/3		73	F	MX	I, Notes	
1957 May 19	210029	12.0	87.3	s	D	5.5/4		61		NI	I	
1957 Jul 08	153036	14.5	91.2	137	M	5.4/5		83	D	GU	I, Notes	
1957 Jul 10	090411	7.7	82.6	n	M	6.2/16	6.7/8	177	F	PA	I, Notes	
1957 Jul 21	060408	14.3	92.1	40	M	5.6/9		83	D	GU	I, Notes	
1957 Jul 27	064909	14.1	91.1	132	M					GU	U	
1957 Aug 15	083256	10.	80.	n	U				F	PA	Notes	
1957 Sep 12	002802	17.0	85.6	s	M	5.8/10	6.1/4	153	F	HO	I, Notes	
1957 Sep 17	050329	13.0	90.0	n	MR			1	F	ES	Notes	

Date	Time (UT) h.m.s.	Epicentre °N	Epicentre °W	Depth km	Ag	M_S	m_B/ m_b	M^* $logM_O$	N	Felt	Reg	Other Agencies, Felt Radii Comments	
1957 Sep 23	123757	13.0	87.5	n	MR				1	F	ES	Notes	
1957 Sep 26	133523	14.4	92.2	156	M					F	MX	Notes	
1957 Oct 15	040214	8.7	84.1	53	M	5.1/1			66		CR	I	
1957 Oct 27	182910	16.0	88.1	56	S					F	HO	U, Notes	
1957 Nov 01	145410	15.6	90.6	s	M						GU	B	
1957 Nov 16	050652	17.0	85.3	n	M						HO	U	
1957 Nov 26	232408	11.6	86.7	90	M						NI	U	
1957 Dec 02	174112	13.0	88.6	67	M					F	ES	U, Notes	
1957 Dec 07	082408	15.6	91.7	n	M	5.1/1					GU	U	
1957 Dec 07	221851	13.6	82.4	s	M	5.1/1					NI	U	
1957 Dec 21	191541	13.9	92.8	s	M						GU		
1957 Dec 23	131929	12.1	86.7	120	M					F	NI	U, Notes	
1958 Jan 08	061650	13.5	89.5	n+	MR				>1	D	ES	Notes	
1958 Jan 13	025246	7.2	82.6	n	M	5.0/1					PA	U	
1958 Jan 15	073742	13.5	89.5	n+	MR				>1	F	ES	Notes	
1958 Feb 01		11.0	86.0	n	MR					F	NI	Notes	
1958 Feb 03	122850	13.6	89.0	n	MR				>1	F	ES	Notes	
1958 Feb 09	041506	7.7	79.3	s	M	5.1/1			64	F	PA	I, Notes	
1958 Mar 01	172134	13.6	89.5	60	M					F	ES	U, Notes	
1958 Mar 03	004446	14.1	91.6	111	M					F	GU	U, Notes	
1958 Mar 11	084728	14.8	90.3	192	M					F	GU	U, Notes	
1958 Mar 21	141504	15.0	92.2	104	M					F	MX	U, Notes	
1958 Mar 27	060549	14.5	92.9	89	M	5.0/1					MX	U	
1958 Apr 08	171706	13.0	88.8	57	M					F	ES	Notes	
1958 Apr 11	091255	8.4	82.5	105	M						PA		
1958 Apr 11	201501	12.9	88.4	90	MR					F	ES	Notes	
1958 Apr 15	021723	12.8	89.0	100	MR						ES		
1958 Apr 15	035234	8.0	84.5	s	M	6.3/16	6.3/7		145	F	CR	I, Notes	3/210
1958 Apr 17*	140617	8.1	84.4	81	M	4.8/1					CR	U	

1958 May 17	103318	13.1	88.9	90	MR				F	ES	Notes	
1958 May 18	224125	13.4	89.7	75	MR					ES		
1958 Jun 05	132357	12.5	86.5	n	U					NI		
1958 Jun 06	091115	7.9	84.5	s	M	6.5/19	6.4/10	203	F	CR	I, Notes	3/230
1958 Jun 06*	143857	7.9	84.4	s	M					CR	U	
1958 Jun 06*	155211	8.0	84.4	s	M	4.9/1				CR		
1958 Jun 06*	224406	8.0	84.2	s	M	5.2/4				CR	U	
1958 Jun 12*	115411	7.7	84.3	45	M	5.4/5		76		CR	I	
1958 Jun 16	143202	12.9	88.4	78	M				F	ES	U, Notes	
1958 Jun 19	040108	15.5	92.0	100	U				F	MX	Notes	
1958 Jun 27	054432	13.1	88.4	64	M	5.6/8		99	D	ES	I, Notes	
1958 Jul 03	190049	13.4	89.6	96	M				F	ES	Notes	
1958 Jul 09	151821	14.0	91.7	96	M	5.2/2		45		GU	I	
1958 Jul 15	012128	8.8	82.5	n	Am			2	F	PA	Notes	
1958 Jul 19	081601	12.5	88.3	50	MR			1	F	ES	Notes	
1958 Jul 21	093908	12.2	88.1	s	M					ES	U	
1958 Jul 23	180356	10.0	89.0	n	U					CR		
1958 Sep 02	200705	14.6	92.9	38	M	5.6/8		61	F	MX	I, Notes	
1958 Sep 03*	041809	14.9	92.7	64	M					MX	B	
1958 Nov 03	135505	12.5	87.5	n	Am				F	NI	Notes	
1958 Nov 14	054630	14.3	91.6	78	M			60	D	GU	I, Notes	
1958 Nov 14	152213	12.5	86.2	n	Am	5.1/1		46	F	NI	I, D, Notes	
1958 Nov 22	055135	13.8	89.7	120	MR				F	ES	Notes	
1958 Dec 04	123434	14.0	91.5	100	U				F	GU	Notes	
1958 Dec 04	191926	11.5	86.6	50	D	5.3/1		49	F	NI	I, Notes	
1958 Dec 09	190312	8.4	82.8	45	M					PA	U	
1959 Jan 09	205208	14.4	90.5	72	M				F	GU	U, Notes	
1959 Jan 11	072242	14.5	90.4	206	M	5.3/1				GU	U	
1959 Jan 13	083408	9.5	84.0	58	M	5.5/6			V	CR	U, Notes	3/170, 4/90, 5/50
1959 Jan 15	164345	9.0	83.5	n	U				F	CR	Notes	
1959 Jan 24	194229	14.6	92.4	83	M			85		MX	I	
1959 Feb 16	175415	13.2	87.4	s	M			32		HO	I	
1959 Feb 20	181619	15.9	90.6	45	M	5.2/2		103	D	GU	I, Notes	3/200, 4/140

Date	Time (UT) h.m.s.	Epicentre °N	Epicentre °W	Depth km	Ag	M_S	m_B/ m_b	M^* $log M_O$	N	Felt	Reg	Other Agencies, Felt Radii Comments
1959 Mar 09	220303	15.2	90.9	168	M						GU	U
1959 Mar 10	224952	14.5	91.7	93	M	4.9/1					GU	U
1959 Mar 11	143136	14.1	92.1	60	M						GU	U
1959 Mar 15	104435	12.	85.	n	U						NI	
1959 Mar 30	071825	7.6	82.2	n	M						PA	U
1959 Apr 09	173614	7.0	82.3	n	M	5.7/10			86		PA	I
1959 Apr 11	092518	9.5	82.4	67	M				42	F	PA	I, Notes 2/130
1959 Apr 18	085100	13.3	89.4	100	MR					F	ES	Notes
1959 Apr 20	042116	8.3	82.9	42	M	5.5/6			71		PA	I
1959 Apr 22*	190145	11.4	86.6	n	D	5.4/9			72	F	NI	I, Notes
1959 Apr 24	093137	11.4	86.5	n	D				69		NI	I
1959 Apr 28	110935	14.5	92.5	58	M	6.5/20	6.4/9		205	F	MX	I, Notes
1959 Apr 29	053627	15.1	92.4	97	M					D	MX	Notes
1959 May 03	044118	12.3	87.5	s	D	5.4/2			87	F	NI	I, Notes
1959 May 12	112827	14.3	91.2	89	M					F	GU	Notes
1959 May 15	074936	14.6	92.2	84	M						MX	
1959 Jun 04	015759	9.6	84.2	62	M				36		CR	I
1959 Jun 05	203715	12.0	86.7	s	D	5.3/6			55		NI	I
1959 Jul 04	030422	13.5	87.8	37	M						ES	
1959 Aug 12	003335	11.3	86.1	51	M						NI	U
1959 Aug 15	225643	12.7	86.8	190	D						NI	U
1959 Sep 11	230847	11.7	87.0	84	M						NI	U
1959 Sep 17	212428	13.3	89.0	51	M				73	D	ES	I, Notes
1959 Nov 14	121330	14.0	89.7	60	MR					D	ES	Notes
1959 Nov 19	140820	11.5	87.1	n	M						NI	U
1959 Dec 01	141924	16.6	82.0	s	M						HO	
1960 Jan 26	181959	12.8	87.8	76	M				38	F	NI	I, Notes
1960 Jan 30	170115	8.3	83.8	n	B						CR	
1960 Feb 15	073615	11.9	87.0	s	D	5.0/1			51		NI	I

1960 Feb 21	021528	14.4	92.0	52	M				GU	U	
1960 Feb 21	233745	13.9	90.0	n	M			D	ES	Notes	
1960 Feb 29	021204	7.1	79.9	s	M			F	PA	U, Notes	
1960 Mar 02	043446	17.	93.	n	U				MX		
1960 Mar 10	185558	14.6	91.6	95	M		41		GU	I	
1960 Mar 18	011455	14.6	90.3	162	M		47		GU	I	
1960 Mar 28	001341	7.6	81.7	n	M	5.9/8	174	F	PA	I, Notes	
1960 Apr 13	123743	15.5	92.1	n	M	5.7/5	113	D	GU	I, Notes	
1960 Apr 25	185312	12.5	88.3	67	M			F	ES	U, Notes	
1960 May 10	231625	14.8	92.8	90	M				MX	U	
1960 May 12	223232	8.2	80.6	s	Am	5.9/10	176	VI	PA	I, M, Notes	3/180, 6/30
1960 May 12*	230043	7.7	80.6	44	M				PA	U	
1960 May 13*	003041	7.5	80.7	41	M				PA	U	
1960 May 23										Ts, Notes	
1960 Jun 05	093458	14.5	91.0	200	U			F	GU	Notes	
1960 Jul 10	133957	12.4	86.6	119	M		43	F	NI	I, Notes	
1960 Jul 19	160323	16.2	92.4	211	M		66	F	MX	I, Notes	
1960 Aug 10	123853	9.7	83.1	40	M				CR	U	
1960 Aug 20	001934	14.6	91.5	105	M		56		GU	I	
1960 Aug 21	035535	8.0	82.5	57	M				PA	U	
1960 Aug 23	075450	14.6	92.4	110	M				MX	U	
1960 Sep 13	175114	13.9	90.3	122	M				GU	U	
1960 Sep 21	034454	13.6	89.8	82	M			D	ES	U, Notes	
1960 Sep 29	185428	15.3	89.8	61	M			F	GU	U, Notes	
1960 Oct 16	093447	14.0	91.0	114	M				GU	U	
1960 Oct 17*	202908	14.3	92.6	119	M				GU		
1960 Oct 17	221527	14.7	92.6	37	M	5.5/1		F	GU	U, Notes	
1960 Oct 18*	063417	15.0	91.7	139	M				GU	U	
1960 Oct 23	214827	13.4	88.6	n	Am			D	ES	U, Notes	
1960 Nov 14	195932	14.6	92.7	73	M	5.2/1			MX	U	
1960 Nov 18	185106	14.8	91.0	136	M				GU	U	
1960 Dec 04	183024	13.3	88.1	62	M			F	ES	U, Notes	
1960 Dec 06	085614	8.6	82.7	46	M			F	PA	U, Notes	2/210, 4/36

Date	Time (UT) h.m.s.	Epicentre °N	Epicentre °W	Depth km	Ag	M_S	m_B/ m_b	M^* $logM_O$	N	Felt	Reg	Other Agencies, Comments	Felt Radii
1960 Dec 14	190926	13.5	88.5	n	Am					F	ES	Notes	
1960 Dec 15	232434	11.9	87.8	39	M						NI	U	
1960 Dec 23	060709	14.4	92.2	91	M						GU	U	
1961 Jan 05	184738	8.4	82.8	91	M						PA	U	
1961 Jan 21	144759	8.5	82.8	48	M						PA	U	
1961 Jan 24	231251	8.4	82.8	79	M					F	PA	U, Notes	
1961 Feb 05	153835	8.1	82.7	35	M	5.5/8			129	F	PA	I, Notes	3/90
1961 Feb 07	032115	13.6	89.9	128	U					F	ES	Notes	
1961 Feb 16	052152	12.9	86.5	40	U					F	NI	Notes	
1961 Feb 23	143155	15.5	92.2	53	U					F	MX	Notes	
1961 Feb 27	154422	9.9	83.9	68	M					D	CR	U, Notes	4/20
1961 Feb 28	182917	15.6	91.5	215	U						GU		
1961 Mar 02	000408	15.2	91.7	86	M					F	MX	U, Notes	
1961 Mar 04	210912	12.8	88.0	62	M					F	NI	U, Notes	
1961 Mar 10	231224	10.5	83.3	44	M						CR	U	
1961 Mar 20	061623	11.4	86.4	48	D				112	F	NI	I, Notes	
1961 Mar 22*	141941	11.8	86.4	75	D						NI		
1961 Apr 12	222033	13.3	88.8	87	M	5.8/14			192	D	ES	I, Notes	
1961 Apr 19	013922	14.1	90.5	98	U						GU		
1961 Apr 25	010930	14.7	90.0	250	M						GU		
1961 Apr 26	121916	15.6	89.9	s	W						GU	U	
1961 May 13	124510	12.6	88.5	40	M					D	ES	U, Notes	
1961 May 16	175721	13.4	88.1	249	M						ES	U	
1961 May 17	015848	15.5	87.0	50	U						HO		
1961 May 17	040930	14.9	85.9	52	U						HO		
1961 May 18	043527	15.9	87.2	n	U						HO		
1961 May 19	092541	12.7	88.2	72	M						NI	U	
1961 May 23	034026	9.8	84.1	100	M	5.3/6	5.9/3		121	V	CR	I, Notes	2/160, 5/40
1961 May 23*	034309	9.8	84.3	112	M	5.3/1					CR	U	

1961 May 23*	035220	10.1	84.0	100	U					CR		
1961 May 23	164459	12.6	87.4	92	D	5.4/8	6.0/3	160	D	NI	I, Notes	
1961 May 26	050640	15.6	91.7	231	M			50		GU	I	
1961 Jun 13	133629	8.7	83.2	43	U					CR		
1961 Jun 17	150733	14.3	92.0	69	M	5.9/11	6.3/3	150	F	GU	I, Notes	
1961 Jun 17*	183950	14.4	91.9	84	M			71		GU	I	
1961 Jun 20*	094902	15.7	87.1	n	M					HO	U	
1961 Jun 20	212151	15.3	87.4	114	M					HO	U	
1961 Jun 21	035740	15.6	87.3	48	M			74		HO	I	
1961 Jun 21	151032	13.6	90.1	100	U				F	ES	Notes	
1961 Jun 21	230400	15.4	86.4	75	U					HO		
1961 Jun 22	200013	12.9	90.1	53	M					ES	U	
1961 Jun 23*	010833	15.2	86.4	67	U					HO		
1961 Jun 23*	011113	15.5	87.2	60	U					HO		
1961 Jun 24	050755	13.5	89.9	71	M	5.1/1				ES	U	
1961 Jun 28	011716	12.	89.	n	U				F	ES	Notes	
1961 Jul 06	051236	15.4	86.8	74	S					HO	U	
1961 Jul 09	045729	15.7	87.1	45	M					HO	U	
1961 Jul 09	063252	15.5	87.5	94	M			61		HO	I	
1961 Jul 09	135057	15.7	86.2	s	S					HO	U	
1961 Jul 22	162432	12.2	88.6	n	U					ES		
1961 Aug 20	151506	15.6	92.4	n	U					MX		
1961 Aug 25	054159	15.6	87.1	s	M					HO	U	
1961 Aug 27	042020	7.5	82.1	n	U					PA		
1961 Sep 01	031607	15.3	87.4	71	M					HO	U	
1961 Sep 01	185035	13.8	92.4	n	M	5.9/11		186		GU	I	
1961 Sep 01	201453	15.3	87.3	44	M				F	HO	U, Notes	
1961 Sep 03	044825	12.3	86.7	116	M				D	NI	U, Notes	
1961 Sep 12	092810	13.3	92.1	98	M					GU	U	
1961 Sep 14	010724	15.0	91.6	152	U					GU		
1961 Sep 14	011138	15.5	97.4	99	M					HO	U	
1961 Sep 19	094617	6.9	82.3	s	M	5.6/11		154	F	PA	I, Notes	2/250
1961 Sep 30	185015	14.1	91.0	140	M					GU	U	

Date	Time (UT) h.m.s.	Epicentre °N	°W	Depth km	Ag	M_S	m_B/ m_b	M^* $log M_O$	N	Felt	Reg	Other Agencies, Comments	Felt Radii
1961 Oct 22	151342	10.5	85.9	143	M						CR	U	
1961 Nov 10	132004	13.6	86.7	38	U						HO		
1961 Nov 11	122709	13.9	90.8	89	M				103	F	GU	I, Notes	
1961 Nov 14	044225	7.4	82.4	s	M	5.9/8			124		PA	I	
1961 Nov 22	123016	15.3	91.6	216	M	5.1/1				D	GU	U, Notes	
1961 Nov 23	103549	10.8	84.9	n	U						CR		
1961 Dec 02	044759	13.9	91.8	105	M						GU	U	
1961 Dec 15	141030	13.6	92.8	160	U						GU		
1961 Dec 17	005143	15.7	88.0	s	M						HO	U	
1961 Dec 20	230028	14.3	90.8	98	U					D	GU	Notes	
1961 Dec 22	153618	11.9	85.5	49	U						NI		
1961 Dec 29	145320	14.1	92.8	59	M	5.0/1					GU	U	
1961 Dec 30	065653	13.4	92.3	n	U						GU		
1962 Jan 08	220855	11.8	85.7	s	D						NI	U	
1962 Jan 17	233419	12.5	87.1	145	M						NI	U	
1962 Jan 28	052243	13.5	92.2	47	M						GU	U	
1962 Jan 30	083425	12.5	87.7	47	D	5.0/1			85	F	NI	I, Notes	
1962 Feb 04	174735	7.2	82.4	s	M	5.0/1					PA	U	
1962 Feb 06	025819	14.4	91.1	s	M						GU	U	
1962 Feb 19	201607	11.9	88.2	n	U					F	ES	Notes	
1962 Feb 21	172216	16.8	92.9	248	I				47		MX		
1962 Feb 24	010325	13.0	88.8	59	M					F	ES	U, Notes	
1962 Mar 11	022554	14.2	91.0	115	M						GU	U	
1962 Mar 12*	094134	8.2	82.8	n	M	5.4/7			122	D	PA	I, Notes	
1962 Mar 12	114012	8.3	83.9	s	M	6.5/13	6.9/11		247	VI	PA	I, Ts, Notes	3/310, 6/50
1962 Mar 12*	134236	8.6	82.9	34	M	5.2/2					PA	U	
1962 Mar 14*	030426	8.2	83.3	68	M						CR	U	
1962 Mar 14*	202017	8.6	82.9	n	U						PA		
1962 Mar 23*	233646	8.0	82.6	n	U						PA		

1962 Apr 04*	140231	8.0	82.8	s	M	5.5/9		85		PA	I
1962 Apr 04*	151605	7.8	83.2	s	M					PA	U
1962 May 03*	031215	8.3	82.8	40	M					PA	U
1962 Jun 01	064831	13.1	88.1	100	M					ES	U
1962 Jun 04	185036	7.5	80.7	s	M				F	PA	U, Notes
1962 Jun 09	195727	13.0	91.1	58	M			89		GU	I
1962 Jun 16	111046	13.3	87.3	n	U					HO	
1962 Jun 18	132204	14.5	83.0	n	U					HO	
1962 Jun 20	225906	14.9	92.7	95	M					MX	U
1962 Jun 21	051316	8.2	83.0	49	M					PA	U
1962 Jun 25	185833	14.5	82.4	s	M					HO	U
1962 Jul 17	041246	11.1	86.7	71	M					NI	U
1962 Jul 17	094046	14.6	93.0	s	M					MX	U
1962 Jul 23	011258	10.9	86.2	72	M	5.1/3				NI	U
1962 Jul 24	035936	11.3	85.6	183	M					NI	U
1962 Jul 24	210822	15.4	92.5	136	M	5.4/7	5.8/4	176	F	MX	I, Notes
1962 Jul 26	081441	7.5	82.7	s	M	7.3/16	7.2/12	296	F	PA	I, Notes
1962 Jul 28	135840	14.3	93.1	67	M	5.3/3				MX	U
1962 Jul 30	101944	15.2	92.4	69	M					GU	
1962 Aug 03	224619	12.8	92.8	n	U					GU	
1962 Aug 04	024945	14.1	93.0	n	U	4.8/1				MX	
1962 Aug 12	051343	12.9	87.9	78	M					ES	U
1962 Aug 13	100926	15.2	92.4	95	M					MX	U
1962 Aug 20	131507	14.7	93.1	41	M					MX	U
1962 Aug 24	012430	16.6	93.0	174	U					MX	
1962 Aug 29	101413	7.3	80.1	53	M				F	PA	U, Notes
1962 Sep 04	151741	15.5	91.8	168	M			71	F	GU	I, Notes
1962 Sep 06	133859	14.4	90.7	53	M					GU	
1962 Sep 10	215230	12.6	86.7	191	M					NI	U
1962 Sep 14	135548	10.5	84.7	n	Am				F	CR	U, Notes
1962 Sep 18	002902	7.5	82.4	s	M	6.7/24	6.8/11	251	D	PA	I, Notes
1962 Sep 18*	023940	7.7	82.1	n	U					PA	
1962 Sep 18*	031657	7.8	82.3	n	U					PA	

Date	Time (UT) h.m.s.	Epicentre °N	°W	Depth km	Ag	M_S	m_B/ m_b	M^* $logM_O$	N	Felt	Reg	Other Agencies, Felt Radii Comments
1962 Sep 18*	032225	7.5	82.2	n	U						PA	
1962 Sep 18*	045515	7.2	82.7	n	U						PA	
1962 Sep 18*	051334	7.5	82.2	s	M				68		PA	I
1962 Sep 19*	014216	7.4	82.3	39	M				54		PA	I
1962 Sep 24*	142245	7.8	82.3	54	M						PA	U
1962 Oct 07	085223	9.4	81.8	59	U						PA	
1962 Oct 25*	155230	8.2	82.6	69	M	5.4/5			87	F	PA	I, Notes
1962 Oct 27	081024	14.1	90.6	94	M				77	F	GU	I, Notes
1962 Oct 27	135255	11.5	86.5	93	D	5.6/1			55		NI	I
1962 Oct 29	001942	7.0	82.7	41	M				101		PA	I
1962 Oct 29	105330	13.0	88.4	43	U						ES	
1962 Oct 30	083151	12.6	87.9	66	M	5.3/2			88	F	NI	I, Notes
1962 Nov 22	013015	14.9	92.8	111	M						GU	U
1962 Dec 16	222123	15.6	92.7	123	M						MX	U
1963 Jan 02	095332	10.3	85.2	73	M						CR	U
1963 Jan 11	143615	12.5	88.1	79	U						ES	
1963 Jan 15	032241	12.0	87.7	n	U						NI	
1963 Jan 15*	035031	11.9	87.7	n	U						NI	
1963 Jan 19	192902	16.6	84.7	37	S						HO	U
1963 Jan 19*	195058	16.8	84.9	35	S						HO	U
1963 Jan 21	073852	9.3	82.8	38	M						CR	U
1963 Feb 03	042318	14.8	91.9	91	U	3.8/2				F	GU	Notes
1963 Feb 11	164746	11.6	87.5	41	U						NI	
1963 Feb 13	002250	13.7	90.9	81	I	3.9/4			63		GU	
1963 Feb 15	181530	12.4	89.5	76	U	3.7/1					ES	
1963 Feb 21	234235	12.7	84.8	n	U	3.2/1					NI	
1963 Feb 24	133416	14.6	91.3	134	I	5.2/6	6.0/3		163	F	GU	Notes
1963 Feb 25	085844	12.3	88.2	58	U	4.2/5					NI	
1963 Feb 26*	163013	12.1	88.1	n	U	3.8/3					NI	

1963 Mar 05	024850	12.6	90.9	98	U					GU	
1963 Mar 08	222626	9.0	84.1	n	U	3.8/3				CR	
1963 Mar 10	060433	14.0	89.8	s	Am	3.5/1	4.1		D	ES	Notes
1963 Mar 29	074752	12.3	90.4	97	U	3.6/1				GU	
1963 Apr 10	141153	14.0	91.0	161	U		4.1			GU	
1963 Apr 13	185318	11.7	87.8	n	U	3.7/2	4.5			NI	
1963 Apr 15	223558	14.3	91.9	88	U	3.7/2	4.5			GU	
1963 Apr 25	072011	12.2	87.3	75	U		4.0			NI	
1963 May 01	163852	14.9	91.8	109	U					GU	
1963 May 03	100137	13.7	89.1	24	M			79		ES	
1963 May 04	161927	13.6	90.9	123	U	3.6/4				GU	
1963 May 05	124443	14.3	91.9	112	U	4.0/1				GU	
1963 May 09	150344	12.5	86.9	s	D	5.2/11		71	D	NI	I, Volc, Notes
1963 May 09*	150627	12.7	86.6	n	U				F	NI	Notes
1963 May 13	124401	14.5	93.2	65	I	5.4/6	5.6	128		MX	
1963 May 20	013531	8.2	83.0	n	U	4.7/4	4.9			PA	
1963 May 20*	024316	8.3	83.0	n	U	4.4/3	4.4			PA	
1963 May 24	061009	12.1	88.5	42	U	3.3/1	4.3			ES	
1963 May 25	145228	13.1	89.9	91	U	3.8/4	4.3			ES	
1963 May 29	224137	13.9	91.8	131	U	3.8/2	4.4			GU	
1963 May 31	215346	11.2	91.2	n	U					GU	
1963 Jun 02	222159	13.2	91.4	100	I	4.8/3	4.9	72		GU	
1963 Jun 03	142557	15.3	91.5	n	U		3.9			GU	
1963 Jun 20	033427	13.9	89.5	43	U	3.7/2	4.1			ES	
1963 Jun 21	033238	10.7	86.0	s	D	3.8/2				CR	U
1963 Jun 21	040254	15.0	91.7	219	U	3.8/2	4.1			GU	
1963 Jun 25	114338	8.5	83.0	s	U	4.2/4	4.5		F	PA	Notes
1963 Jun 26	174240	7.1	82.4	s	I	5.7/12	5.9	167		PA	
1963 Jul 02	180534	9.0	83.0	64	U	3.5/3	4.1			CR	
1963 Jul 02	220049	13.7	88.9	173	U	3.3/2	4.2			ES	
1963 Jul 04	014321	12.8	88.7	55	U					ES	
1963 Jul 05*	144055	12.8	88.9	85	U	3.8/2	4.2			ES	
1963 Jul 09	092434	8.4	83.2	39	I	5.7/12	5.8	141	F	CR	Notes

Date	Time (UT) h.m.s.	Epicentre °N	Epicentre °W	Depth km	Ag	M_S	$m_B/$ m_b	M^* $logM_O$	N	Felt	Reg	Other Agencies, Felt Radii Comments
1963 Jul 10*	182248	8.8	82.9	51	U	3.6/2	4.2				CR	
1963 Jul 10*	224937	8.5	83.0	48	U						CR	
1963 Jul 11	070450	16.8	91.3	n	U	3.0/2	3.7				MX	
1963 Jul 15	163219	14.4	90.2	n	U						GU	
1963 Jul 22	001633	13.6	89.9	34	U						ES	
1963 Jul 24	140604	12.3	88.3	47	U						ES	
1963 Jul 30	122539	11.4	87.3	n	U	4.0/1	4.4				NI	
1963 Jul 31	183329	9.2	82.4	n	U	3.3/1					PA	
1963 Aug 02	142836	11.8	89.5	n	U	3.2/1					ES	
1963 Aug 07	093538	14.6	91.3	235	U						GU	
1963 Aug 07	183646	13.6	90.9	67	U	5.0/7					GU	
1963 Aug 10	183745	8.8	82.9	n	U	4.1/1					CR	
1963 Aug 11	181537	10.3	83.5	s	M	3.7/1					CR	
1963 Aug 12	171905	9.8	84.2	114	U	3.3/1					CR	
1963 Aug 26	013009	14.4	91.3	n	U	3.0/1					GU	
1963 Aug 28	055551	12.6	87.2	115	U						NI	
1963 Aug 31	130846	11.9	87.0	48	U	3.7/3					NI	
1963 Sep 01	225734	11.3	85.5	120	U	3.8/4					NI	
1963 Sep 08	052838	12.3	87.9	n	U	3.1/1					NI	
1963 Sep 14	094352	15.0	91.0	130	U						GU	
1963 Sep 17	034810	12.6	87.1	144	U						NI	
1963 Sep 25	004738	16.5	86.6	n	U	3.6/4					HO	
1963 Sep 25*	010940	16.6	87.1	n	U	4.0/4					HO	
1963 Sep 28	052503	10.2	86.0	n	U	2.9/1					CR	
1963 Sep 29	224413	14.7	91.7	134	I	5.0/3			90		GU	
1963 Sep 29*	230522	14.3	92.0	n	U						GU	
1963 Sep 29*	230821	14.5	92.0	42	U						GU	
1963 Sep 30*	030538	14.1	91.9	86	U	3.8/3					GU	
1963 Sep 30	140121	12.9	89.0	65	U						ES	

1963 Oct 01	103031	10.2	84.5	46	U	4.4/3			CR		
1963 Oct 05	061626	11.7	85.1	156	U				NI		
1963 Oct 07	035954	11.6	86.9	50	U	4.0/3			NI		
1963 Oct 13	051012	9.8	84.1	63	U	4.2/2			CR		
1963 Oct 19	212225	15.2	89.2	n	U				GU		
1963 Oct 28	022508	14.8	90.8	n	U	3.2/1			GU		
1963 Nov 01	113456	13.6	90.1	128	U	4.0/4			ES		
1963 Nov 06	214958	12.2	88.0	37	U	3.4/1			NI		
1963 Nov 10	071808	12.9	88.5	45	U				ES		
1963 Dec 13	195727	14.5	91.9	139	U				GU		
1963 Dec 16	062320	12.2	88.4	34	U	3.5/1			ES		
1963 Dec 16	104952	13.9	90.9	59	U	3.9/1			GU		
1963 Dec 20	222831	13.2	88.0	64	U				ES		
1963 Dec 27	035710	14.5	90.7	n	U				GU		
1963 Dec 31	142207	12.4	87.9	77	U				NI		
1964 Apr 09	041521	13.13	89.73	86	E		60		ES	I	
1964 Apr 24	144030	13.47	88.70	146	E	5.3/3	112		ES	I	
1964 Jul 06	0515	9.91	83.78	n	AM			F	CR	Notes	
1964 Jul 30	051606	11.14	86.23	52	E	5.9/10	127	V	NI	I, D, Notes	5/50
1964 Aug 08	154515	12.55	88.02	81	E	5.2/8	103		NI	I	
1964 Aug 10	0715	9.9	83.8	n	Am			F	CR	Notes	2/19, 4/6
1964 Dec 15	121324	14.64	91.80	87	E	5.6/3	117		GU	I	
1965 Jan 23	200729	8.77	83.25	9	E		52	F	CR	I, Notes	2/30
1965 Jan 30	024543	9.87	83.93	s	AM		9	F	CR	I, Notes	3/12
1965 Feb 24	080918	13.96	92.18	48	E	5.7/3	99		GU	I	
1965 Mar 01	213214	15.36	92.57	104	E	5.9/14	208		MX	I	
1965 Mar 21	094243	11.57	86.42	48	E	5.7/1	95		NI	I, D	
1965 May 03	100139	13.70	89.17		AM	5.9/16	135	VIII	ES	E, I, Notes	3/180, 6/20, 7/8, 8/5
1965 May 26*	045847	13.74	90.58	98	E		102		ES	I	
1965 Jul 19	221422	10.7	85.3	81	I		21	F	CR	Notes	2/48
1965 Aug 29	014556	14.04	90.69	81	E	5.4/7	116		GU	I	
1965 Oct 04	062306	8.79	82.85	34	E		73	F	CR	I, Notes	2/130
1965 Oct 16	142254	8.98	83.48	24	E		89		CR	I, GD	

Date	Time (UT) h.m.s.	Epicentre °N	°W	Depth km	Ag	M_S	m_B/ m_b	M^* $log M_O$	N	Felt	Reg	Other Agencies, Comments	Felt Radii
1965 Oct 20	235432	12.44	87.40	74	E	5.6/7			107		NI	I, D	
1965 Oct 20	235444	12.62	86.79	19	E				37		NI	I	
1965 Nov 12	085957	10.60	84.44	37	E				51	F	CR	I, Notes	3/45
1965 Dec 15	230524	7.55	82.19	26	E	6.4/14			207		PA	I, GD	
1966 Mar 13	214623	13.15	88.61	88	E				30	F	ES	I, Notes	2/100
1966 Mar 23	051134	16.76	85.93	32	E				85		HO	I	
1966 Mar 27	185341	8.79	83.54	26	E	5.6/2			142	V	CR	I, GD, Notes	5/13
1966 Apr 09*	023429	9.47	84.19	34	E				143	F	CR	I, GD, Notes	
1966 Apr 09	024210	9.42	84.20	30	E	5.6/8			166	V	CR	I, GD, Notes	3/120, 5/45
1966 May 04	181358	12.61	87.68	74	E				65		NI	I	
1966 May 11	143535	8.7	83.5	n	Am				21	D	CR	I, Notes	
1966 Jul 10	191858	9.9	83.8	n	Am				9	F	CR	I, Notes	
1966 Jul 13	082100	12.57	87.72	53	E	5.1/2			147		NI	I	
1966 Aug 09	111242	9.25	84.01	53	E	5.0/1			67	F	CR	I, Notes	
1966 Aug 18	103316	14.44	91.84	67	E	6.0/7			222		GU	I	
1966 Oct 12	202010	11.13	86.31	67	E	5.4/4			95		CR	I, D	
1966 Dec 10	130634	14.26	92.10	66	E	6.6/8			239		GU	I	
1966 Dec 11	172200	10.17	84.18	104	E				31	F	CR	I, Notes	2/40
1966 Dec 27	212217	13.18	88.92	73	E				129	V	ES	I, Notes	5/50
1967 Jan 26	161034	14.9	93.0	65	I	5.7/6			154		MX		
1967 Apr 22	144320	8.24	82.81	17	E	5.0/2			118	F	PA	I, GD, Notes	
1967 May 16	125808	13.46	90.78	61	E	5.3/4			113		GU	I	
1967 Aug 01	11	9.9	84.1	s	Am				1	F	CR	Notes	2/13
1967 Aug 27	130856	12.15	86.31	175	E	5.2/4			210		NI	I, D	
1967 Oct 03	181607	10.94	5.90	37	E	6.1/8			245	F	CR	I, GD, D, Notes	4/90
1967 Oct 04	060218	10.73	86.02	28	E	5.3/1			111		CR	I, GD, D	
1967 Oct 15	080052	11.90	85.95	160	E	6.5/11			317	F	NI	I, D, Notes	3/240
1967 Nov 08	031055	16.81	85.92	33	E	5.1/4			147		HO	I	
1967 Nov 18	121656	13.14	89.44	69	E				86	F	ES	I, Notes	

1967 Dec 06	025307	12.35	87.33	76	E		111		NI	D	
1968 Jan 04	100406	11.84	86.66	77	E		33	VIII	NI	I, Notes	3/66, 4/27, 6/7, 7/3, 8/1
1968 Jan 21	234446	15.64	92.33	146	E	4.9/1	159		MX	I	
1968 Jan 21	234518	16.73	92.29	73	E		72		MX	I	
1968 Mar 01	163453	11.71	85.62	261	E		53	F	NI	I, Notes	2/180
1968 Mar 03	084326	10.1	83.1	n	I		11	F	CR	Notes	
1968 Mar 24	171323	12.08	86.75	105	E	5.3/2	77		NI	I, D	
1968 Mar 28	010737	15.09	92.08	90	E	5.5/3	137	V	MX	I, Notes	2/360, 5/45
1968 May 06	143749	14.42	91.07	110	E		117	F	GU	I, Notes	2/215
1968 Jun 15	051120	14.42	92.90	35	E	5.4/7	143		MX	I	
1968 Jun 15	070849	5.6	82.6	s	I	6.1/9	216	F	PA	Notes	2/340
1968 Jul 02	161544	8.5	83.1	n	I		11	F	CR	Notes	
1968 Jul 04	151559	10.28	84.29	95	E		27	F	CR	I, Notes	
1968 Jul 17	062314	10.37	83.40	25	E	5.1/1	101	F	CR	I, GD, Notes	2/110
1968 Sep 25	103838	15.54	92.67	122	E	6.3/23	214	VII	MX	I, Ts, Notes	3/285, 4/210, 5/160, 6/116, 7/50
1968 Oct 02	024559	14.24	90.98	93	E	3.6/1	42		GU	I	
1968 Oct 09	030942	14.4	93.0	63	I	4.6/1	34		MX		
1968 Oct 22	064204	13.08	88.43	69	E	3.7/2	53		ES	I	
1968 Nov 03	195324	7.86	85.90	14	E	3.9/1	45		CR	I, W	
1968 Nov 11	031108	15.06	92.73	74	E	4.2/1	60		MX	I	
1968 Nov 21	071942	11.5	87.0	82	I	4.0/1	21		NI		
1968 Nov 28	134435	12.35	88.23	38	E	4.9/1	47		ES	I	
1968 Dec 08	031226	13.7	91.0	85	I	4.1/1	14		GU		
1968 Dec 10	180731	12.9	89.7	63	I	3.9/2	21		ES		
1968 Dec 16	030727	7.11	82.30	20	E	4.7/2	108		PA	I	
1968 Dec 27	114523	8.55	82.81	27	E		49	F	PA	I, Notes	
1968 Dec 29	173632	14.49	92.36	64	E	5.5/7	156		MX	I	
1969 Jan 03	111916	14.47	91.66	96	E		82		GU	I	
1969 Feb 25	073902	15.23	87.44	9	E	5.1/2	150	F	HO	I, Notes	3/130
1969 Mar 10	081510	12.36	87.46	57	E		119	F	NI	I, D, Notes	2/240, 3/100
1969 Mar 14	084717	12.72	86.87	178	E	5.6/5	221	F	NI	I, D, Notes	

Date	Time (UT) h.m.s.	Epicentre °N	Epicentre °W	Depth km	Ag	M_S	$m_B/$ m_b	M^* $logM_O$	N	Felt	Reg	Other Agencies, Comments	Felt Radii
1969 Apr 01	211954	12.75	88.20	51	E				88	F	ES	I, Notes	
1969 Apr 21	021908	14.09	91.06	72	E	5.9/9			207	F	GU	I, Notes	3/205
1969 Apr 25	033420	7.45	82.03	27	E	6.6/9			184		PA	I, GD	
1969 May 13	141654	11.47	86.36	71	E	6.5/11			204	V	NI	I, D, Notes	3/250, 4/150, 5/95
1969 Jun 24	003511	11.51	85.85	137	E				148		NI	I, D	
1969 Jun 28	043442	12.73	89.42	47	E	5.1/4			124	F	ES	I, Notes	3/105
1969 Jul 04	111603	7.48	82.69	30	E	5.1/2			120		PA	I, GD	
1969 Jul 11	075105	10.39	85.35	30	E	5.2/2			84	F	CR	I, Notes	
1969 Aug 15	014315	9.72	83.96	15	E				56	F	CR	I, Notes	3/70
1969 Nov 06	130118	15.52	92.68	104	E				77		MX	I	
1970 Jan 01	014346	8.51	83.39	9	E	5.5/6			141	F	CR	I, GD, Notes	3/80
1970 Jan 15	165247	11.84	86.69	63	E	5.3/3			92		NI	I, D	
1970 Jan 20	082752	11.37	86.52	71	E	4.9/4			104		NI	I, D	
1970 Jan 22	151431	14.40	92.32	62	E	5.6/4			128		MX	I	
1970 Jan 31	222339	15.75	88.68	42	E				98		HO	I	
1970 Mar 27	183921	12.70	86.92	177	E				72		NI	I, D	
1970 Apr 09	162430	13.21	92.29	15	E	5.5/6			167		MX	I	
1970 Apr 21	053709	12.84	87.89	77	E				65		HO	I	
1970 Apr 29*	112235	14.46	92.76	18	E	6.5/16			215	D	MX	I, Notes	
1970 Apr 29*	120731	14.27	92.72	17	E				52		MX	I	
1970 Apr 29*	140122	14.48	92.77	50	E				168		MX	I	
1970 Apr 29	140136	14.46	92.68	50	E	7.3/14			233	V	MX	I, Notes	3/850, 4/380, 5/230
1970 May 02*	051546	14.5	93.0	55	I	5.2/3			108		MX		
1970 May 15*	094446	14.23	92.97	36	E	5.5/10			140		MX	I	
1970 Jun 27*	225328	14.48	92.92	51	E	5.3/4			86		MX	I	
1970 Aug 12	092417	11.89	86.75	72	E	6.5/9			271		NI	I, D	
1970 Aug 12*	102431	11.97	86.75	77	E	5.4/7			154		NI	I, D	
1970 Sep 29	044248	11.53	85.46	190	E	5.0/2			197		NI	I, D, GD	
1970 Nov 29	200055	15.29	92.84	95	E				154		MX	I	

1971 Jan 20	044600	8.79	79.09	22	E	5.5/22	217		PA	I	
1971 Feb 19	062759	14.17	91.69	59	E	4.7/4	99		GU	I	
1971 Feb 19	165944	13.85	88.83	176	E		67		ES	I	
1971 Feb 25	041544	12.12	87.55	58	E	5.4/16	110		NI	I, D	
1971 Mar 24	161828	14.06	92.44	33	E		66		GU	I	
1971 May 01	143213	13.23	88.50	83	E	5.0/2	93		ES	I	
1971 May 28	141208	14.72	91.44	117	E	4.7/2	178		GU	I	
1971 Jun 05	142044	9.34	84.23	27	E	5.4/8	189		CR	I, GD	
1971 Jun 28	024410	14.41	91.73	100	E		65		GU	I	
1971 Aug 05	225812	9.56	84.19	55	E	3.8/2	76		CR	I, GD	
1971 Aug 20	213610	13.28	92.44	30	E	5.8/12	245		GU	I	
1971 Aug 27	092204	12.13	88.91	55	E	4.2/2	81		ES	I	
1971 Sep 06	095850	14.01	92.51	52	E	4.5/2	60		GU	I	
1971 Sep 28	054721	9.80	86.25	56	E		45		CR	I, GD, D	
1971 Oct 12	094858	15.80	91.31		AM	5.4/9	238	VIII	GU	E, I, Notes	3/140, 7/15, 8/8
1972 Jan 05	115230	12.08	86.37		AM	3.9/2	26	D	NI	E, I, Notes	
1972 Jan 22	130848	14.01	91.09	71	E	5.9/27	236		GU	I	
1972 Feb 07	191450	8.50	83.97	20	E	5.6/20	195		CR	I, GD	
1972 Mar 11	035047	13.37	90.06	77	E	5.6/14	212		GU	I	
1972 Jun 11	182119	13.44	90.65	49	E		70		GU	I	
1972 Jun 30	124524	14.57	92.81	55	E	4.8/6	78		GU	I	
1972 Aug 31	022412	13.02	88.69	68	E	5.1/10	151		ES	I	
1972 Oct 05	104818	13.82	91.19	65	E	5.5/14	155		GU	I	
1972 Nov 19	043525	12.98	88.57	64	E	5.2/4	113		ES	I	
1972 Dec 23	062945	12.15	86.27		AM	6.3/29	192	IX	NI	E, I, D, Notes	4/130, 6/60, 6/37, 7/22, 8/11, 9/6
1972 Dec 23*	071742	12.08	86.24	31	E		44		NI	I, D	
1972 Dec 23*	071952	12.22	86.01	6	E		89		NI	I, D	
1973 Feb 15	001601	11.50	86.63	84	E	5.0/10	86		NI	I	
1973 Feb 22	003021	14.46	91.65	95	E	4.9/6	164		GU	I	
1973 Apr 14	083402	10.47	84.97		AM	6.5/37	244	VIII	CR	E, I, GD, Notes	5/115, 6/75, 7/40, 8/9
1973 Apr 24	211643	7.36	82.37	3	E		126		PA	I, GD	

Date	Time (UT) h.m.s.	Epicentre °N	°W	Depth km	Ag	M_S	m_B/ m_b	M^* $log M_O$	N	Felt	Reg	Other Agencies, Felt Radii Comments
1973 May 24	193414	14.59	91.41	95	E	4.8/1			182		GU	I
1973 May 31	053921	13.96	90.94	107	E	5.1/11			238		GU	I
1973 Jun 07	183244	14.31	91.98	66	E	6.4/13			258	F	GU	I, Notes
1973 Jun 07	183444	14.41	91.79	28	E				177		GU	I
1973 Jun 07*	194543	14.35	92.01	83	E				91		GU	I
1973 Jul 27	194249	12.76	86.73	195	E	5.4/6			171		NI	I
1973 Jul 28	204827	12.33	87.13	85	E	4.3/2			117		NI	I
1973 Aug 04	004446	9.85	84.66	43	E	5.4/13			202		CR	I, GD
1973 Aug 19	221753	12.94	88.49	70	E	4.2/2			48		ES	I
1973 Oct 02	210507	10.44	85.18	87	E	4.2/2			92		CR	I, GD
1973 Oct 18	010908	9.15	84.12	58	E	5.1/8			163		CR	I, GD
1973 Nov 21	133136	14.51	92.78	80	E	5.6/17			154		GU	I
1973 Dec 05	040149	14.58	91.81	85	E	4.6/2			100		GU	I
1974 Feb 02	070738	13.95	91.90	43	E	4.6/4			64		GU	I
1974 Feb 28*	201536	9.42	83.98	25	E				132		CR	I, GD
1974 Feb 28	202009	9.34	84.05	22	E	6.1/21			254		CR	I, GD
1974 Mar 06	014029	12.30	86.45	119	E	6.0/7			339	D	NI	I, Notes
1974 Mar 31	211349	11.76	87.94	60	E	5.0/3			103		NI	I
1974 Apr 10	224300	14.55	91.64	88	E	5.1/11			236	F	GU	I, Notes
1974 Apr 14	065500	14.67	91.46	110	E	4.3/2			179		GU	I
1974 May 04	175408	7.65	82.74	27	E	4.1/2			112		PA	I, GD
1974 May 16	230750	11.26	86.20	47	E	5.6/26			189		NI	I
1974 Jul 05	200936	14.67	81.82	18	E	4.6/2			108		HO	I
1974 Jul 21	082835	14.28	92.03	56	E	5.6/24			174		GU	I
1974 Jul 21	223339	11.23	86.29	77	E	4.9/11			98		NI	I
1974 Aug 16	120913	8.46	82.82	42	E	4.1/4			64		PA	I, GD
1974 Aug 19	195444	12.29	88.94	40	E	5.0/8			148		ES	I
1974 Aug 26	062742	15.90	91.05	32	E	4.7/6			127		GU	I
1974 Oct 07	171036	11.58	85.49	204	E				135		NI	I, GD

1974 Dec 02	123646	14.59	91.59	97	E	5.3/7		159		GU	I	
1974 Dec 06	135840	8.10	82.90		AM	5.5/14		143	VI	PA	E, I, Notes	2/220, 3/170, 4/90, 5/40, 6/20
1974 Dec 10	2330	9.76	84.00	s	AM			1	F	CR		2/50
1974 Dec 24	021026	14.24	90.11	143	E	4.9/4		196		GU	I	
1974 Dec 31	201533	14.19	91.88	64	E	5.7/8		139		GU	I	
1974 Dec 31	202116	14.18	91.87	84	E	6.1/11		168		GU	I	
1975 Jan 28	022107	9.80	84.34	s	AM	3.4/1		19	F	CR	I	3/25
1975 Feb 03	010327	15.66	91.67	216	E	5.2/3		228	F	GU	I, Notes	
1975 Mar 13	234012	16.26	87.02	28	E	5.1/7		140		HO	I	
1975 Mar 25	023127	13.68	90.82	60	E	5.4/13		109		GU	I	
1975 Apr 16	044625	10.51	85.74	89	E	4.9/6		143		CR	I, GD	
1975 Jul 15	155357	7.75	82.88	33	E	5.0/9		122		PA	I, GD	
1975 Jul 17	175201	13.64	88.90	s	AM	3.9/2		10	VI	ES	I, Notes	2/55, 3/27, 4/10, 6/4
1975 Jul 19	234847	8.37	82.97	39	E	4.2/2		90		PA	I, GD	
1975 Sep 01	215042	8.64	83.15	53	E	3.7/2		88		CR	I, GD	
1975 Dec 15	142757	14.31	92.71	56	E	4.3/1		56		GU	I	
1975 Dec 15	173631	14.42	92.70	53	E	4.2/1		71		GU	I	
1976 Jan 04	113455	13.01	88.45	70	E	4.2/1		113		ES	I	
1976 Feb 04	090144	15.16	89.18	10	AM	7.5/20	20.41	433	VIII	GU	E, I, Ts, Notes	2/800, 4/330, 5/190, 6/120, 7/60, 8/40
1976 Feb 04∗	093039	14.78	90.43	73	E			85		GU	I	
1976 Feb 06∗	181925	14.75	90.55		AM	5.6/11		246	D	GU	E, I, Notes	
1976 Feb 08	081350	15.60	88.55	7	E	5.8/19		199		HO	I	
1976 Feb 09	114451	15.28	89.14	21	E	4.9/9		168		GU	I	
1976 Feb 20	142045	13.71	92.26	40	E	4.7/1		137		GU	I	
1976 Feb 25	162902	10.36	85.29	64	E	5.6/9		167		CR	I, GD	
1976 Mar 07∗	025407	14.67	91.02	11	E	4.9/5		117		GU	I	
1976 Mar 07∗	031549	14.81	90.90	52	E	5.0/5		129		GU	I	
1976 Mar 09∗	074243	14.77	91.08	37	E	4.5/3		112		GU	I	
1976 Mar 13∗	163044	14.77	90.98		AM	5.5/16		159	D	GU	E, I, Notes	
1976 May 04	084324	12.97	87.78	4	E	4.8/1		140		ES	I	
1976 May 19	174401	16.90	85.58	32	E	4.8/3		191		HO	I	

Date	Time (UT) h.m.s.	Epicentre °N	°W	Depth km	Ag	M_S	m_B/ m_b	M^* $logM_O$	N	Felt	Reg	Other Agencies, Comments	Felt Radii
1976 May 24	105456	11.48	86.90	44	E	5.2/5			141		NI	I	
1976 Aug 05	184811	12.77	88.21	50	E	4.8/1			133		ES	I	
1976 Aug 09	170151	12.85	88.20	54	E	4.4/1			99		ES	I	
1976 Sep 19	122336	7.39	82.25	27	E	4.8/4			144		PA	I	
1976 Oct 09	123116	10.74	86.05	75	E	5.4/11			162		CR	I, GD	
1976 Oct 29	044854	14.29	92.06	36	E				116		MX	I	
1976 Nov 25	064523	9.78	84.71	34	E	4.9/3			150		CR	I, GD	
1976 Dec 01	141538	9.75	84.71	32	E	5.3/14			207		CR	I, GD	
1976 Dec 01	174449	13.28	89.49	78	E				34		ES	I	
1976 Dec 01	232457	14.6	91.1	n	I				10		GU		
1976 Dec 20	101857	9.35	83.85	48	E	5.2/6			198		CR	I, GD	
1977 Jan 23	170038	13.96	91.80	53	E	4.4/1			138		GU	I	
1977 Feb 19	031908	11.54	87.01	41	E	4.5/2			100		NI	I	
1977 Apr 11	100126	13.13	88.18	15	E	4.5/2			61		ES	I	
1977 Aug 04	132057	12.27	87.45	59	E	5.2/12			178		NI	I	
1977 Aug 07	070807	8.49	82.82	28	E	5.7/18		18.08	256		PA	I, GD	
1977 Aug 20	024613	16.69	86.78	11	E	5.8/23		18.20	255		HO	I	
1977 Aug 20	035156	16.68	86.67	31	E	6.0/17		18.58	319		HO	I	
1977 Sep 03	223307	12.54	87.62	63	E	4.9/6		17.59	179		NI	I	
1977 Sep 10	102107	13.97	91.77	25	E	4.9/8		17.32	193		GU	I	
1977 Sep 30	071027	11.19	85.92	68	E	5.0/6		17.75	187	D	CR	I, GD, Notes	2/250, 3/140, 4/70
1977 Oct 20	092037	11.92	86.95	54	E	4.4/3			112		NI	I	
1977 Oct 31	005318	13.08	88.96	76	E	4.1/1			88		ES	I	
1978 Jan 20	021241	14.24	92.26	67	E	4.6/1			119		MX	I	
1978 Feb 22	060735	14.29	91.46	66	E	6.2/45		18.54	382	D	GU	I, Notes	4/250
1978 Feb 22*	193140	13.99	90.82	68	E	4.8/3			125		GU	I	
1978 Mar 04	173509	14.42	91.32	74	E	4.7/1			174		GU	I	
1978 Mar 30	193013	14.93	90.45	24	E	4.2/1			57		GU	I	
1978 Apr 12	165110	12.58	87.86	63	E				104		NI	I	

1978 Apr 15	171841	14.18	91.39	113	E	5.0/2		227		GU	I	
1978 Apr 26	210057	11.38	86.67	62	E			70		NI	I	
1978 May 15	211647	11.4	86.6	92	I	3.9/1		23		NI		
1978 May 31	010721	12.76	87.29	49	E	6.0/63	18.86	387	V	NI	I, Notes	2/520, 3/250, 5/140
1978 Jul 03	041803	10.60	86.10	56	E			115		CR	I, GD	
1978 Jul 20	093446	12.18	86.70	95	E	5.2/5	17.91	264	F	NI	I, Notes	
1978 Jul 29	143733	14.67	91.01		AM	4.5/1		124	D	GU	E, I, Notes	3/50
1978 Jul 31	201851	14.09	89.98	49	E	4.2/1		53		GU	I	
1978 Aug 18	153518	14.37	91.51	80	E	5.9/34	18.38	370	V	GU	I, Notes	3/260, 5/100
1978 Aug 23	003830	10.19	85.30	25	E	7.0/68	19.52	433	D	CR	I, GD, Notes	
1978 Aug 23*	005029	10.29	85.14	54	E			240		CR	I, GD	
1978 Sep 10	232415	14.28	91.57	76	E	5.8/49	18.20	378	D	GU	I, Notes	3/265, 4/120
1978 Oct 06	124545	12.72	88.10	64	E	4.1/1		51		ES	I	
1978 Oct 17	201650	13.76	89.78	98	E	4.1/1		95		ES	I	
1978 Oct 30	182336	13.83	91.13	91	E	6.0/53	18.51	296	F	GU	I, Notes	3/210, 4/210
1978 Oct 30*	190943	13.82	91.18	59	E			84		GU	I	
1978 Oct 31*	063036	13.81	91.12	58	E	5.1/11		115		GU	I	
1978 Dec 06	115341	13.31	89.74	70	E	6.0/46	18.26	369	V	ES	I, Notes	4/170, 5/75
1978 Dec 06	142655	13.60	92.31	44	E			146		GU	I	
1978 Dec 14	175652	11.5	87.4	n	I			11		NI		
1978 Dec 18	023127	12.94	88.39	67	E		17.46	223	F	ES	I, Notes	4/125
1978 Dec 19	085729	11.71	87.33	47	E	5.6/15	17.73	138		NI	I	
1978 Dec 19*	163402	11.72	87.36	30	E	3.8/1		28		NI	I	
1978 Dec 20*	025507	11.80	87.24	26	E	3.6/1		50		NI	I	
1979 Jan 12	035902	14.30	91.59	76	E	5.8/47	18.26	348	F	GU	I, Notes	
1979 Mar 15	034556	13.6	89.2	s	Am			13	F	ES	I, Notes	4/9
1979 May 26	001623	13.60	92.26	27	E	4.6/3	16.90	208		GU	I	
1979 Jun 27	095005	7.11	82.30	8	E	6.1/56	18.63	346		PA	I	
1979 Jun 30*	083300	7.05	82.38	6	E	5.1/6	17.32	119		PA	I	
1979 Jul 01	203807	8.34	82.94	39	E	6.4/53	18.36	334	VI	PA	I, GD, Notes	3/230, 5/90, 6/45
1979 Jul 02*	061812	8.34	83.01	46	E	5.1/19	17.32	161		PA	I, GD	
1979 Jul 07	093302	14.10	91.09	70	E	5.2/22	17.52	183		GU	I	
1979 Aug 03*	170749	8.21	82.96	1	E	4.4/3		62		PA	I, GD	

Date	Time (UT) h.m.s.	Epicentre °N	Epicentre °W	Depth km	Ag	M_S	m_B/ m_b	M^* $log M_O$	N	Felt	Reg	Other Agencies, Comments	Felt Radii
1979 Aug 13	114225	8.7	88.8	n	I				5		ES		
1979 Aug 24	042654	8.92	83.50	30	E	6.4/47		18.76	425	VI	CR	I, GD, Notes	5/95, 6/45
1979 Oct 01	141412	15.77	92.26	163	E	5.4/15		18.32	322		MX	I	
1979 Oct 09	074927	14.14	90.24		AM	5.0/4			141	VII	GU	E, I, Notes	2/120, 5/40, 6/18, 7/9
1979 Oct 27	143600	13.86	90.86	64	E	6.8/62		19.36	441	V	GU	I, Notes	3/440, 5/180
1979 Oct 27*	214326	13.78	90.75	65	E	6.6/52		19.28	406	F	GU	I, Notes	4/180
1979 Dec 09	154618	13.2	91.0	100	I				15		GU		
1979 Dec 12	213955	11.8	88.0	56	I				16		NI		
1979 Dec 16	125000	15.24	92.27	115	E	4.2/1			93		MX	I	
1979 Dec 18	103759	11.40	86.47	64	E	5.6/27		17.92	197		NI	I	
1980 Mar 04	062541	13.00	88.42	63	E	4.6/2			76		ES	I	
1980 Mar 06	005251	7.55	82.96	26	E	4.9/7			100		PA	I, GD	
1980 Mar 20	165421	16.86	85.67	31	E	4.4/1			87		HO	I	
1980 May 18	070939	14.73	91.30	101	E	4.9/11		17.85	264	D	GU	I, Notes	
1980 Jun 07	134733	13.14	88.66	53	E	3.8/1			28		ES	I	
1980 Jul 20	002405	16.19	91.93	61	E				95		MX	I	
1980 Aug 01	081613	12.54	87.66	25	E	4.8/7		17.41	199		NI	I	
1980 Aug 09	054511	15.88	88.52		AM	6.5/32		18.83	362	VII	GU	E, I, Notes	7/29
1980 Sep 02	102812	15.93	88.30	34	E	4.7/7		16.98	211	F	GU	I, Notes	
1980 Nov 04	162116	13.89	90.99	73	E	5.0/6		17.38	156	F	GU	I, Notes	2/190
1981 Jan 27	015313	11.38	86.20	44	E	4.6/2		17.18	92		NI	I	
1981 Apr 05	102622	14.48	92.71	12	E	4.7/5		17.36	155		MX	I	
1981 Apr 20	055021	14.37	91.68	79	E	4.8/5		17.45	190	F	GU	I	
1981 May 01	055032	12.51	87.72	57	E	4.6/2			68		NI	I	
1981 May 01	061719	9.75	84.89	25	E	4.7/2			97		CR	I, GD	
1981 Jun 08	054755	12.82	88.28	60	E	3.7/1			56		ES	I	
1981 Jun 11	183424	16.67	86.13	33	E	5.4/23		17.95	230	F	HO	I, Notes	
1981 Aug 11	060034	11.8	87.3	n	I				10		CR		

1981 Aug 29	181752	12.72	87.75	78	E	4.4/1		46		NI	I	
1981 Sep 03	080903	11.94	87.38	45	E	3.9/2		57		NI	I	
1981 Sep 27	133235	14.42	91.64	89	E	4.5/2	17.46	182	F	GU	I, Notes	
1981 Nov 07	220249	14.59	90.45	188	E	4.7/6	17.90	304		GU	I	
1981 Dec 21	103214	14.82	92.37	80	E	5.4/8	18.04	223	F	GU	I	
1982 Jan 12	054820	13.18	87.49	8	E	6.1/25	18.26	298	VII	HO	I, Notes	4/200, 7/20
1982 Mar 04	114103	13.27	88.93	59	E	4.8/5		208	V	ES	I, Notes	5/50
1982 Mar 06	212443	12.77	88.04	62	E			84	F	ES	I,	2/160
1982 Mar 17	135617	11.74	86.72	59	E	5.0/9		140		NI	I	
1982 Mar 26	224946	10.70	85.97	42	E	4.7/2		129		CR	I, GD	
1982 Apr 06*	154947	14.22	91.98	25	E	5.3/16		285	F	GU	I	
1982 Apr 06	195653	14.28	92.07	45	E	6.6/61	19.15	445	F	GU	I, Notes	
1982 Apr 07	192051	10.66	85.76	35	E	5.4/18		297		CR	I, GD	
1982 Apr 18	104210	14.28	91.61	65	E	5.4/11		330	D	GU	I	
1982 Jun 19	062159	13.35	89.37	74	E	7.3/57	20.00	567	VII	ES	I, Notes	3/600, 5/200, 7/70
1982 Jun 21*	194918	13.3	89.3	92	I			8	F	ES		
1982 Jul 07	015241	11.37	86.49	53	E	5.6/18		208	F	NI	I	
1982 Aug 17	182404	9.52	84.10	28	E	5.6/23	17.83	245	F	CR	I, GD	
1982 Aug 19	155904	6.71	82.71	11	E	6.7/52	19.26	480	V	PA	I, Notes	3/430, 5/140
1982 Aug 20	212805	13.99	90.89	64	E	4.8/5		190	F	GU	I	
1982 Sep 05	023636	14.12	91.81	53	E	5.4/16		285		GU	I	
1982 Sep 29*	034801	14.55	89.13	13	E			175	F	GU	I	2/210, 3/93
1982 Sep 29	055034	14.62	89.21		AM	5.2/11	17.49	231	VI	GU	E, I, Notes	2/190, 3/95, 6/10
1982 Oct 18	125659	13.80	90.84	71	E	4.4/3		113	F	GU	I	
1982 Oct 27	074106	13.46	89.41	79	E	5.2/8		196	F	ES	I, Notes	
1982 Oct 28	122905	14.26	91.69	50	E	4.4/2		89		GU	I	
1982 Oct 28	124524	14.25	91.80	51	E	4.4/2		71		GU	I	
1982 Oct 31	060306	14.03	90.34	96	E	5.6/9	18.40	313	F	GU	I	
1982 Nov 14	011936	14.48	92.34	56	E	5.1/15	17.69	249	F	GU	I	
1982 Nov 19	232207	10.85	86.00	52	E	4.6/3		56		CR	I, GD	
1982 Nov 24	102301	12.80	90.96	25	E	5.0/10	17.44	262	F	GU	I	
1982 Dec 09	082549	7.51	81.80	19	E	5.1/10		145		PA	I	
1983 Feb 09	210537	7.93	82.86	30	E	4.7/4		117	F	PA	I, GD	

Date	Time (UT) h.m.s.	Epicentre °N	°W	Depth km	Ag	M_S	$m_B/$ m_b	M^* $log M_O$	N	Felt	Reg	Other Agencies, Comments	Felt Radii
1983 Apr 03	025003	8.67	83.23		AM	7.2/24		20.06	574	VII	CR	E, I, GD, Notes	3/410, 5/210, 6/130 7/55
1983 Apr 03*	030035	9.03	83.20	8	E				85		CR	I, GD	
1983 Apr 03*	030422	8.76	83.32	89	E				107		CR	I, GD	
1983 Apr 03*	031339	8.61	83.19	18	E				121		CR	I	
1983 Apr 03*	050453	8.75	83.68	18	E				145	F	CR	I, GD	
1983 Apr 03*	054930	8.77	83.50	18	E	5.3/4			187	F	CR	I, GD	
1983 Apr 03*	144650	9.25	84.03	18	E	5.1/14			145		CR	I, GD	
1983 Apr 04*	131111	8.87	84.16	13	E	5.2/11			89		CR	I, GD	
1983 Apr 07*	192918	8.02	82.73	46	E	5.6/32		18.00	305	D	PA	I, GD, Notes	
1983 Apr 07*	204642	7.99	82.79	24	E	4.7/2			188	D	PA	I, GD	
1983 Apr 14	071930	14.16	91.44	78	E	5.6		17.73	314	F	GU	I, Notes	2/245, 3/110
1983 Apr 29	010221	13.41	89.21	81	E	4.9/9			180	F	ES	I, Notes	
1983 May 09*	155305	8.26	82.94	38	E	6.1/45		18.46	312		PA	I, GD	
1983 May 16*	155202	8.96	83.48	56	E				61		PA	I, GD	
1983 Jul 03*	171340	9.48	83.67		AM	5.5			66	F	CR	E, I, GD, Notes	3/140
1983 Jul 03	171422	9.52	83.66		AM	6.3/42		18.69	347	VIII	CR	E, I, GD, Notes	3/230, 4/170, 5/110 6/55, 7/12, 8/5
1983 Jul 18	125245	12.89	87.21		AM	6.0/18		18.74	389	D	NI	E, I, Notes	2/440, 3/220, 4/130
1983 Sep 23	234432	8.40	83.42	42	E			17.52	316	F	CR	I, GD, Notes	3/100
1983 Sep 29	152117	10.89	85.62	84	E	4.6/9			160	F	CR	I, GD, Notes	3/55
1983 Oct 12	033942	8.06	82.70	27	E	5.6/21		18.04	289	D	PA	I, GD, Notes	
1983 Oct 12	131209	13.69	90.98	64	E	5.4/21			215	F	GU	I, Notes	2/150
1983 Oct 13	114219	13.80	91.46	59	E	4.8/7			100		GU	I	
1983 Oct 17	231233	12.62	89.97	29	E				101	F	ES	I, Notes	2/120
1983 Oct 19	093312	12.60	87.75	76	E	4.1/5			114	F	NI	I, Notes	2/200
1983 Nov 01	115346	8.84	83.62	17	E	4.5/2			82		CR	I, GD	
1983 Nov 26	201826	7.33	82.31	20	E	6.0/38		18.52	290		PA	I	
1983 Nov 30	035000	13.65	91.03	74	E	4.7/7			120	F	GU	I, Notes	2/200

1983 Dec 02	030904	14.03	91.96	35	E	6.9/40	19.57	467	V	GU	I, Notes	5/160
1983 Dec 02*	115525	14.13	91.90	60	E	5.2/10		170		GU	I	
1983 Dec 10*	202239	14.29	91.79	78	E	5.6/29	18.08	296	F	GU	I, Notes	2/220
1983 Dec 11*	061936	14.08	91.90	69	E	4.2/4		67		GU	I	
1984 Jan 28	225244	9.14	83.96	32	E	4.3/3		86	F	CR	I, GD	
1984 Mar 26	202333	12.5	88.3	n	I			18	F	ES	Notes	
1984 Apr 09	124931	8.42	83.01	10	E	4.7/2	16.91	104	F	CR	I, Notes	2/200, 4/43
1984 Apr 13	091719	12.72	87.80	81	E	3.9/3		64	F	NI	I, Notes	2/185
1984 May 18	170109	11.62	86.78	77	E	4.6/3		156		NI	I	
1984 Jul 30	213300	13.28	90.24	65	E	4.7/12	17.30	126	F	GU	I	
1984 Aug 03	143100	14.39	92.30	39	E	5.5/27	17.86	295		GU	I	
1984 Aug 22*	181326	11.44	86.58	61	E	5.1/12	17.75	187	F	NI	I, Notes	3/100
1984 Aug 22*	235302	11.50	86.57	61	E	4.8/10		146	F	NI	I, Notes	2/115
1984 Aug 23*	003018	11.40	86.57	62	E	4.6/7		126	F	NI	I, Notes	2/320, 3/115
1984 Aug 23	194919	11.53	86.55	65	E	6.1/52	18.28	319	F	NI	I, Notes	3/120, 4/77
1984 Aug 31	044258	12.89	87.17	7	E	5.2/16	17.51	256	F	NI	I, Notes	2/180, 4/60
1984 Sep 13	073659	11.36	86.74	54	E	5.6/26		193		NI	I	
1984 Nov 02	034802	13.18	89.95	75	E	4.9/13	17.15	264	F	ES	I	2/100
1984 Dec 20	162617	11.25	86.43	74	E	5.6/23	18.00	248	F	NI	I, Notes	2/134, 3/84
1985 Jan 05	111133	10.13	80.06	41	E	5.9/31	18.27	472	V	PA	I, Notes	3/320, 5/97
1985 Feb 08	042940	8.7	84.4	s	I			16	F	CR	Notes	2/150, 3/100
1985 Mar 10	193312	13.49	88.99	50	E	5.0/22	17.69	277	V	ES	I, Notes	5/36
1985 Mar 27	222918	13.35	90.11	70	E			36		ES	I	
1985 Apr 04	235224	12.91	91.24	24	E	4.6/4		130		GU	I	
1985 Apr 19	174313	11.73	86.80	77	E	5.9/40	18.26	384	F	NI	I, Notes	
1985 Apr 19*	175536	11.72	86.87	62	E	5.2/5		187	F	NI	I	
1985 Apr 23	151625	13.50	88.45		AM	4.0/1		61	VI	ES	E, I, Notes, Volc.	2/85, 5/6, 6/4
1985 May 03	070248	11.96	86.82	20	E	5.6/42	17.85	286		NI	I	
1985 May 18	031257	8.33	82.83	17	E			107	F	PA	I, GD	
1985 Jun 03	024530	13.19	90.12	29	E	5.9/31	18.45	371		ES	I	
1985 Jun 04*	110422	13.34	90.00	62	E	5.0/14	17.26	203		ES	I	
1985 Jun 05	042143	14.11	92.10	64	E	4.3/4		166		GU	I	

Date	Time (UT) h.m.s.	Epicentre °N	°W	Depth km	Ag	M_S	$m_B/$ m_b	M^* $log M_O$	N	Felt	Reg	Other Agencies, Comments	Felt Radii
1985 Jul 30	215138	13.35	89.19	74	E				100	F	ES	I, Notes	4/38
1985 Sep 25	120956	9.12	84.16	22	E	4.9/14		17.28	207	F	CR	I, GD, Notes	3/100, 4/40
1985 Oct 11	033913	15.34	90.87		AM				83	D	GU	E, I, Notes	3/90
1985 Oct 12	202923	13.16	89.76	49	E	5.5/16		17.74	351	F	ES	I	2/180, 4/82
1985 Nov 15	053913	13.05	88.75	73	E	3.8/2			122		ES	I	
1985 Nov 16	181613	13.39	90.07	71	E				25		ES	I	
1985 Nov 27	150859	13.93	91.03	68	E	4.6/8		17.38	244	F	GU	I, Notes	2/170, 3/90
1985 Dec 16	024436	11.72	85.85		AM	6.1/34		18.42	371	VII	NI	E, I, Notes	2/300, 3/240, 4/130, 6/35, 7/15
1986 Jan 31	054721	13.6	91.3	60	I				18		GU		
1986 Feb 14	200028	14.22	92.14	42	E				84		GU	I	
1986 Feb 15	213748	14.78	91.44	120	E	5.0/8		17.83	286	F	GU	I, Notes	2/270, 3/90
1986 Mar 24	103359	14.13	91.26	63	E	4.5/4		17.08	147	F	GU	I, Notes	2/350, 3/90
1986 Mar 29	030830	14.2	90.6	162	I				25		GU		
1986 Jun 30	225214	11.15	86.15	65	E	5.7/54		17.93	351	F	NI	I, Notes	2/260, 3/110
1986 Jul 05	220939	15.48	92.59	114	E	5.2/18		18.08	438	F	MX	I, Notes	3/160
1986 Aug 02	174925	13.07	88.85	61	E	5.0/23		17.51	275	F	ES	I, Notes	4/80
1986 Aug 07	223254	7.44	81.31	17	E	5.4/28		17.59	249	F	PA	I, Notes	2/250, 3/170
1986 Oct 10	174926	13.70	89.18		AM	5.4/17		17.67	182	VIII	ES	E, I, Notes	2/200, 7/5, 8/3
1986 Oct 21	210937	13.37	90.15	35	E	5.2/7		17.53	184	F	GU	I, Notes	2/150
1986 Dec 19	103914	12.47	88.48	46	E	4.6/3		17.15	86		NI	I	
1986 Dec 19	115548	12.49	88.56	49	E	4.6/4			83		NI	I	
1986 Dec 21	033538	12.82	88.25	51	E				54		ES	I	
1987 Jan 15	164341	13.75	89.20	s	AM				17	D	ES	I, Notes	
1987 Feb 03	002216	12.91	88.53	70	E	4.5/2			189	F	ES	I	
1987 Mar 06	102327	14.59	92.38	81	E			17.46	249		GU	I	
1987 Mar 13	065634	12.25	87.45	59	E	5.2/4		17.41	239		NI	I	
1987 Mar 21	120900	8.85	83.43	25	E	5.2/5		17.75	221	V	CR	I, Notes	2/140, 3/90, 5/15
1987 Apr 02	054608	11.70	87.31	42	E	4.7/5		17.15	124		NI	I	

1987 Apr 08	174239	11.64	86.51	62	E	6.3/38	18.70	445	F	NI	I, Notes	2/350, 3/230, 4/160
1987 Jun 02	031152	9.09	83.89	33	E	5.1/13	17.36	225	V	CR	I, Notes	3/140, 4/80, 5/35
1987 Jun 22	192334	7.13	82.31	13	E	5.2/11	17.60	243	F	PA	I, Notes	2/350, 3/150
1987 Jul 15	113622	9.50	84.20		AM	4.5/3		86	F	CR	E,.I, Notes	2/150, 3/75, 4/32
1987 Jul 15*	143127	9.50	84.20		AM			84	F	CR	E, I, Notes	2/130, 3/60
1987 Sep 13	112054	14.20	90.02	127	E	4.6/6	17.72	308	F	GU	I, Notes	3/100
1987 Sep 27	211907	10.75	86.39	43	E	5.1/3	17.18	88	F	CR	I	
1987 Sep 27	212913	10.5	86.8	n	I			24		CR		
1987 Oct 04	081521	10.74	85.97	55	E	6.0/34	18.30	407	V	CR	I, Notes	2/300, 3/170, 4/110, 5/60
1987 Oct 14	190602	13.39	89.66	77	E	5.3/20	17.66	245	F	ES	I, Notes	4/55
1987 Oct 20	184705	9.66	84.55	s	AM	3.8/2			F	CR	I, Notes	2/30, 3/14
1987 Nov 17	034010	12.50	87.00		AM	6.1/30	18.76	547	VII	NI	E, I, Notes	3/370, 4/260, 5/90, 7/21
1987 Nov 19	160220	9.69	83.71	14	E			73	F	CR	I, Notes	2/130, 3/60, 4/20
1988 Jan 06	013508	12.25	87.82	50	E	4.3/1		42		NI	I	
1988 Jan 31	233147	9.82	83.70	12	E	4.3/3	16.60	91	VI	CR	I, Notes	3/105, 4/64, 5/30, 6/12
1988 Feb 21	190323	11.54	86.85	50	E	4.8/5	16.86	97		NI	I	
1988 Mar 02	071316	9.60	84.84	28	E	5.2/14	17.30	153	V	CR	I, Notes	4/50, 5/17
1988 Mar 11	034458	8.90	83.20		AM	6.0/36	18.11	307	VI	CR	E, I, Notes	2/400, 3/240, 4/180, 5/100, 6/40
1988 Mar 13	232849	13.18	89.40	67	E	5.2/6		159	F	ES	I	3/55
1988 May 06	144619	11.44	85.95	99	E	6.3/55	18.63	589	VI	NI	I, Notes	2/400, 3/250, 4/170, 5/100, 6/45
1988 May 19	141022	12.78	88.75	58	E	4.0/2		77		ES	I	
1988 Jun 18	184204	13.64	91.09	49	E	5.6/42	17.65	368	F	GU	I, Notes	2/205
1988 Jul 27	201803	13.96	92.96	24	E	4.6/2	16.83	202		MX	I	
1988 Sep 11	010535	14.74	92.56	63	E	5.2/10	17.36	333	F	MX	I	
1988 Sep 13	194531	13.81	90.76	61	E	4.9/7	17.45	157	F	GU	I	
1988 Sep 16	081158	13.96	91.56	43	E	4.4/1	16.85	109		GU	I	
1988 Oct 12	092736	15.99	92.64	163	E	3.6/2	16.85	99		MX	I	
1988 Nov 03	144713	13.91	90.56	74	E	6.4/78	18.93	528	D	GU	I, Notes	2/800, 4/190

Date	Time (UT) h.m.s.	Epicentre °N	°W	Depth km	Ag	M_S	$m_B/$ m_b	M^* $log M_O$	N	Felt	Reg	Other Agencies, Felt Radii Comments	
1988 Nov 04	024313	13.86	90.62	67	E	5.1/7		17.53	223	F	GU	I	
1988 Nov 05	013107	11.96	86.59	35	E				26		NI	I	
1988 Dec 09	093106	14.51	92.50	52	E	4.3/1			86		MX	I	
1989 Feb 26	122112	9.67	84.17	14	E	4.8/6		17.11	125	VI	CR	I, Notes	2/140, 3/72, 5/25, 6/12
1989 Mar 24	153131	11.27	86.50	62	E	4.8/5		16.98	175	F	NI	I	3/110
1989 Apr 02	103558	10.53	85.92		SJ	4.5/6		17.11	146	V	CR	E, I, Notes	2/190, 3/105, 4/57, 5/27
1989 Apr 25	061135	11.19	86.65	50	E	4.5/5		17.08	69		NI	I	
1989 Apr 28	074821	13.28	89.78	56	E	4.8/9		16.92	216	F	ES	I	
1989 May 12	222528	12.81	88.00	62	E	4.6/3		17.04	63		NI	I	
1989 May 25	213622	13.95	91.83	63	E	5.0/6		17.08	162		GU	I	
1989 Jul 26	160847	14.26	91.61	68	E	4.6/3		17.08	134	F	GU	I	
1989 Oct 23	205513	11.78	86.36	70	E	4.8/4		17.26	129	F	NI	I, Notes	
1989 Dec 06	112949	15.58	92.86	19	E	3.9/2			106	F	MX	I, Notes	
1990 Jan 12	032458	12.64	87.46	55	E	4.4/5		17.43	133	F	NI	I, Notes	2/220, 3/142
1990 Feb 16	074927	8.71	83.20	2	E				156		CR	I	
1990 Feb 22	185949	11.67	86.71	48	E	5.2/10		17.26	129	F	NI	I, Notes	2/185
1990 Mar 25	131608	9.82	84.79	25	E	7.0/18			535	F	CR	I, Notes	3/280
1990 Mar 25	132257	9.94	84.77	22	E	7.1/47		19.96	674	VII	CR	I, Notes, Ts	3/350, 5/180, 6/105, 7/50
1990 Mar 25*	213523	9.54	84.70	25	E	5.3/14		17.60	352	F	CR	I	3/110
1990 Apr 03	225701	11.47	86.32	40	E	6.5/74		19.28	542	VI	NI	I, Notes	2/400, 3/300, 4/220, 5/120, 6/50
1990 Apr 04*	041926	11.55	86.47	60	E	5.6/34		17.89	303	F	NI	I	
1990 Apr 28	012313	8.92	83.49	21	E	6.4/68		18.66	581	V	CR	I, Notes	3/280, 4/130, 5/85
1990 May 01	114438	14.17	91.70	53	E	5.4/20		17.75	209	F	GU	I, Notes	3/140
1990 May 10	211537	9.60	84.73	18	E	4.4/3			121	F	CR	I	
1990 May 16	133236	10.83	85.60	65	E	4.9/15		17.28	214		CR	I	

1990 Jun 22	211654	11.62	86.07	30	E	4.8/8	17.20	105	F	NI	I	
1990 Jun 30	145109	9.83	84.38		AM	5.1/15	17.30	199	VII	CR	E, I, Notes	2/160, 3/95, 4/40, 5/40, 6/20, 7/7
1990 Jul 13	135025	12.96	87.70	7	E			65	V	ES	I, Notes	5/40
1990 Jul 23	052709	9.49	84.62	21	E	5.2/20	17.28	237		CR	I	
1990 Jul 27	005458	16.07	86.23	14	E	4.9/20	17.20	342	F	HO	I, Notes	
1990 Aug 30	183302	6.20	82.60	7	E	5.1/16	17.28	182		PA		
1990 Aug 30	222051	9.79	84.49	39	E	5.0/7		87	F	CR	I	
1990 Dec 11	085739	13.17	89.95	60	E	4.7/13	17.08	118		ES	I	
1990 Dec 18	183619	15.6	87.7	n	I			8	F	HO	Notes	3/30
1990 Dec 22	172756	9.88	84.29		AM	5.7/34	18.00	326	VIII	CR	E, I, Notes	3/170, 4/80, 5/50, 6/27, 7/14, 8/7
1990 Dec 22*	172845	9.9	84.3	s	I			4	F	CR	Notes	
1991 Jan 23	151642	12.66	86.29	15	E	4.9/5		83	F	NI	I, Notes	2/180, 3/58
1991 Jan 29	224641	12.90	90.87	39	E	4.7/15	17.22	242		GU	I	
1991 Feb 16	1324	10.0	84.2	s	SJ	3.5			D	CR	Notes	
1991 Mar 01	173028	10.86	84.69	205	E	5.5/31	18.30	679	F	CR	I, Notes	2/130
1991 Mar 13	010917	13.25	87.95	12	E			46	F	NI	I	
1991 Mar 16	060214	9.90	85.70		AM	6.3/69	18.49	422	VI	CR	E, I, Notes	3/270, 4/190, 5/120, 6/40
1991 Apr 12	084133	13.10	88.43	65	E	4.0/2	17.26	93	F	ES	I	
1991 Apr 22	215655	9.63	83.16		AM	7.6/55	20.34	721	VIII	CR	E, I, Notes, Ts	2/710, 3/450, 4/320, 5/230, 6/150, 7/100, 8/57
1991 Apr 22*	220837	9.95	83.41	29	E			111	F	CR	I	
1991 Apr 22*	221928	9.94	83.35	13	E			146	F	CR	I	
1991 Apr 22*	224155	10.10	82.97	21	E			99	F	CR	I	
1991 Apr 22*	225139	9.73	82.53	26	E			117	F	CR	I	
1991 Apr 22*	230052	10.0	83.5	n	I			7		CR		
1991 Apr 23*	053712	9.51	82.70	47	E	5.7/20		181		CR	I	
1991 Apr 23	063407	14.06	91.66	62	E	5.6/40	17.82	312	F	GU	I, Notes	2/260, 3/110
1991 Apr 24*	191303	9.7	83.5		Am	6.1/46	18.35	403	F	CR	E, I, Notes	3/200, 4/120
1991 Apr 27*	054245	10.28	83.30	18	E	4.9/17	17.01	210	F	CR	I	

Date	Time (UT) h.m.s.	Epicentre °N	Epicentre °W	Depth km	Ag	M_S	m_B/ m_b	M^* $logM_O$	N	Felt	Reg	Other Agencies, Comments	Felt Radii
1991 May 02	070532	9.1	81.9	s	I				10	F	PA		
1991 May 04*	034256	9.56	82.40	7	E	6.2/58		18.41	368	VII	PA	I, Notes	3/230, 5/70, 7/9
1991 May 15*	123425	9.2	82.4	n	Am	4.6/5			143	F	PA	I, Notes, Ts	3/86, 4/33
1991 May 27*	184033	9.59	82.53	23	E	5.0/7		17.58	78	VI	PA	I, Notes	3/100, 4/56, 6/17
1991 Jun 21	062740	13.37	89.70	66	E	5.5/35		17.91	353	F	ES	I, Notes	2/160, 4/60
1991 Aug 09	093351	9.77	84.02		AM	4.7/10		16.93	92	VII	CR	E, I, Notes	2/105, 3/68, 4/38, 5/24, 6/10, 7/4
1991 Sep 18	094815	14.65	90.98		AM	6.2/48		18.41	350	VIII	GU	E, I, Notes	3/240, 5/53, 7,15, 8/8
1991 Nov 10	082024	10.01	83.26	48	E	4.8/7		17.08	198	V	CR	I	2/150, 3/100, 4/50, 5/20
1991 Dec 20	181047	9.74	83.57	35	E	4.5/2			53	F	CR	I	
1991 Dec 21	220212	8.23	82.92	25	E	4.6/5		16.75	129	F	PA	I	
1992 Jan 21	125417	14.09	91.12	66	E				45		GU	I	
1992 Feb 21	220957	14.04	92.17	49	E	5.6/21		17.96	199		GU	I	
1992 Mar 07	015339	10.16	84.31	80	E	5.9/76		18.79	716	V	CR	I, Notes	3/275, 5/50
1992 Mar 17	064401	14.49	92.74	34	E	5.1/13		17.26	223	F	MX	I	
1992 Mar 20	171627	13.82	90.83	71	E	4.5/7		17.11	249	F	GU	I, Notes	2/130
1992 Apr 05	141345	11.12	86.51	68	E	5.6/30		17.74	248	F	NI	I	
1992 Apr 27	082956	12.23	87.15	84	E	5.3/34		17.62	312	F	NI	I	
1992 May 18	231921	7.31	82.33	12	E	6.1/67		18.52	551	V	PA	I, Notes	2/350, 5/110
1992 May 23	103037	13.50	90.06	75	E	5.7/25		18.04	417	F	ES	I, Notes	4/93
1992 May 30	163004	14.39	92.99	58	E	6.2/76		18.63	478	F	MX	I, Notes	2/410, 3/260
1992 Jun 06	155143	12.85	88.30	66	E	5.2/37		17.58	266	F	NI	I	3/160
1993 Jun 29	022508	11.79	87.53	50	E	5.4/30		17.30	167		NI	I	
1992 Aug 10	060921	11.81	87.29	72	E	5.6/62		17.78	218		NI	I	
1992 Aug 11	052352	11.77	87.45	39	E	5.3/26		17.30	167	F	NI	I	
1992 Aug 28	095130	12.81	89.79	24	E	4.3/10			134		ES	I	
1992 Sep 02	001603	11.70	87.40	40	E	7.2/65		20.38	477	D	NI	I, Notes, Ts	4/400

1992 Sep 02*	002115	11.26	87.44	76	E			113		NI	I	
1992 Sep 02*	002820	10.96	87.23	6	E			55		NI	I	
1992 Sep 02*	003107	12.06	87.96	45	E			87		NI	I	
1992 Sep 02*	005737	11.51	87.86	20	E			167		NI	I	
1992 Sep 02*	014247	12.02	87.90	46	E			169		NI	I	
1992 Sep 02*	020656	11.85	87.61	22	E	5.4.6		156		NI	I	
1992 Sep 02*	091906	11.23	87.75	34	E	5.0/14		187		NI	I	
1992 Sep 02*	110813	11.61	87.89	29	E	4.6/8		187		NI	I	
1992 Sep 02	165049	10.40	86.77	20	E	5.3/29	17.64	334	F	CR	I	
1992 Sep 02*	182901	11.26	86.74	30	E	5.9/56	18.11	309	F	NI	I	
1992 Sep 02*	202948	12.00	87.84	27	E	4.0/1		39		NI	I	
1992 Sep 03*	235756	11.5	87.4	40	I			17		NI		
1992 Sep 04*	002126	11.47	87.94	32	E	4.6/14	16.89	200		NI	I	
1992 Sep 04*	082543	11.8	87.3	n	I	3.6/1		15		NI		
1992 Sep 05*	214820	12.06	87.43	58	E	6.0/78	18.19	336	F	NI	I	
1992 Sep 06*	131246	11.86	87.55	56	E	5.6/50	17.57	264	F	NI	I	
1992 Sep 10	150321	10.20	86.53	35	E	5.5/44	17.92	314	F	CR	I	
1992 Sep 13*	105337	11.53	86.88	52	E	5.5/34	17.40	249		NI	I	
1992 Sep 15*	184034	11.01	86.82	36	E	5.4/50	17.66	370	F	NI	I	
1992 Sep 16*	235153	12.13	87.58	58	E	5.1/19	17.43	169	F	NI	I	
1992 Sep 28	074128	13.40	90.85	54	E	6.1/93	18.36	355	F	GU	I	
1992 Oct 09*	020816	11.7	87.3	n	I			25		NI		
1992 Oct 31	181253	14.4	90.7	173	I			75		GU		
1992 Nov 05	212806	11.2	87.7	52	I			25		NI		
1992 Nov 18	004851	12.42	88.44	44	E	5.2/7		105		ES	I	
1993 Feb 12	182657	11.23	86.78	52	E	4.5/15		161	F	NI	I	
1993 Mar 06	113035	12.56	87.64	63	E	5.8/9		442	VI	NI	I, Notes	3/210, 4/110
1993 Mar 28	180134	10.25	86.46	43	E	5.1/1		94		CR	I	
1993 Apr 30	122937	14.02	91.55	60	E	4.2/16	16.73	148		GU	I	
1993 May 20	045559	13.73	91.10	59	E		17.72	266	F	GU	I	
1993 May 23	014439	9.06	85.52	16	E		17.04	132		CR	I	
1993 Jun 12	111509	13.27	87.54	216	E		17.84	473	F	HO	I	
1993 Jul 08*	231900	9.74	83.62	10	E			106	F	CR	I, Notes	2/150, 3/88, 4/43

Date	Time (UT) h.m.s.	Epicentre °N	Epicentre °W	Depth km	Ag	M_S	m_B/ m_b	M^* $logM_O$	N	Felt	Reg	Other Agencies, Felt Radii Comments
1993 Jul 10	204100	9.77	83.62	16	E			17.82	345	VII	CR	I, Notes 2/200, 3/130, 4/82, 5/44, 6/20, 7/6
1993 Sep 03*	123505	14.64	92.80	47	E			19.19	559	F	MX	I
1993 Sep 10	191257	14.80	92.69	34	E			19.94	717	D	MX	I, Notes
1993 Sep 11*	192948	14.41	92.44	63	E			17.67	427	F	MX	I
1993 Sep 12	032241	13.92	90.58	75	E			18.04	394	D	GU	I, Notes
1993 Sep 13	225818	11.31	86.47	56	E			17.72	226	F	NI	I
1993 Sep 18*	020147	14.42	92.83	63	E			17.23	182		MX	I
1994 Mar 12	230031	11.93	86.89	35	E			17.49	198		NI	I
1994 Mar 14	205126	15.98	92.43	160	E			19.34	696	D	MX	I, Notes
1994 Mar 15	033621	11.16	88.11	13	E			18.15	509	F	NI	I
1994 Jun 08	205328	16.83	85.90	7	E			24.30	149		HO	I
1994 Jun 29	120101	11.53	86.24	56	E			17.51	122		NI	I
1994 Jul 04	145558	7.73	82.30	22	E			17.34	206		PA	I
1994 Aug 16	211552	14.39	92.77	58	E			16.85	139		GU	I
1994 Oct 31	225928	8.99	83.32	57	E			17.41	325		CR	I
1994 Dec 26	213827	11.47	86.81	78	E				90		NI	I
1994 Dec 28	212236	9.64	84.26	18	E				163		CR	I
1995 Jan 20	112914	14.06	91.82	64	E			17.58	205		GU	I
1995 Jan 27	162452	14.28	91.82	68	E			17.26	278	F	GU	I
1995 Feb 05	203712	6.79	82.64	13	E			18.04	391		PA	I
1995 Feb 11	224536	12.58	81.65	18	E			17.58	281	D	NI	I, Notes
1995 Feb 28	211212	6.90	82.56	7	E			18.11	258	F	PA	I
1995 Jun 14*	054342	11.71	88.89	30	E			17.30	221	F	NI	I
1995 Jun 14	111149	12.17	88.37	27	E			18.88	510	F	NI	I
1995 Jun 14*	111508	12.17	88.35	45	E				130	F	NI	I
1995 Aug 23	074548	9.34	83.96	22	E				125		CR	I
1995 Sep 19	154010	13.78	90.62	68	E			16.98	173	F	GU	I
1995 Dec 19	205609	15.23	90.18	15	E			17.11	139	D	GU	I, Notes

References

Abe K. (1981), Magnitudes of large shallow earthquakes from 1904 to 1980, *Phys. Earth & Plan. Inter.*, **27**, 72-93.

Abe K. (1994), Instrumental magnitudes of historical earthquakes, 1892 to 1898, *Bull. Seism. Soc. Am.*, **84**, 415-425.

Abe K. and Noguchi S. (1983a), Determination of magnitude of large shallow earthquakes 1898-1917, *Phys. Earth & Plan. Inter.*, **32**, 45-59.

Abe K. and Noguchi S. (1983b), Revision of magnitude of shallow earthquakes 1897-1912, *Phys. Earth & Plan. Inter.*, **33**, 1-11.

Acharya H.K. (1965), Seismicity of the Galapagos islands and vicinity, *Bull. Seism. Soc. Am.*, **55**, 609-617.

Adams R.D., Hughes A.A. and McGregor D.M. (1982), Analysis procedures at the International Seismological Centre, *Phys. Earth & Plan. Inter.*, **30**, 85-93.

Adams R.D. and Richardson W.P. (1996), A view of South and Central America from the International Seismological Centre, *Geofísica Inter.*, **35**, 193-203.

Alfaro A., Michaud G. and Biolley C. (1912), Informe sobre el terremoto del Toro Amarillo, Grecias., *Anales Centro Estud. Sism. Costa Rica, año 1911*, San José, 35-41.

Algermissen S.T., Dewey J.W., Langer C.J. and Dillinger W.H. (1974), The Managua, Nicaragua earthquake of December 23, 1972; location, focal mechanism and intensity distribution, *Bull. Seism. Soc. Am.*, **64**, 993-1004.

Alvarado G., Morales L., Climente A. and Rojas W. (1988), Aspectos sismológicos y morfotectónicos en el extremo occidental de la cordillera volcánica Central de Costa Rica, *Revista. Geol. Am. Centr.*, **9**, 75-98.

Amador J., Chacon R. and Lizano O. (1994), Estudio de efectos geofísicos del terremoto de Limón mediante percepción remota y análisis hidro meteorológico, *Revista Geol. Am. Centr.*, **Vol. Especial**, San José, 153-170.

Ambraseys N. N. (1973), The earthquake of Managua, Nicaragua, 1972, *Nature*, **244**, 427-428.

Ambraseys N. N. (1990), Uniform magnitude re-evaluation of European earthquakes associated with strong-motion records, *J. Earthq. Eng. Struct. Dyn.*, **19**, 1-20.

Ambraseys N. N. (1995), Magnitudes of Central American earthquakes 1898-1930, *Geophys. J. Int.*, **121**, 545-556.

Ambraseys N. N. and Adams R. D. (1986), Seismicity of West Africa, *Annales Geophys.*, **4B**, 679-702.

Ambraseys N. N. and Adams R. D. (1991), Reappraisal of major African earthquakes south of 20°N, 1900-1930, *Natural Hazards*, **4**, 389-419.

Ambraseys N. N. and Adams R. D. (1998), The Rhodes earthquake of 26 June 1926, *J. Seismology*, **2**, 267-292.

Ambraseys N. N. and Douglas J. (1999), "Surface-wave magnitude reappraisal 10°–44°N, 18°–70°E", *ESEE Report no.6/99*, Dept. Civil Eng., Imperial College of Science, London.

Ambraseys N. N. and Douglas J. (2000), Reappraisal of surface wave magnitudes in the Eastern Mediterranean region and the Middle East, *Geophys. J. Int.*, **141**, 357-373.

Ambraseys N. N. and Finkel C. (1987), Seismicity of Turkey and neighbouring regions, 1899-1915, *Annales Geophys.*, **5B**, 701-726.

Ambraseys N. N. and Free, M.W. (1997), Surface-wave magnitude calibration for European region earthquakes, *J. Earthquake Engineering*, **1**, 1-22.

Ambraseys N. N. and Melville C. P. (1982), "A history of Persian earthquakes", Camb. Univ. Press. pp.119-126.

Ambraseys N. N., Melville C. P. and Adams R. D. (1994), "The seismicity of Egypt, Arabia and the Red Sea", Camb. Univ. Press, 181pp.

Anderson R.W. (1987), The San Salvador earthquake of October 10, 1986; review of building damage, *Earthquake Spectra*, **3**, 497-541.

Anderson T. (1908), The volcanoes of Guatemala, *Geogr. J.*, **31**, 473-489.

Anonymous (1911), Il terremoto catastrófico del 4 Maggio 1910 in Costarica, *Annali della Soc. del. Ingegn. Rome*, **11**, 348-354.

Anonymous (1917), *Zeitschr. d. Ges f. Erdkunde*, Berlin, 575.

Anonymous (1918), The Guatemala earthquakes, *Pan-American Mag.*, **26**, 305-317.

Anonymous (1954), Scientific intelligence, *Am. J. Sci.*, **18**, 277-284.

Anonymous (1979), *Nature*, **297**, 159.

Arenales R. (1961), El terremoto de San Salvador, Corpus-Christi-Junio 7 de 1917, Publ. Depto. Editor. Minist. Educat., San Salvador (Harvard SA 4961-79.20).

Ballore, F. M. de. (1888), "Tremblements de terre et éruptions volcaniques au centre-Amerique", Publ. Soc. Sci. Naur. Saône-et-Loire, Dijon.

Barquero R. and Alvarado G. (1987), Actividad sísmica de la zona de Quepos, *Rep. Inst. Costaric. Electric.*, San José.

Barquero R., Baschini I. and Rojas W. (1991), "Actividad sísmica relacionada con el sismo de Cobano", La Crisis sísmica del Golfo de Nicoya, Publ. Inst. Costaric. Elecricidad, 45-73, San José.

Barquero R. and Rojas W. (1994a), Sismicidad inducida por el terremoto de Limón, *Revista Geol. Am. Centr.*, **Vol. Especial**, San José, 111-120.

Barquero R. and Rojas W. (1994b), Catálogo de mapas de isosistas de temblores y terremotos importantes de Costa Rica, *Publ. Inst. Costaric. Electr.* OSV.94.03, San José.

Barquero R., Rojas W. and Fernandez M. (1991), "El enjambre de Puriscal y Mora de 1990", La Crisis sísmica del Golfo de Nicoya, Publ. Inst. Costaric. Elecricidad, 74-94, San José.

Båth M. (1969), "Handbook on earthquake magnitude determination", VESIAC Spec. Rep. 7885-36-X, Seism. Inst., 158pp, Uppsala.

Belt T. (1874), "The naturalist in Nicaragua [1868–72]", J.Murray, London.

Bengstsson J.and Söderberg J. (1991), "Landslides in Costa Rica", Report 91/4 Inst. Jord & Bergmekan. Kungl. Tekn. Högskol., Stockholm.

Bermudez M. (1994), El manejo institucional y percepción de la población en el terremoto de Limón, *Revista Geol. Am. Centr.*, **Vol. Especial**, San José, 211-224.

Boddam-Whetham J.E. (1877), "Across Central America", Hurst and Blackett, London.

Bois C. (1934), Chronique sismologique, *Matériaux pour l'Étude des Calamit. Genève*, **34**, 152-155.

Bohnenenberger O. and Bonis S. (1976), "Fracture map; Earthquakes of February 1976, Guatemala Valley", Publ. Inst. Geograf. Nacional, Guatemala City.

Boschini I.M. (1987), "Sismos sentidos en Costa Rica durante el año 1987", Rep. Instit. Costaric. Electr.

Boschini I.M. (1989), "Incidencia de las fuentes sísmicas en la region caribe de Costa Rica", Tessis Licent. Univ. de Costa Rica, 97pp.

Boschini I.M. and Montero W. (1974), Sismicidad histórica e instrumental del caribe de Costa Rica, *Revista Geolog. Am. Centr.*, **Vol. Especial**, San José, 65-71.

Brown R.D. (1968), "Managua, Nicaragua, earthquake of January 4, 1968", Proj. Rep., Nicaragua Invest. U.S. Geol. Surv. Internal Rep. NI.1, Menlo Park.

Brown R.D., Ward P.L. and Plafker G. (1973), "Geologic and seismologic aspects of the Managua, Nicaragua, earthquake of December 23, 1973", Profess. Paper 838, U.S.Geol. Surv., Washington.

Bucknam R., Plafker G. and Sharp R. (1978), Fault movement following the Guatemala earthquake of February 4, 1976, *Geology*, **6**, 170-173.

Calvert A. and Calvert P. (1917), "A year of Costa Rican natural history", Macmillan, New York.

Camacho E. (1991), The Puerto Armuelles earthquake, southwestern Panama, of July 18, 1934, *Revista. Geol. Am. Centr.*, **13**, 1-13.

Camacho E. (1993), The tsunami of April 22 1991 in Central America, *Tsunami Newsletter*, **25**, 6-7, Honolulu.

Camacho E. (1994), El tsunami del 22 de Abril de 1991 en Bocas del Toro, Panama, *Revista Geol. Am. Centr.*, **Vol. Especial**, San José, 61-64.

Camacho E. and Viquez V. (1992), Historical seismicity of the north Panama deformed belt, *Inst. Geociencas Univ. Panama*, 50pp, and *Revista Geol. Am. Centr.*, **15**, 49-64, 1993.

Camacho E. and Viquez V. (1994), Liquefacción y hundimientos costeros en el nordeste de Panama durante el terremoto de Limón, *Revista Geol. Am. Centr.*, **Vol. Especial**, San José, 133-138.

Carden L.E. (1911), "Dispatches, and reports of the British Minister in Guatemala and Honduras", PRO/FO, Kew.

Carr M. and Stoiber R. (1977), "Destructive earthquakes in Central America", Geol. Soc. Am. Data Repository, Item no.7701, 14pp.

Chacón B. (1909), Los temblores del año pasado en Copan, *Revista de la Universidad, Tegucigalpa*, **1(4)**, 369-371.

Chavez M..and Castro R. (1988), Attenuation of *MM* intensity with distance in Mexico, *Bull. seism. Soc. Amer.*, **78**, 1875-1884.

Chieruzzi R. (1987), The San Salvador earthquake of October 10, 1986; geotechnical effects, *Earthquake Spectra*, **3**, 483-489.

Christiansen A. and Ziemendorff G. (1909), Les tremblements de terre pendant l'année 1905, *Publ. Bureau Centr. Assoc. Intern. Seism., Ser.B*, Strasbourg.

Christoskov L., Kondorskaya N.V. and Vaněk J. (1983), Earthquake magnitude in seismological practice: *PH*, *S* and *L* waves, *Rada Ceskolovenske Akad. Ved, Matem.-Prirod. ved.*, **93(1)**, 113pp, Prague.

Climente A. (1991a), "Observatiónes macrosísmicas y aceleraciónes", La Crisis sísmica del Golfo de Nicoya, Publ. Inst. Costaric. Elecricidad, San José, 15-33.

Climente A. (1971b), "Aceleraciónes producidas por los sismos de Puriscal", La Crisis sísmica del Golfo de Nicoya, Publ. Inst. Costaric. Elecricidad, San José, 105-113.

Cortes J., Soto R. and Jimenez C. (1994), Efectos ecológicos del Terremoto de Limón, *Revista Geol. Am. Centr.*, **Vol. Especial**, San José, 187-192.

Crawford J. (1898), Recent severe seismic disturbances in Nicaragua, *Am. Geologist*, **22**, 56-58.

Crawford J. (1902a), Earthquakes in Nicaragua, *Am. Geologist*, **29**, 323.

Crawford J. (1902b), List of the most important volcanic eruptions and earthquakes in western Nicaragua, *Am. Geologist*, **30**, 111-113.

Cruz G. and Wyss M. (1983), Large earthquakes, mean sea level, and tsunamis along the Pacific coast of Mexico and Central America, *Bull. Seism. Soc. Am.*, **73**, 553-570.

Davis I. (1976), "Guatemala; shelter and housing policy", Publ. Dept. Archit., Oxford Polytechnic, Oxford.

Dengo G., Fiedler G., Knudsen C., Lomnitz C. and Rosenblueth E. (1965), "Misión de reconocimiento sismológico", UNESCO Rep. no.WS/0665.131 AVS, Paris.

Denyer P., Arias O. and Personius S. (1994a), Efecto tectónico del terremoto de Limón, *Revista Geol. Am. Centr.*, **Vol. Especial** San José, 39-52.

Denyer P., Arias O. and Arias M. (1994b), Esfuerzos y paleo-esfuerzos de la cuenca de Limón, *Revista Geol. Am. Centr.*, **Vol. Especial**, San José, 53-60.

Denyer P., Personius S. and Arias O. (1994c), Generalidades sobre los efectos geologicos del terremoto de Limón, *Revista Geol. Am. Centr.*, **Vol. Especial**, San José, 29-38.

Dewey J.W., Algermissen S.T., Langer C.J., Dillinger W.H. and Hopper M.G. (1973), The Managua earthquake of 23 December 1972, , *Proc. Managua Nicaragua earthquake*, Publ. Earthq. Eng. Res. Inst., San Francisco, **1**, 66-88.

Dewey J.W. and Algermissen S.T. (1974), Seismicity of the Middle America arc-trench system near Managua, Nicaragua, *Bull. Seism. Soc. Am.*, **64**, 1033-1048.

Dewey J.W. and Julian B. (1976), "Main event source parameters from teleseismic data", USGS Prof. Paper 1002, 19-23.

Diaz V.M. (1918), "Narraciones", Guatemala City.

Duda S. J. (1965), Secular seismic energy release in the circum-Pacific belt, *Tectonophysics*, **2**, 409-452.

Durham H. (1931), Managua: its reconstruction and utilities, *Eng. News Rec.*, April 23, 696-700.

Durkin M. and Hopkins J. (1987), The San Salvador earthquake of October 10, 1986; architecture and urban planning, *Earthquake Spectra*, **3**, 609-620.

Dziewonski A.M. and Woodhouse J.H. (1983), An experiment in the systematic studies of global seismicity: centroid-moment tensor solutions for 201 moderate and large earthquakes of 1981, *J. Geoph. Res.*, **88**, 3247-3271.

Eisen G. (1903), The earthquake and volcanic eruption in Guatemala in 1902, *Bull. Am. Geogr. Soc.*, **35**, 325-352.

Engdahl E.R., van der Hilst R. and Buland R. (1998), Global teleseismic earthquake relocation with improved travel times, and procedures for depth determination, *Bull. Seism. Soc. Am.*, **88**, 722-743.

Espinoza A.F. (1976), ed. "The Guatemalan earthquake of February 4, 1976, a preliminary report", U.S. Geol. Surv. Prof. Paper 1002, Washington.

Estrada S. J. and Flores E. (1934), Los terremotos de Occidente 1934, *Renacimiento*, no.147, 30 Deciembre, Tegucigalpa, and in *Anales Archivo Nacional*, **8(4)**, 84-86, Tegucigalpa.

Evernden J. F. (1971), Variation of Rayleigh-wave amplitude with distance, *Bull. Seism. Soc. Am.*, **61**, 231-240.

Feldman L.H. (1984), "Catalogue of historic documents pertaining to the earthquake damage in Panama and Costa Rica, collected in San José and Washington", (unpublished report).

Feldman L.H. (1993), "Mountains of fire; lands that shake", Pub. Labyrinthos, Culver City, Cal.

Freeman J.R. (1932), "Earthquake damage and earthquake insurance", McGraw-Hill, N.Y., 589-595.

Figueroa J. (1973), "Sismicidad en Chiapas", Publ. Inst. Ingen. UNAM, no.318, Mexico City.

Gann T. (1928), "Discoveries in Central America", Duckworth, London.

Grases J. (1974), "Sismicidad de la región Centroamericana asociada a la cadena volcánica del cuaternario", 4 vols, Caracas.

Guardia F.L. and Marin C.A. (1910), "The Cartago earthquake of May 4, 1910", Ed. A.Lehman, San José, 51pp.

Güendel F.D. (1982), The foreshock-mainshock-aftershock sequence of the 1978 Samara, Costa Rica earthquake; a unique dataset, *Earthquake Notes*, **53**, 81-82.

Güendel F.D. (1986), "Seismotectonics of Costa Rica", Ph.D. Thesis, Earth Sci., Univ. Cal. Santa Cruz.

Gutenberg B. (1945), Amplitudes of surface waves and magnitudes of shallow earthquakes, *Bull. Seism. Soc. Amer.*, **35**, 3-12.

Gutenberg B. (1958), Unpublished Work-notes, Millikan Memorial Library, Cal. Inst. Tech.

Gutenberg B. and Richter C.F. (1936), Magnitude and energy of earthquakes, *Science*, **83**, 183-185.

Gutenberg B. and Richter C.F. (1954), "Seismicity of the earth and associated phenomena", Princeton Univ. Press, New Jersey.

Gutenberg B. and Richter C.F. (1956), Magnitude and energy of earthquakes, *Ann. Geofis.*, **9**, 1-15.

Gutenberg B. and Richter C.F. (1965), "Seismicity of the earth and associated phenomena", New York.

Habermann R.E. (1987), Man-made changes of seismicity rates, *Bull. Seism. Soc. Am.*, **77**, 141-159.

Hafström P. and Skogsberg J. (1994), "Liquefaction during the Limón-Telire earthquake, 22 April 1991", Report no.94/7, Inst. Jord & Bergmek., Kungl. Tekn. Högskolan, Stockholm.

Hansen F. (1975), "Seismicity of Nicaragua", *in* Engineering Rep. on the Managua Earthquake of 23 December 1972, Publ. Natl. Acad. Sci., Washington.

Hansen F. and Chavez V. (1973), Isoseismal maps of the Managua December 23, 1972 earthquake, *Proc. Managua, Nicaragua earthquake*, Publ. Earthq. Eng. Res. Inst., San Francisco, **1**, 104-114.

Harlow D.H. and White R.A. (1980), "Preliminary catalogue of seismicity prior to the Guatemala earthquake of February 4, 1976", Open File Rep. 80-60, U.S. Geol. Surv., Menlo Park.

Harlow D.H., White R.A., Rymer M.J. and Alvarez S. (1993), The San Salvador earthquake of 10 October 1986 and its historical context, *Bull. Seism. Soc. Am.*, **83**, 1143-1154.

Hayes C.W. (1899), "Geology and physiography of the Nicaragua Canal route", Report of the Nicaragua Canal Commission, Appendix 2, 132-137, Baltimore.

Herak M. and Herak D. (1993), Distance dependence of M_S and calibrating function for 20 second Rayleigh waves, *Bull. Seism. Soc. Am.*, **83**, 1881-1892.

Herak M., Panza G.F. and Costa G. (2000), Theoretical and observed depth correction for M_S, *Pure Appl. Geophys.*, (in press).

IASPEI, (1967), Recommendations of the committee on magnitudes, *Comptes Rendus*, **15**, 65.

Jaggar. (1911a), The Costa Rica volcanoes and earthquakes, *J. Am. Assoc. Eng. Soc.*, **46(2)**.

Jaggar. (1911b), The earthquakes of Costa Rica, *Sci. Conspectus*, **1(2)**.

Jeffreys H. and Bullen K.E. (1940), "Seismological Tables", British Association, London.

Jimenez C..and Cortez J. (1994), Colonización vegetal y actividades humanas en las plataformas arrecifales expuestas por el terremoto de Limón, *Revista Geol. Am. Centr.*, **Vol. Especial**, San José, 193-200.

Johansson L.O. (1988), "Seismic hazard analysis of Managua, Nicaragua", Examensarbete Inst. Jord-och Bergmech. no.3.88, Kungl. Tekn. Högsk, Stockholm.

Jones J.O. (1903), "On the effect of the recent seismic disturbances in Guatemala, Costa Rica and Nicaragua upon the level of the waters in Lakes Nicaragua and Managua",U.S. Senate 57th Congress, 2nd Session, Document no.131, 33pp, Washington.

Jordan J. and Martines M. (1980), "Seismic history of El Salvador", Publ. Centro Invest. Geotechnicas, Depto. Sismologia, San Salvador.

Jorgensen C. (1933), "Earthquake history of the vicinity of Panama Canal and adjacent regions", manuscript notes.

Jorgensen K. (1966), "Seismological history of the Canal Zone and Panama", Balboa Heights, Panama, Feb. 1966.

Kanamori H. and Abe K. (1979), Reevaluation of the turn-of-the-century seismicity peak, *J. Geoph. Res.*, **84**, 6131-6139.

Kanamori H. and Stewart G. (1978), Seismological aspects of the Guatemala earthquake of February 4, 1976, *J. Geoph. Res.*, **83**, 3427-3434.

Karim A., Ruiz T., Padilla S. and Santos S. (1968), "Informe sobre el terremoto de 4 de enero 1968", Publ. Assoc. Nicaraguense Ingen. y Arquit. no.14, Managua.

Kárník V. (1968), "Seismicity of the European area", Academia, Prague.

Kárník V. (1977), Comparison of empirical A/T curves in near epicentral distances, *Publ. Inst. Geoph. Polish Acad. Sci.*, **A5**, 61-64.

Kárník V. *et al.* (1962), "Standardisation of the earthquake magnitude scale, *Studia Geoph. et Geod.*, **6** 41-47.

Kárník V. and Christokov L. (1977), Magnitude determination at short epicentral distances in Europe, *Publ. Inst. Geoph. Polish Acad. Sci.*, **A5**, 51-60.

Kellar J.G. (1935), Raz de marée à Panama, *Annales des Raz de Marée*, **5**, 28, Paris.

Kelleher J., Sykes L.R. and Oliver J. (1973), Possible criteria for predicting earthquake locations, *J. Geoph. Res.*, **78**, 2547-2585.

Kennett B.L.N., Engdahl E.R. and Buland R. (1995), Constraints on seismic velocities in the earth from travel times, *Geophys. J. Int.*, **122**, 108-124.

Kirkpatrick R.Z. (1920), Earthquakes in Panama up to January 1, 1920, *Bull. Seism. Soc. Am.*, **10**, 120-128.

Kirkpatrick R.Z. (1936), "The Panama earthquakes of July and August 1934", United States Earthquakes 1934, no.593, 51-56, Washington.

Knudson C.F., Perez V. and Matthiesen R.B. (1974), Strong-motion instrumental records of the Managua earthquake of December 23, 1972, *Bull. Seism. Soc. Am.*, **64**, 1049-1067.

Kowalle G. and Thürmer C. (1991), "Magnitudenbestimmung für die Wiechert Seismogramme der Station Potsdam", Veröff. Zentralinst. Phys. Erde, no.119, Potsdam.

Krinitzsky E.L. and Bonis S.B.(1976), "Notes on earthquake shaking in soils, Guatemala earthquake of 4 February 1976", Rep. U.S. Eng. Waterways Exper. Stat. Pavements Lab., Vicksburg.

Lais R. (1913), Catalogue général de tremblements de terre de l'année 1907, *Publ. Bureau Centr. Assoc. Int. Seism., Ser. B*, Strasbourg.

Lais R. and Scheu E. (1912), Catalogue régional des tremblements de terre en 1907, *Publ. Bureau Centr. Assoc. Int. Seism., Ser. B*, Strasbourg.

Langer C.J. and Bollinger G.A. (1979), Secondary faulting near the terminus of a seismogenic strike-slip fault: aftershocks of the 1978 Guatemala earthquake, *Bull. Seism. Soc. Am.*, **69**, 427-444.

Langer C.J., Hopper M.G., Algermissen S.T. and Dewey J.W. (1974), Aftershocks of the Managua, Nicaragua, earthquake of December 23, 1972, *Bull. Seism. Soc. Am.*, **64**, 1005-1016.

Lara M.A.(1987a), The San Salvador earthquake of October 10, 1986; history of construction practices in San Salvador, *Earthquake Spectra*, **3**, 491-496.

Lara M.A. (1987b), The San Salvador earthquake of October 10, 1986; detailed evaluation of the performance of eight engineered structures, *Earthquake Spectra*, **3**, 543-562.

Lara V.C. (1964), "Fechas de la historia de Honduras", Tipogr. Natl., 392-394, Tegucigalpa.

Larde J. (1916), ed. "El terremoto del 6 de Setiembre de 1915", Revista de la Enseneza, San Salvador.

Larde J. (1960), "Orbas completas" Dept. Editor. Minist. Cultura, San Salvador

Ledbetter S.R. and Bommer J.J. (1987), "The San Salvador earthquake of 10th October 1986", EEFIT Publ., Inst. Civil Eng., London.

Leeds D.J. (1973), Destructive earthquakes of Nicaragua, *Proc. Managua, Nicaragua earthquake*, Publ. Earthq. Eng. Res. Inst., San Francisco, **1**, 26-51.

Leeds D.J. (1974), Catalogue of Nicaraguan earthquakes, *Bull. Seism. Soc. Am.*, **64**, 1135-1158.

Leonardo L. (1918), "Reconocimiento del terreno donde estan las grietas: zona nordeste o Yampuc", Archivo General de Centro America, Gobern. no.29405.4.2.1918, Guatemala.

Levin S.B. (1940), The Salvador earthquakes of December 1936, *Bull. Seism. Soc. Am.*, **30**, 377-407.

Lienkaemper J.J. (1984), Comparison of two surface-wave magnitude scales: M of Gutenberg and Richter (1954) and M_S of Preliminary Determination of Epicentres, *Bull. Seism. Soc. Am.*, **74**, 2357-2378.

Lisowski M. and Thatcher W. (1981), Geodetic determination of horizontal deformation associated with the Guatemala earthquake of 4 February 1976, *Bull. Seism. Soc. Am.*, **71**, 845-856.

List K. (1903), Erdbeben an der Küste Guatemalas im Jahre 1902 und deren Folgeerscheinungen, *Ann. Hydrograph. und Marit. Meteorolog.*, Feb.1903, 52-54, Hamburg.

Lomnitz C. and Schultz R. (1966), The San Salvador earthquake of May 3, 1965, *Bull. Seism. Soc. Am.*, **56**, 561-575.

Lutz O. (1915), Beiträge zur Geologie des Panamakanals, *Petterm. Mitteil.*, 394.

MacDonald D.F. and Johnston W.C. (1913), Isthmian earthquakes, *Canal Record*, **7**, 144-149.

Margottini C. (1993), "La magnitudo dei terremoti Italiani del XX secolo", Publ. Ente Nuove Tecn. Energ. Ambiente, Rome.

Marshall P. D. and Basham P W. (1973), Rayleigh wave magnitude scale, M_S, *Pure Appl. Geophys.*, **103**, 406-414.

Martin P.F. (1911), "Salvador of the 20th century", Arnold, London.

Martinez H.M. (1977), "En nuevo enfoque par el sismo de la 3 de Mayo de 1965", Revista de ASIA, no.44, 23-36, San Salvador.

Martinez H.M. (1978), "Cronología sísmica y eruptiva de la república de El Salvador a partir de 1520", Publ. Centro Invest. Geotecnicas, Minist. Orbas Publ., San Salvador.

Masters V.F. (1916), in *Bull. Seism. Soc. Am.*, **6**, 48-49.

Matthiesen R.B. (1975), "Strong-motion measurements of Managua earthquake", Engineering Rep. on the Managua Earthquake of 23 December 1972, Publ. Natl. Acad. Sci., 13-20, Washington.

Matumoto T. and Latham G. (1973), Aftershock and intensity of the Managua earthquake of 23 December 1972, *Proc. Managua, Nicaragua earthquake*, Publ. Earthq. Eng. Res. Inst., San Francisco, **1**, 97-103.

McNally K.C. and Minster J.B. (1981), Nonuniform seismic slip rates along the Middle American trench, *J. Geoph. Res.*, **86**, 4949-4959.

Meyer-Abich H. (1952), Das Erdbeben von Jucuapa in El Salvador vom 6 und 7 Mai 1951, *Neues Jahrb. Geol. u. Paläont.*, **95**, 311-336, and "Terremoto de Jucuapa en el Salvador", Comunicaciónes Inst. Tropical Investig. Cientif., no.2, 35pp, San Salvador.

Meyer-Abich H. (1956), Los volcános activos de Guatemala y El Salvador, *Anales Serv. Geol. Nacl. El Salvador*, **3**, 1-102, San Salvador.

Michaud G. (1910), The Cartago earthquake, *Sci. Amer.*, **102**, 440-441.

Michaud G. and Biolly P. (1912), Informe sobre el terremoto de Toro Amarillo, Grecia, *Annales Centro Estud. Sismol. Costa Rica*, 34-41.

Milne J. (1911a), "Double and multiple earthquakes", British Assn Adv. Sci., 16th Rep. Seism. Investig., 3-5, Portsmouth.

Milne J (1911b), "A catalogue of destructive earthquakes AD 7 to AD 1899", Pub. British Assn. Adv. Sci., London.

Mitchell C. (1918), Seismic disturbances connected with the Guatemala earthquake, *Nature*, **100**, 514-515.

Miyamura S. (1975), Recent crustal movements in Costa Rica disclosed by relevelling surveys, *Tectonophysics*, **29**, 191-198.

Miyamura S. (1976), Provisional magnitudes of Middle American earthquakes not listed in the magnitude catalogue of Gutenberg-Richter, *Bull. Int. Inst. Seism. Earthq. Eng.*, **14**, 41-46, Tokyo.

Miyamura S. (1980), "Sismicidad de Costa Rica", Pub. Univ. de Costa Rica.

Molnar P.H. and Sykes L.R. (1969), Tectonics of the Caribbean and Middle America regions from focal mechanisms and seismicity, *Bull. Geol. Soc. Am.*, **80**, 1639-1684.

Montero W. and Miyamura S. (1981), Distribución de intensidades y estimatión de los parámetros focales de los terremotos de Cartago de 1910, *Rivist. Inst. Geograf. National, San.José*, January, 9-34.

Montero W. and Mora S. (1985), "Estudio de las intensidades efectos y distribución del terremoto de Osa del 3 abril 1983", Proc. 3rd Semin. Nac. Geotechn., San José.

Montero W. and Alvarado G. (1988), Los terremotos de Bagaces de 1935 y 1941, *Ciencia y Tecnologia*, **12**, 67-87, San José.

Montero W. and Climente A. (1990), Los terremotos de Grecia de 1882 y de Quepos de 1952; distribución de intensidades y origen tectónico, *Revist. Ciencia y Tecnol.*, **14**, 81-105, San José.

Montero W., Boschini I. and Rojas W. (1991a), "El terremoto de Cobano, Costa Rica, 25 de Marzo de 1990", La Crisis sísmica del Golfo de Nicoya, Pub. Inst. Costaric. Elecricidad, 1-14, San José.

Montero W. and Boschini I.(1991b), "Interpretación sismotectonica de la región del Golfo de Nicoya", La Crisis sísmica del Golfo de Nicoya, Pub. Inst. Costaric. Elecricidad, 34-44, San José.

Montero W., Boschini I. and Rojas W. (1991c), "Interpretatión neotectónica", La sísmica del Golfo de Nicoya, Pub. Inst. Costaric. Elecricidad, 114-130, San José.

Montero W., Camacho E., Espinoza A. and Boschini I. (1994a), Sismicidad y macro neotectónica de Costa Rica y Panama, *Revista Geol. Am. Centr.*, **Vol. Especial**, San José, 73-82.

Montero W., Pardo M., Ponce L., Rojas W. and Fernandez M. (1994b), Evento principal y replicas importantes del terremoto de Limón, *Revista Geol. Am. Centr.*, **Vol. Especial**, San José, 93-102.

Mora S. and Mora R. (1994), Los deslizamientos causados por el terremoto de Limón, *Revista Geol. Am. Centr.*, **Vol. Especial**, San José, 139-152.

Mora S. and Yasuda S. (1994), Liquefacción de suelos y fenómenos asociados durante el terremoto de Limón, *Revista Geol. Am. Centr.*, **Vol. Especial**, San José, 121-132.

Morales L.D. (1994), Danos causados por el terremoto de Limón: perdidas y medidas de mitigación, *Revista Geol. Am. Centr.*, **Vol. Especial**, San José, 201-210.

Morales L.D. and Leandro G.(1985), "Riesgos geológicos asociados al terremoto de Buenavista, Perez Zeladon", Proc. 3rd Natl. Geotechn. Seminar, Asociat. Costarr. Mech. Suelos, San José.

Morgan J and Swan S. (1987), The San Salvador earthquake of October 10, 1986; performance of lifelines, *Earthquake Spectra*, **3**, 585-607.

Morley S.G. (1918), The Guatemala earthquake, *Am. Museum J.*, **18**, 200-210.

Neumann F. (1936), United States earthquakes 1934, Ser.no.593, p.41.

Nishenko S. and Singh S. (1987), Relocation of the great Mexican earthquake of 14 January 1903, *Bull. Seism. Soc. Am.*, **77**, 256-259.

Nogues A.F. (1892), Tremblements de terre du Salvador, *La Nature*, **20**, 78-79, Paris.

Nuttli, O. (1973), Seismic wave attenuation and magnitude relations for Eastern North America, *J. Geoph. Res.*, **78**, 876-885.

Oddone E. (1907), Tremblements de terre ressentis pendant l'année 1904, *Publ. Bureau Centr. Assoc. Int. Sism., Ser. B*, Strassburg.

Oldham R.D. (1905), The rate of transmission of the Guatemala earthquake of April 19, 1902, *Proc. Royal Soc.*, **76**, 102-111.

Osegueda F. (1932), "El macrosismo del sismo centroamericano del 21 de mayo 1932", Rep. Obsrv. Nacional, San Salvador.

Pacheco J.F. and Sykes L. R. (1992), Seismic moment catalogue of large shallow earthquakes, 1900 to 1989, *Bull. Seism. Soc. Am.*, **82**, 1306-1349.

Page R.A. (1976), ed. "Interim report on the Guatemala earthquake of 4 February 1976", U.S. Geol. Surv. Open-File Rep.76-295, Menlo Park.

Panza G.F., Duda S.J. and Herak M. (1989), Gutenberg's surface wave magnitude calibrating function: theoretical basis from synthetic seismograms, *Tectonophysics*, **166**, 35-43.

Perez O. J. (1999), Revised world seismicity catalogue 1950-1997 for strong $M_S > 6$ shallow $h < 70$ earthquakes, *Bull. Seism. Soc. Am.*, **89**, 335-341.

Perez O. J. and Scholtz C. H. (1984), Heterogeneities of the instrumental seismicity catalogue 1904-1980 for strong shallow earthquakes, *Bull. Seism. Soc. Am.*, **74**, 669-686.

Perrey A. (1855), Note sur les tremblements de terre en 1854, avec suppléments pour les années anterieures, *Bull. Acad. R. Sci. Bruxelles.*, **22**, 526-572.

Pittier H. (1910), Costa Rica; Vulcan's smithy, *Natl. Geogr. Mag.*, **21**, 494-525.

Plafker G. (1973), Field reconnaissance of the effects of the earthquake of April 13[sic] 1973, near Laguna de Arenal, Costa Rica, *Bull. Seism. Soc. Am.*, **63**, 1847-1856.

Plafker G. (1976), Tectonic aspects of the Guatemala earthquake of 4 Feb. 1976, *Science*, **193**, 1201-1208.

Plafker G. and Brown R. (1973), The surface geologic effects of the Managua earthquake of December 23, 1972, *Proc. Managua, Nicaragua earthquake*, Publ. Earthq. Eng. Res. Inst., San Francisco, **1**, 115-142.

Pollog K. (1931), Augenzeugenberichte über das Erdbeben von Managua, *Peterm. Mitteil.*, **77**, 184-185.

Ponce L., Pardo M., Dominguez J., Montero W., Rojas W., Boschini I., Suarez G. and Camacho E. (1994), Estudio de replicas del terremoto de Limón usando datos locales, *Revista Geol. Am. Centr.*, **Vol. Especial**, San José, 103-110.

Poppe B. Naab D. and Derr J. (1978), "Seismograph station codes and characteristics", U.S. Geol. Surv. Circ., no.791.

Proceedings, Managua, Nicaragua earthquake of December 23, 1972 (1973), Publ. Earthq. Eng. Res. Inst., San Francisco, 2 vols.

Ramirez R. and Laurent R. (1974), Determinación de la magnitud local a partir de acelerogramas: terremoto de Limón, *Revista Geol. Am. Centr.*, **Vol. Especial**, San José, 83-92.

Reid H.F. (1917), Note on the earthquakes at Almirante, *Bull. Seism. Soc. Am.*, **7**, 27-30.

Rezapour M. and Pearce R.G. (1998), Bias in surface-wave magnitude M_S due to inadequate distance correction, *Bull. Seism. Soc. Am.*, **88**, 43-61.

Richter C.F. (1935), An instrumental earthquake magnitude scale, *Bull. Seism. Soc. Am.*, **25**, 1-32.

Rios R., Grases J., Ponce L., Weidmann N. and Zegarra L. (1986), "El terremoto de San Salvador 10 de Octubre 1986", UNESCO Publ. no.SC/RP 233090.6, Paris.

Roberts M. (1924), "On the earthquake line", Arrowsmith, London.

Rockstroh E. (1902), Earthquake in Guatemala, *Nature*, **66**, 150.

Rockstroh E. (1903), Recent earthquakes in Guatemala, *Nature*, **67**, 271-272.

Rudolph E. (1905), Katalog der im Jahre 1903 bekannt gewordenen Erdbeben, *Beitr. z. Geophys.*, **Suppl. 3**, 674pp, Leipzig.

Rojas W. and Barquero R. (1991), "El temblor de Piedras Negras 22 de Dicembre de 1990", La Crisis sísmica del Golfo de Nicoya, Publ. Inst. Costaric. Elecricidad, 95-104, San José.

Rymer M.J. (1987), The San Salvador earthquake of October 10, 1986; geologic aspects, *Earthquake Spectra*, **3**, 435-463.

Santana G. (1989), Current seismic design regulations for Costa Rica, *Bull. Int. Inst. Seism. Earthq. Eng.*, **23**, 95-122, Tokyo.

Santana G., Altamirano J. and Vargas W. (1987), "Accelerogramas registrados durante los sismos de Quepos de 15 Julio de 1987", Informe INII-39-87 Inst. Invest. Ingen., Univ. de Costa Rica, San José.

Sapper K. (1899), Das nicaraguensische Erdbeben vom 29 April 1898 und die Maribios-Vulkane, *Globus*, **75**, 201-208,222-227, Braunschweig.

Sapper K. (1902a), Das Erdbeben in Guatemala vom 18 April 1902, *Peterm. Mitteil.*, **9**, 193-195.

Sapper K. (1902b), Mittelamerikanische Reisen aus des Jahren 1888 bis 1900, 104-126, Braunschweig

Sapper K. (1912), Das Erdbeben von Sarchi, Costarika am 6 Juni 1912, *Peterm. Mitteil.*, **58(2)**, 340-341.

Sapper K. (1913), Die mittelamericanischen Vulkane, *Peterm. Mitteil., Suppl. 178*, Gotha.

Sauter F. (1987), The San Salvador earthquake of October 10, 1989; structural aspects of damage, *Earthquake Spectra*, **3**, 563-584.

Sauter F. (1994), Evaluatión de danos en puentes y otras estructuras civiles causado por el terremoto de Limón, *Revista Geol. Am. Centr.*, **Vol. Especial**, San José, 171-186.

Satake K., Bourgeois J. *et al.* (1993), Tsunami field survey of the 1992 Nicaragua earthquake, *EOS*, **74**, 145,156-157, and *Tsunami Newsletter*, **25**, 1-2.

Saville M.H. (1918), The Guatemala earthquake of December 1917 and January 1918, *Geograph. Rev.*, **5**, 459-469.

Scherzer C. (1857), "Travels in the free states of Central America; Nicaragua Honduras and San Salvador", 2 vols, London.

Scheu E. (1911), Catalogue régional de tremblements de terre pendant l'année 1906, *Publ. Bureau Centr. Assoc. Int. Seism., Ser. B*, Strasbourg.

Schneider R. (1914), Seismische Registierungen in Wien, im Jahre 1912, *Mitteil. Erdbeben-Komm. Akadem. Wiss. Wien*, **47**, 7-56, Vienna.

Schultz R. (1963), Estudio sobre la sismicidad en la región Centro Americana, *Boll. Bibliogr. Geofis. y Oceanorg. Americ., Parte Geofis. 1960-2*, **3**, 134-144, Mexico.

Severit W. (1933), Die Anthropogeographische Bedeutung der Erdbeben, *Matér. Étude Calam.*, **30**, 101-218.

Shakal A., Huang M. and Linares R. (1987), The San Salvador earthquake of October 10, 1986; processed strong motion data, *Earthquake Spectra*, **3**, 465-481.

Shea G.H. (1971), ed. "Costa Rica earthquake of April 22, 1991; reconnaissance report", *Earthquake Spectra*, **7 (Suppl. B)**, 127pp.

Sieberg A. (1913), Zerstörendes Erdbeben in Panama, *Beitr. z. Geophys.* 121-124.

Sieberg A. (1917a), Catalogue général des tremblements de terre de l'année 1908, *Publ. Bureau Centr. Assoc. Int. Seism., Ser. B*, Strasbourg

Sieberg A. (1917b), Catalogue régional des tremblements de terre 1908, *Publ. Bureau Centr. Assoc. Int. Seism., Ser. B*, Strasbourg.

Sieberg A. (1924), "Bemerkenswerte Erdbeben und Vulkanausbrüche des Jahres 1923", Mitteil. Reichsanstalt Erdbebenforschung in Jena, nr.1, Jena.

Sieberg A. (1932), "Die Erdbeben", in *Handbuch der Geophysik*, **4**, ed. B. Gutenberg, Berlin.

Soloviev S.L. (1955), Classification of earthquakes in order of energy, *Trudy Geofiz. Inst. AN SSSR*, **30(157)**, 3-31 (in Russian).

Soloviev S. L. and Shebalin N. (1957), " Opredelenie intensivnosti zemletriasenii po smeshchenia pochvi v poverkhnostikh volnakh", Ser. Geofis. Izvest. Aka. Nauk, no.7.

Sozen M.A. (1975), "A general description of structural damage" Engineering Rep. on the Managua Earthquake of 23 December 1972, Pub. Natl. Acad. Sci., 21-111, Washington.

Sozen M.A. and Roesset J. (1976), "Structural damage caused by the 1976 Guatemala earthquake", Struct. Res. Series, no.426, Univ. of Illinois, Urbana.

Spofford C.M. (1910), Earthquake effects on structures at Cartago, Costa Rica, *J. Am. Assoc. Eng. Soc.*, **46**, 63-65.

Stoiber R.E. and Carr M.J. (1973), Quaternary volcanic and tectonic segmentation of Central America, *Bull. Volcan.*, **37**, 304-325.

Sultan D. (1932), An army engineer explores Nicaragua, *Natl. Geogr. Mag.*, **61**, 593-625.

Sutch P.L. (1979), "Historic seismicity of Honduras", MSc Thesis, Branner Earth Sci. Library, Stanford University, Stanford.

Sutch-Osiecki P. (1981), Estimated intensities and probable tectonic sources of historic pre-1898 Honduran earthquakes, *Bull. Seism. Soc. Am.*, **71**, 865-881.

Sykes L R. and Ewing M. (1965), The seismicity of the Caribbean region, *J. Geoph. Res.*, **70**, 5065-5074.

Szirtes S. (1909), Katalog der im Jahre 1905 registrierten seismischen Störungen, *Publ. Zentral-Bureaus Int. Seism. Assoz., Ser.B*, Strassburg.

Szirtes S. (1910), Katalog der im Jahre 1906 registrierten seismischen Störungen, *Publ. Zentral-Bureaus Int. Seism. Assoz., Ser. B*, Strassburg.

Szirtes S. (1912a), Registrierungen des Jahres 1907, *Publ. Zentral-Bureaus Int. Seism. Assoz.*, Strassburg

Szirtes S. (1912b), Katalog der in Jahre 1907 registrierten seismischen Störungen, *Publ. Zentral-Bureaus Int. Seism. Assoz., Ser. B*, Strassburg.

Termer F. (1930), Die Erdbeben in der Republik Guatemala im Juni und Juli 1930, *Peterm. Mitteil.*, **76**, 302-304.

Tristan J. (1912a), Apuntes sobre el temblor del 25 de agosto, *Anales Centr. Estud. Sismol. Costa Rica, año 1911*, 43-45, San José.

Tristan J. (1912b), Notas sobre el terremoto de Guatuso 10 de octubre de 1911, *Anales Centr. Estud. Sismol. Costa Rica, año 1911*, 47-51, San José.

Tristan J. (1912c), El temblor del 21 de junio de 1900, *Anales Centr. Estud. Sismol. Costa Rica, año 1911*, 61-62, San José.

Tristan J. (1916), The Costa Rica earthquake of February 27, 1916, *Bull. Seism. Soc. Am.*, **6**, 232-235.

Tristan J. and Biolley P. (1912), Registro de temblores 1911, *Anales Centr. Estud. Sismol. Costa Rica, año 1911*, 18-32, San José.

Tristan J., Biolley P. and Cots C. (1912), The Sarchi earthquake, Costa Rica, *Bull. Seism. Soc. Am.*, **2**, 201-208.

Tristan J. (1925), Calamités causées au Costa Rica par des phénomènes géologiques et volcaniques, *Matér. Étude Calamit.*, **5**, 62-67.

Turner H.H. (1930), Shocks felt by ships, *Int. Seism. Summary*, Oct.-Dec. 1926, 315-316.

Uribe G.E. (1973), "Managua antisísmica", Fondo Colombiano de Invest. Cient. F.J. de Caldes, Bogota, 230pp.

Utsu T. (1982), Catalogue of large earthquakes in the region of Japan from 1885 through 1980, *Bull. Earthq. Res. Inst.*, **57**, 401-463.

Van de Putte J. (1924), "Phénomènes séismiques et volcaniques au Guatemala", anex aux *Ciel et Terre* pour 1923-24, 1-138, Bruxelles.

Vaněk J.(1995), Comment on "Distance dependence of M_S calibration function for 20 second Rayleigh waves" by M. Herak and D. Herak, *Bull. Seism. Soc. Am.*, **85**, 961.

Vaněk J., Kárník V., Zátopek A. *et al.* (1962), Standardization of magnitude scales, *Engl. transl. Izvest. Geophys. Ser.*, **2**, 108-111.

Van Everdingen E. (1915), Seismische Registrierungen in De Bilt, *Publ. Kon. Nederl. Meteor. Inst.*, no.108, Utrecht.

Vassaux J. (1969), "Cinquenta años de sismología en Guatemala", Publ. Observ. Natl. de Guatemala, Guatemala City.

Veith K.F. and Clawson G.E. (1972), Magnitudes from short-period *P*-wave data, *Bull. Seism. Soc. Am.*, **62**, 435-452.

Villacorta J.A. (1926), Monografía del departamento de Guatemala, Tipogr. Nacional, Guatemala.

Viquez C.G. (1910), "Temblores, terremotos, inundaciónes y erupciónes en Costa Rica", Publ. A. Alsina, San José.

Von Hagen V.W. (1945), "The jungle in the clouds", R.H. Hale, London.

Von Seggern D.H. (1977), Amplitude-distance relation for 20-second Rayleigh waves, *Bull. Seism. Soc. Am.*, **67**, 405-411.

Wacher C.J. (1917), The earthquake and eruption at San Salvador, 7 June 1917, *Geogr. J.*, **50(ii)**, 230-231.

Ward P.L., Gibbs J., Harlow D.H. and Aburto A. (1974), Aftershocks of the Managua, Nicaragua, earthquake and the tectonic significance of the Tiscapa fault, *Bull. Seism. Soc. Am.*, **64**, 1017-1029.

Walker J.W. (1899), "Report on the western division" Report of the Nicaragua Canal Commission, Appendix 4, Baltimore.

Wallström T. (1953), "Andarin, en resa i Central Amerika", Norstedt, Stockholm.

Ward P., Harlow D.H., Gibbs J. and Aburto A. (1973), Location of the main fault that slipped during the Managua earthquake, *Proc. Managua, Nicaragua earthquake*, Pub. Earthq. Eng. Res. Inst., San Francisco, **1**, 89-96.

Wells, W. (1857), "Exploration and adventures in Honduras 1854", Harper, NewYork.

White R.A. (1984), "Catalogue of historical seismicity in the vicinity of the Chixoy-Polochic and Montagua faults", Open File Report 84-88, U.S. Geol. Surv., Menlo Park.

White R.A. and Harlow D.H. (1993), Destructive upper-crust earthquakes of Central America, *Bull. Seism. Soc. Am.*, **83**, 1115-1142.

White R.A., Harlow D.H. and Alvarez S.(1987), The San Salvador earthquake of October 10, 1986; Seismological aspects, *Earthquake Spectra*, **3**, 419-434.

White R.A. and Sanchez E. (1980), "Preliminary report to the Government of Guatemala on the on-going earthquake swarm in the Department of Santa Rosa, Guatemala", Open File Rep. 80-800, U.S. Geol. Surv., Menlo Park.

Williams R.L. (1972), "The geology of western Nicaragua", Report Natural Resources Inventory Project, **4**, Managua.

Willmore P.L. (1979), "Manual of seismological observatory practice", Report SE-20 World Data Center A for Solid Earth Geoph., U.S Dept. Comm., Boulder.

Wolge, H. (1991), "Analysis of the Puriscal landslide", Report 91/2, Inst. Jord & Bergmekan. Kungl. Tekn. Högskol., Stockholm.

Wyllie L., *et al.* (1974), Effects on structures of the Managua earthquake of December 23, 1972, *Bull. Seism. Soc. Am.*, **64**, 1069-1133.

Wysession M.E., Okal E.A. and Miller K.L. (1991), Intraplate seismicity of the Pacific Basin, *Pure Appl. Geophys.*, **135**, 261-359.

Young C.J., Lay T. and Lynnes C.S. (1989), Rupture of the 4 February 1976 Guatemalan earthquake, *Bull. Seism. Soc. Am.*, **79**, 670-689.

Zamacois E. (1918), "Alegría de andar 1916-1920", Guatemala City.

Archives:

AE Archives des Affaires Étrangéres, Correspondance Consulaire et Commerciale, Paris. Costa Rica. 1898-1901, Guatemala, 1898-1901

AEG Archivo Episcopal General, Guatemala City

AGCA Archivo General de Centro America, Guatemala City

LC Library of Congress

-MS Catalogue 1918, p.57 (1902 Guatemala)

-Dispatches from US Consuls:

Costa Rica F.818.48 (EQ 12.2.1934)
Panama F.819.48 (3.6.1945)
Panama Microfilm #607 R.33 (EQ 21.7.1934)
San José #607 R.33(13.4.1910)
San José (5.5.1910)
San José (21.2.1912)
San José (10.4.1920)
San José (4.3.1924)

ODC Official documents and registers in British, French and German consular archives. 1898-1920

PRO Foreign Office Public Record Office, Kew. El Salvador, 802.1-4; Guatemala, 802; Nicaragua, 804.

USNA: US National Archives:
Record Group 84, Amer. Legation Corresp. C8.8 vol.7 (EQ 29.12.1937)
RG 59 El Salvador, Box 5169 (26.1.1947)
RG 59 El Salvador, Box 4526 (6.5.51)
Foreign Service Dispatch #614,818.49
Costa Rica (30.12.1952)
RG 59 El Salvador, Box 5514 (21.5.1932)

Periodical Publications:

BAAS (1899) "Seismological investigations", 4th Report, British Assn. Adv. Sci., Dover Meeting, London

BAAS (1911) "Seismic activity 1899-1903 inclusive", British Assn. Adv. Sci., 16th Rep. Seism. Invest., pp.26-36, Portsmouth

BAAS (1912) "Seismic activity 1904 to 1909 inclusive", British Assn. Adv. Sci., 17th Rep. Seism. Invest., pp.2-20, Dundee

BAAS (1913a) "Seismic activity in 1910" British Assn. Adv. Sci., 18th Rep. Seism. Invest., pp.2-7, Birmingham

BAAS (1913b) "Shide monthly bulletin 1913", British Assn. Adv. Sci., Shide

BAAS (1917a) "Seismological Committee: Bulletin for June and July 1917", British Assn. Adv. Sci., Oxford

BAAS (1917b) "Seismological Committee: Bulletin for October to December 1917", British Assn. Adv. Sci., Oxford

BCIS Annales de l'Institut de Physique du Globe, Univ. de Strasbourg, Strasbourg. 1919-1945

BPH Balboa Heights

BSS Boletin Sismologico de El Salvador. 1951-1966

BSSA Bulletin Seismological Society of America

CNR The Canal Record, Ancon. 1907-1917

DES Seismological despatches received at the Seismological Station, Georgetown University, Washington. 1915-1941

ERD Die Erdbebenwarte Laibach. 1901-1908

ISC Bulletin, International Seismological Centre. 1964-1995

ISS International Seismological Summary. 1918-1963

LLL Lloyds List, Weather & Navigation, London. 1898-1955

MES Matériaux pour l'Etude des Calamités, Geneva. 1923 -1936

NEN Neueste Erdbebennachrichten, Laibach. 1903-1908

REC Revue pour l'Étude des Calamités, 1938-1961, Revue de l'Union Intern. de Secours, 1964-1966, Annual Summary on Natural disasters, 1966-1974 (published), and Fichier de l'Union Internationale de Secours, 1899-1971 (unpublished), Geneva / Strasbourg

SCB Seismological Committee Monthly Bulletins, British Assn. Adv. Sci., Shide/Oxford. 1913-1917

Shide Circulars, British Assn. Adv. Sci., Seism. Comm. Circulars nos 1-27, Shide, Isle of Wight. 1899-1912

STR Monatsberichte & Seismische Aufzeichnungen der K.Hauptst. Erdbebenforschung. z. Strassburg. 1900-1916

TAC Information from Station Bulletins and Reports issued by Tacubaya Observatory. 1907-1920

USEQ United States Earthquakes, U.S.Coast & Geodetic Survey, Washington. 1928-1955

USSR Seismological Reports, U.S. Coast & Geodetic Survey. 1924-1928

Press reports: (with abbreviations where used)

PAVS El Aviso de San Salvador (ES) 1898

PBPB El Bien Público (Quezaltenango GU) 1898

PBDN Boletín de Noticias, (GU) 1917-1918

PCNT El Centroamericano, Leon (NI) 1924

PECR El Cronista (Tegucigalpa HO) 1917

PDRC El Derecho (CR) 1905

PDCA Diario de Centro-America (GU) 1898, 1899, 1902, 1906, 1907, 1912, 1913, 1915, 1916, 1917, 1918, 1919, 1921, 1929, 1930, 1934

PDCO Diario del Comercio, (CR) 1924

PDCR Diario de Costa Rica, (San José CR) 1920, 1921, 1924, 1931, 1935, 1939, 1940, 1945, 1948, 1951, 1952, 1953

PDRL El Diario Latino (ES) 1919

PDNI Diario Nicaraguence (Granada NI) 1947

PDOC Diario de Occidente (Quezaltenango GU) 1898

PDOF Diario Oficial (San Salvador ES) 1899

PDS Diario del Salvador (El Salvador) 1919
PEPC La Época (Tegucigalpa HO) 1937
PEPO La Época (CR) 1912, 1916
El Guanacaste (CR) 1910
PHRL El Heraldo de Costa Rica (San José CR) 1898
PHRC La Hora (San.José CR) 1933, 1934, 1939, 1952, 1953
PHRG La Hora (Guatemala GU) 1951
PIMC El Imparcial (San José CR) 1916, 1934
PIMG El Imparcial (Guatemala) 1929, 1930, 1932, 1951
PIND El Independente (CR) 1910
PINF La Información (CR) 1909-1913, 1916
PNCN La Nación (CR) 1950, 1952, 1953, 1955
PNTC El Noticiero (CR) 1904, 1909, 1911, 1913
PNUP La Nueva Prensa (CR) 1924
PNVP La Nueva Prensa (Managua NI) 1947
PPCF El Pacifico (San José CR) 1898
La Prensa (CR) 1919
PPRN La Prensa (Managua NI) 1951, 1955
PPLB La Prensa Libre (San José CR) 1904, 1905, 1911, 1912, 1916, 1931, 1934, 1935, 1939, 1940, 1941, 1948, 1950, 1952, 1953
La Razón (CR) 1941
PRPB La República (CR) 1905, 1910, 1912, 1913, 1914, 1952, 1955
PRPN La Repúblicana (CR) 1910
Star and Herald, (PA) 1905
PTIM The Times 1898-1930
PTRB La Tribuna, (San José CR) 1920, 1924, 1931, 1935, 1939, 1940, 1941, 1945
PTRL La Tribuna Libre (ES) 1917, 1919

Index